Springer Proceedings in Physics 400

Indexed by Scopus

The series Springer Proceedings in Physics, founded in 1984, is devoted to timely reports of state-of-the-art developments in physics and related sciences. Typically based on material presented at conferences, workshops and similar scientific meetings, volumes published in this series will constitute a comprehensive up to date source of reference on a field or subfield of relevance in contemporary physics. Proposals must include the following:

- Name, place and date of the scientific meeting
- A link to the committees (local organization, international advisors etc.)
- Scientific description of the meeting
- List of invited/plenary speakers
- An estimate of the planned proceedings book parameters (number of pages/articles, requested number of bulk copies, submission deadline).

Please contact:

For Americas and Europe: Dr. Zachary Evenson; zachary.evenson@springer.com
For Asia, Australia and New Zealand: Dr. Loyola DSilva; loyola.dsilva@springer.com

Chao Chang · Yaxin Zhang · Ziran Zhao ·
Yiming Zhu
Editors

Proceedings of the 5th China and International Young Scientist Terahertz Conference, Volume 1

YTHZ 2024, 22–24 March, Chengdu, China

Springer

Editors
Chao Chang
Peking University
Beijing, China

Ziran Zhao
Tsinghua University
Beijing, China

Yaxin Zhang
University of Electronic Science
and Technology of China
Chengdu, China

Yiming Zhu
University of Shanghai for Science
and Technology
Shanghai, China

ISSN 0930-8989 ISSN 1867-4941 (electronic)
Springer Proceedings in Physics
ISBN 978-981-97-3785-7 ISBN 978-981-97-3786-4 (eBook)
https://doi.org/10.1007/978-981-97-3786-4

This Springer imprint is published by the registered company Springer Nature Singapore Pte Ltd.
The registered company address is: 152 Beach Road, #21-01/04 Gateway East, Singapore 189721, Singapore

Contents

THz Sensors and Detection

THz Sensors and Detection

Explosives Detection by Terahertz Spectroscopy and Artificial Neural Network

Sijia Li[1,2], Meng Chen[2], Yingxin Wang[1,2(✉)], and Ziran Zhao[1,2]

[1] Department of Engineering Physics, Tsinghua University, Beijing 100084, China
wangyingxin@tsinghua.edu.cn

[2] National Engineering Research Center for Dangerous Articles and Explosives Detection Technologies, Beijing 100084, China

Abstract. Terahertz spectroscopy can reveal the spectral fingerprint characteristics of different materials in the terahertz range, which allows distinguishing explosives from common materials. In this work, by using the terahertz spectrum as a feature vector to establish an artificial neural network recognition model, we successfully achieve the identification of explosives and conventional materials, as well as the determination of different explosive classes.

Keywords: Terahertz Spectroscopy · Explosives · Artificial Neural Network

1 Introduction

Thanks to the fingerprint characteristics of terahertz spectroscopy, qualitative identification and even quantitative detection of prohibited items have become possible [1–5]. The first key issue involved is what identification method to use to determine "whether" and "what" prohibited items are. The simplest way is to directly compare the spectrum of the tested material with the standard spectrum in the database. However, for practical applications, especially in the detection of hidden prohibited or packaged products, interference from obstructions and noise can complicate the spectrum and increase uncertainty. A more effective approach is to establish an identification model that contains as much spectral variation information as possible. The specific process is to design a classifier by training (or learning) a certain number of samples, and then make classification decisions for the identified samples. There are multiple pattern recognition techniques that can be selected as classification tools, among which artificial neural networks (ANNs) have strong nonlinear mapping ability and have been used for the recognition of terahertz spectra [2]. Here we present to use ANNs for explosives detection based on terahertz spectroscopy.

2 Sample Preparation

Four typical explosives, TNT, RDX, HMX, and Composition B (CB), as well as ten conventional materials, were selected for qualitative identification of prohibited items. The parameters and properties of each material are shown in Tables 1 and 2. In the experiment, the explosive samples were compressed into uniformly thick pellets to reduce

C. Chang et al. (Eds.): YTHZ 2024, SPPHY 400, pp. 3–7, 2024.
https://doi.org/10.1007/978-981-97-3786-4_1

the influence of scattering effects. All other materials were in their normal state and had not been pre-treated. A terahertz time-domain spectroscopy (THz-TDS) system was employed to test the terahertz spectra of various materials [6, 7]. To reduce the impact of water vapor absorption in the air, the THz-TDS system was filled with dry nitrogen, and the final relative humidity inside the system was around 4%.

Table 1. Explosive Samples and Their Properties

Explosives	Molecular formula	Thickness (mm)	Properties
TNT	$C_7H_5N_3O_6$	1.6	Light yellow scaly crystalline flakes, the most widely used military explosive
RDX	$C_3H_6N_6O_6$	1.0	White crystalline powder, commonly mixed with other explosive materials and plasticizers to make plastic explosives
HMX	$C_4H_8N_8O_8$	0.8	Homolog of RDX, white crystalline powder, commonly used as solid fuel for rocket propulsion and military warheads
CB	$C_7H_5N_3O_6/C_3H_6N_6O_6$	2.9	Mixed explosive formed by adding solid particles of RDX into molten TNT, widely used in military applications

Table 2. Non-Explosive Samples and Their Properties

Materials	Cloth	Towel	Leather	Plastic	Envelope
Thickness (mm)	0.6	~1.5	1.0	1.1	0.3
Color	Blue	Yellow	Black	Black	Yellow
Materials	**Polyethylene**	**Foam**	**Paperboard**	**Flour**	**Soda ash**
Thickness (mm)	2.1	3.2	0.5	~1.5	~1.5
Color	White	White	Gray	White	White

3 Experimental Results

ANN is a complex network formed by a large number of processing units (neurons) connected to each other. Its model is based on instance learning and adopts parallel reasoning methods, with features such as association, memory, and induction. The feed-forward neural network is one of the most widely and successfully applied networks in the field of pattern recognition, represented by multi-layer perceptrons (BP networks) that use error backpropagation learning algorithms (BP algorithms). The BP network completes the nonlinear mapping from input signals to output modes through the mutual correlation between layers. The weight of the network is adjusted by the feedback of the error signal between the actual output and the expected output. The continuous correction of the weight gradually makes the actual output of the network closer to the expected output, thus completing the training process.

The feasibility of qualitatively identifying prohibited items through terahertz spectroscopy analysis is verified using BP network below. There are two tasks that need to be addressed here: one is to distinguish prohibited items from conventional materials, and the other is to identify the classes of prohibited items.

3.1 Distinguishing Between Explosives and Non-explosives

The training set was constructed by the spectra of RDX, HMX, TNT obtained from 5 measurements, as well as the spectra of ten non-explosive samples, containing a total of 25 samples. The test set includes 2 spectra of RDX, HMX, TNT, and CB measured at another time, 1 spectrum of RDX concealed in envelopes and fabrics (C-RDX), and 20 spectra of non-explosive samples collected at that time. To eliminate the influence of thickness, all spectra were normalized.

The BP network structure consists of three layers. The input layer has 124 nodes, corresponding to the spectral length within the effective frequency range. The hidden layer has 15 nodes, and the output layer has 1 node, corresponding to whether it is an explosive. The excitation function of the hidden layer in the network is the Sigmoid function $f(x) = 1/(1 + e^{-x})$, and the output layer is a linear function $f(x) = x$. . The training and testing of the network were implemented by the Neural Network Toolbox of Matlab (Mathworks Inc.). During training, set the output value of the network to 1 for explosive samples and 0 for non-explosive samples. The training algorithm adopted the Fletcher Reeves conjugate gradient algorithm, and the mean square error between the actual output and the expected output of the final network reached the predetermined accuracy (10^{-7}).

After training, the test set samples were used as input data to test the recognition performance of the network, as shown in Fig. 1. It can be seen that the output values of explosive and non-explosive samples are distributed around the target values of 1 and 0, respectively, which preliminarily verifies the recognition ability of the network. In order to calculate the recognition rate, a quantitative standard is needed. The deviation between the network output value y and the target output y_0 is defined as $\delta = |y - y_0|$, and a recognition threshold δ_0 is selected, that is, the recognition is considered correct when $\delta < \delta_0$, otherwise it is considered incorrect or the decision is rejected. At this point, the recognition rate can be defined as

$$\eta = \frac{\sum_{i=1}^{N} u(\delta_0 - \delta_i)}{N} \times 100\% \tag{1}$$

where N is the number of test samples and u is the threshold function. When $x < 0$, $u(x) = 0$. In cases of $\delta_0 = 5\%$ and $\delta_0 = 10\%$, the recognition rates are 80% and 95%, respectively. The test results indicate that the use of terahertz spectroscopy and neural network methods can effectively distinguish explosives from conventional items to a certain extent.

Note that the training set does not include the spectra of C-RDX and CB, but the network still correctly recognizes them. Due to the lack of absorption characteristics in the spectra of conventional materials, the spectral characteristics of RDX can still be displayed in hidden situations, while CB is composed of a mixture of RDX and TNT, and its absorption spectrum reflects the characteristics of RDX. Therefore, the network successfully identifies it. This fully demonstrates that terahertz spectroscopy technology can effectively detect specific components in hidden prohibited substances and mixtures.

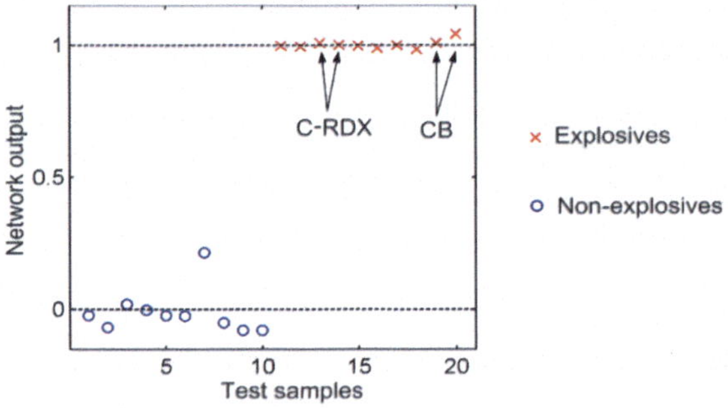

Fig. 1. BP network outputs for distinguishing between explosives and non-explosives.

3.2 Identification of Explosive Classes

On the basis of distinguishing between explosives and non-explosives, it is necessary to further distinguish the explosive classes. To achieve this purpose, a classification network was re-established using the spectra of explosive samples. The training set includes spectra of RDX, HMX, and TNT samples measured 10 times, and contains a total of 30 samples. The test set includes 5 spectra of RDX, HMX, TNT, and CB measured at another time, and 1 spectrum of RDX hidden in envelopes and fabrics, containing a total of 22 samples. Thc network structure was still divided into three layers, and the number of hidden nodes, excitation function, and training algorithm were the same as the previously constructed network. The difference is that the output layer has three neurons, corresponding to RDX, HMX, and TNT. When the output value is 1, it indicates the presence of the sample, while 0 indicates the absence. For example, [1 0 0], [0 1 0], and [0 0 1] represent RDX, HMX, and TNT, respectively.

The network response to the test set is shown in Fig. 2, which shows good discrimination among various explosives, and the output vectors of the same sample are tightly clustered together. According to the class decision method of multi output networks, samples of unknown classes were determined as the class corresponding to the node with the highest output value. Finally, all test samples of RDX, HMX, TNT, and C-RDX were correctly identified. In addition, due to the significant similarity in the spectra of

CB and RDX, while the TNT spectral features are not obvious, the network recognizes CB as RDX, which is understandable.

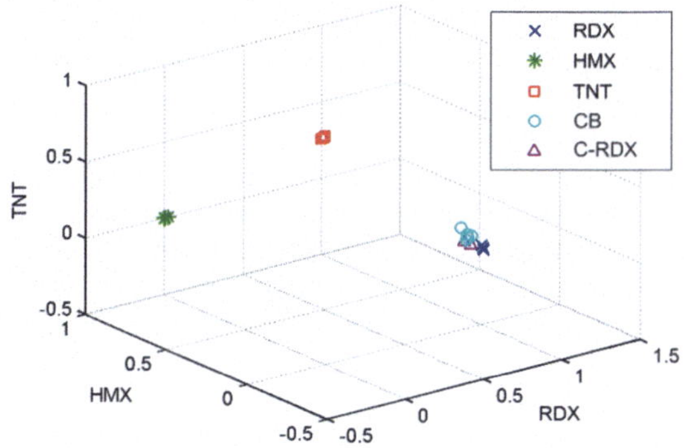

Fig. 2. BP network outputs for identification of explosive classes.

Through the above experiments, it can be seen that ANN can fully utilize all the information of terahertz spectrum for learning and quickly identify unknown samples. Its nonlinear prediction ability reduces the influence of system noise, which is conducive to improving the detection efficiency and recognition rate of the system.

Acknowledgment. This work was supported by the National Key R&D Program of China (No. 2023YFF0715000).

References

1. Kemp, M.: Explosives detection by terahertz spectroscopy—a bridge too far? IEEE Trans. Terahertz Sci. Technol. **1**(1), 282–292 (2011). https://doi.org/10.1109/TTHZ.2011.2159647
2. Wang, C., et al.: Convolutional neural network-based terahertz spectral classification of liquid contraband for security inspection. IEEE Sens. J. **21**(17), 18955–18963 (2021)
3. Liu, G., et al.: Myelin sheath as a dielectric waveguide for signal propagation in the mid-infrared to terahertz spectral range. Adv. Funct. Mater. **29**(7), 1807862 (2019)
4. Li, Y., et al.: Terahertz wave enhances permeability of the voltage-gated calcium channel. J. Am. Chem. Soc. **143**(11), 4311–4318 (2021)
5. Xiang, Z., et al.: A primary model of THz and far-infrared signal generation and conduction in neuron systems based on the hypothesis of the ordered phase of water molecules on the neuron surface. Sci. Bull. **65**, 308 (2020)
6. Wang, Y., et al.: Terahertz absorbance spectrum fitting method for quantitative detection of concealed contraband. J. Appl. Phys. **102**, 113108 (2007)
7. Wang, Y., et al.: THz technology using in exploring the recognition of explosives concealed on personnel. In: Proceedings of the 13th national academic annual conference nuclear electric detective Technology, vol. 397. China (2006)

High-Performance Infrared to Terahertz Detector Based on Topological Semimetal $ZrTe_3$

Guanchu Ding[1], Aosheng Zheng[1], Fan Yang[1], Zhengyi Zhao[1], Tan Qi[1], Yanyan Liu[2(✉)], Yating Zhang[1(✉)], and Jianquan Yao[1]

[1] Key Laboratory of Opto-Electronics Information Technology (Tianjin University), Ministry of Education, School of Precision Instruments and Opto-Electronics Engineering, Tianjin University, Tianjin 300072, China
yating@tju.edu.cn

[2] National Key Laboratory of Electromagnetic Space Security, Tianjin 300308, China
lyy13@126.com

Abstract. Terahertz radiation with applications in the field of medical diagnostics, astronomical remote sensing, imaging, biomedical research, communication etc. has been widely investigated recently. As a vital part of these applications, the development of terahertz photodetectors (PDs) is urgently needed. However, the widely investigated graphene and other two-dimensional materials for terahertz PD still facing optimal combination between bandgap, dark current, and absorption capabilities. Here, adopting the zero-bandgap and high carrier density characteristics of topological semimetals (TSMs), we applied the TSM $ZrTe_3$ for constructing high-performance terahertz PD. The PD achieved high responsivity of 0.11 A/W at 0.22 THz with response time of 4.74/4.97 ms, which exhibits great potentials for terahertz applications. Moreover, taking the zero-bandgap advantage of $ZrTe_3$, the PD was able to respond across ultra broad spectrum range spanning from near infrared to terahertz. The responsivity of the PD was recorded from near infrared to terahertz, confirming the broad band photodetection capability.

Keywords: Terahertz Detection · Photodetector · Topological Semimetals

1 Introduction

Terahertz radiation, refer to the electromagnetic wave within the wavelength of 0.03–3 mm (corresponding frequencies situated at 0.1–10 THz), possess significant application potentials in medical diagnostics, astronomical remote sensing, imaging, biomedical research, communication etc. [1–8]. Terahertz photodetectors (PDs) denote the optoelectronic device that coverts terahertz signals into electric signals have received great attentions these years. Suffering from low photon energy, the development of terahertz PDs still facing significant challenges. Represented by graphene [9], black phosphorus [10], and transition metal dichalcogenides, numerous two dimensional materials have been investigated for building terahertz PDs [11–14]. However, achieving high performance PD in the terahertz range is still challenging due to the lack of the optimal combination between bandgap, dark current, and absorption capabilities. Continuous exploration of other two-dimensional materials in this field is still required.

C. Chang et al. (Eds.): YTHZ 2024, SPPHY 400, pp. 8–12, 2024.
https://doi.org/10.1007/978-981-97-3786-4_2

Adopting the characteristic of zero bandgap and high carrier density, topological semimetals (TSMs) have attracted great attentions [15]. When utilized in the fabrication of PDs, TSMs demonstrate numerous advantages, including lower dark current, ultrafast response, and low noise equivalent power. These characteristics equip them with potential applications in the far infrared and terahertz wavelength range. Among which, TSM $ZrTe_3$ have not yet been investigated.

Herein, we developed a terahertz PD adopting TSM $ZrTe_3$, utilizing the advantages of TSMs, the terahertz PD achieved high responsivity of 0.11 A/W at 0.22 THz with response time of 4.74/4.97 ms, which shows great potentials in terahertz applications. Moreover, the response range of the PD was confirmed from near infrared to terahertz, indicating the broad-band capability.

2 Results and Discussions

We designed and fabricated a $ZrTe_3$ based photodetector (PD) adopting the symmetric electrode structure. The electrode pattern was prepared through a lift-off photolithography following a thermal evaporation and deposition of 5 nm Cr and 30 nm Au on a SiO_2/Si substrate, after which the substrate was washed to remove excess metal. The $ZrTe_3$ working layer was prepared through mechanical exfoliation. The bulk crystal was exfoliated and dry transferred to the electrodes with a polydimethylsiloxane (PDMS) film. The exfoliated nanoribbon was about 100 nm thickness confirmed by atomic force microscopy (AFM) as depicted in Fig. 1a. X-ray diffraction (XRD) pattern confirms the high crystal quality of our $ZrTe_3$ as shown in Fig. 1b. Raman spectrum was also induced as shown in Fig. 1c to confirm the crystal quality, all Raman peaks corresponds well with the previous reports. In this work, the device tunnel was chosen to be 30 μm as depicted in Fig. 1d.

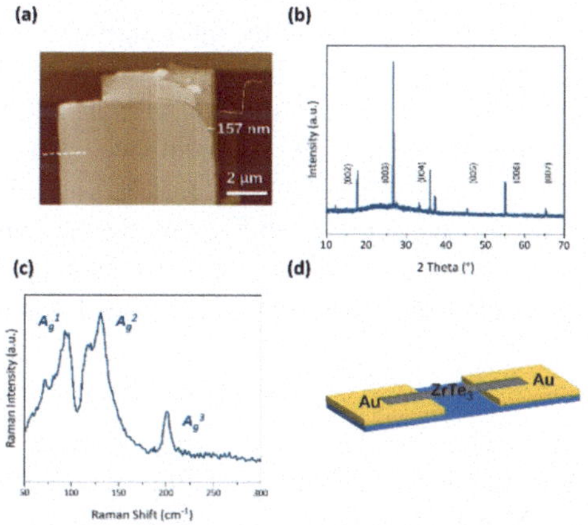

Fig. 1. Material characterization of $ZrTe_3$ (a) Atomic force microscopy (AFM) of exfoliated $ZrTe_3$ (b) X-ray diffraction (XRD) pattern (c) Raman spectrum (d) Device illustration.

We then performed basic electrical characterization on our device using a source meter (Keithly 2636B). As can be inferred from the liner *I-V* curve in Fig. 2a, our device exhibits good ohmic contact, with the resistance of about 100 Ω. Furthermore, the noise power density spectrum were collected as Fig. 2b, at the frequency of 1 Hz, our device exhibit a relatively low noise current density of 3.73×10^{-13} A/Hz$^{1/2}$, showing great potential for building high-performance detector.

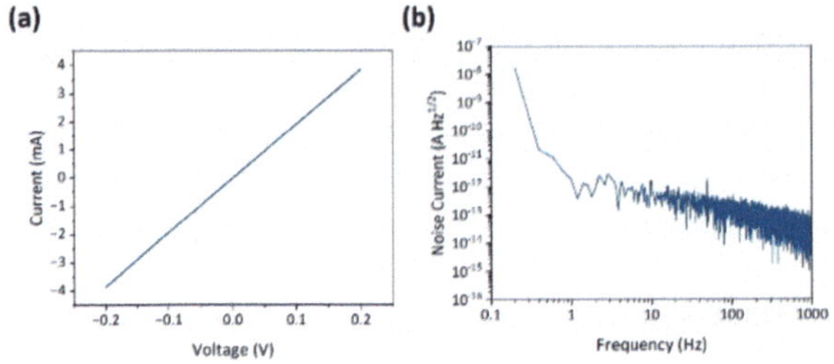

Fig. 2. Electrical characterization of ZrTe3 PD (a) I-V curve (b) noise power density spectra

Benefiting from the zero bandgap characteristics of topological semimetals, $ZrTe_3$ based PD is expected to show photon absorption capabilities across broadband even in the terahertz range with low photon energy. As demonstrated in Fig. 3a, a terahertz signal of 220 GHz (0.22 THz) with 1 Hz square wave modulation was adopted to radiate our PD directly. Under self-powered mode, our PD exhibit periodic switching of photocurrent with stability, indicating the potential of applications in terahertz range. Besides, our self-powered PD can respond to terahertz signal without antenna integration. Responsivity (R) is a figure of merit evaluating PD performance, which can be expressed as:

$$R = I_{ph}/P_{in} \tag{1}$$

where I_{ph} denotes the photocurrent generate by light signal, P_{in} denotes the efficient power of incident light, respectively. To quantitively evaluate the PD performance, we calculated the corresponding R value which is 0.11 A/W at 220 GHz. Response time is also a key figure of merit evaluating a PD performance, which defined as the time required for the photocurrent rise from 10% to 90% (t_{rise}) and decrease from 90% to 10%(t_{fall}) of peak photocurrent. As depicted in Fig. 3b, the t_{rise} and t_{fall} were 4.74 and 4.97 ms respectively, which reached a fast level in terahertz band. Noise equivalent power (*NEP*) is a figure of merit evaluating PD noise level, and is defined as:

$$NEP = i_n/R \tag{2}$$

where i_n is the noise current density. The lower the *NEP* value is, the higher the PD sensitivity is. According to the noise current spectrum recorded in Fig. 2b along with the R value, the *NEP* is 3.07 pW/Hz. Specific detectivity ($D*$) is also a key parameter of PDs, which is defined as:

$$\mathrm{D}^* = (\mathrm{A} \times \Delta\mathrm{f})^{1/2}/\mathrm{NEP} \tag{3}$$

where A is the device effective area, Δf is the device bandwidth, respectively. Under self-powered mode at 220 GHz, our device reached a D^* of 1.49×10^9 Jones. The

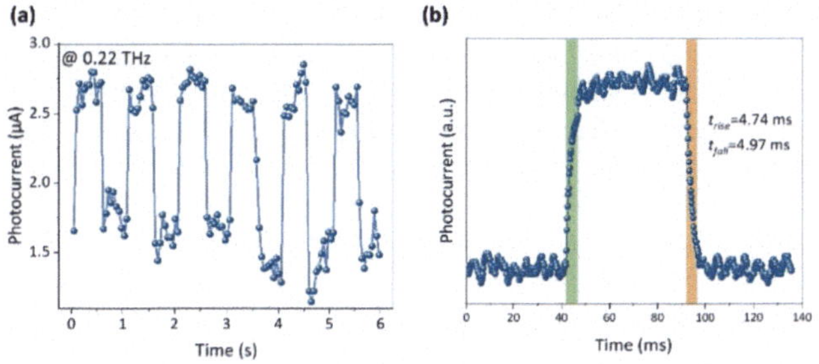

Fig. 3. Device performance of PD under 0.22 THz irradiation (a)I-t curve (b)Response time

high performance of our PD at terahertz band indicates the great potential of terahertz applications such as high-resolution terahertz imaging etc.

Adopting the zero-bandgap characteristic of ZrTe3, our PD is expected to response to broadband spectrum range. Therefore, we evaluated the photoresponse behavior of the PD from near infrared to terahertz range (0.808 μm, 1.55 μm, 4.6 μm, 10.6 μm, 12–30 μm and 1.364 μm). The corresponding *R* value are illustrated in Fig. 4, our PD exhibit peak responsivity of 0.86 A/W at 1.55 μm. Notably, we used blackbody as the light source in 12–30 μm range. The results reveal the broadband response capability of our PD, indicating the huge application potential from near infrared to terahertz.

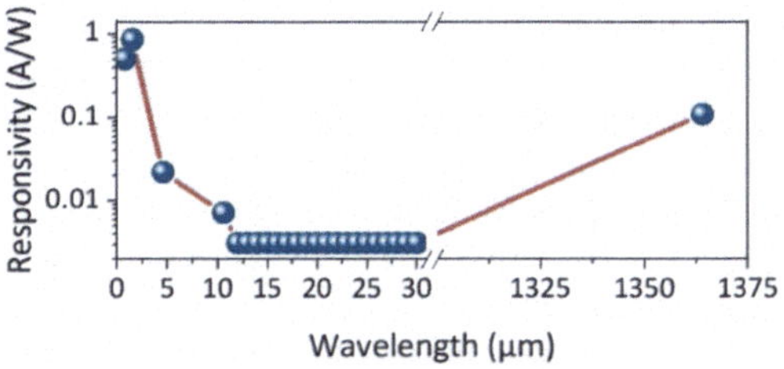

Fig. 4. Broad band detection capability: wavelength dependent responsivity of $ZrTe_3$ PD.

3 Conclusion

In summary, we designed and fabricated a topological semimetal $ZrTe_3$ based PD. The PD achieved *R* of 0.11 A/W, response time of 4.74/4.97 ms, *NEP* of 3.07 $pW/Hz^{1/2}$ and D^* of 1.49×10^9 Jones under self-powered mode without antenna integration at terahertz range of 220 GHz. Benefiting from the zero-bandgap characteristic, our PD exhibit broadband photoresponse capability ranging from near infrared to terahertz, reached peak responsivity of 0.86 A/W at 1.55 μm. Our finding proposed great broadband especially in terahertz range application potential of semi-metal $ZrTe_3$ based PDs.

Acknowledgement. This work was supported by the National Key Research and Development Program of China (2021YFB2800703), the National Natural Science Foundation of China (U22A2008, 62175181), the Opening Funding of National Key Laboratory of Electromagnetic Space Security, and Laoshan Laboratory Science and Technology Innovation Project (No.LSKJ202200801).

References

1. Harter, T., Füllner, C., Kemal, J.N., et al.: Generalized Kramers–Kronig receiver for coherent terahertz communications
2. Ferguson, B., Zhang, X.C.: Materials for terahertz science and technology. Nat. Mater. **1**(1), 26–32 (2012)
3. Horiuchi, N.: Terahertz technology: endless applications. Nat. Photonics **4**(3), 140 (2010)
4. Liu, G., et al.: Myelin sheath as a dielectric waveguide for signal propagation in the mid-infrared to terahertz spectral range. Adv. Funct. Mater. **29**(7), 1807862 (2019)
5. Li, Y., et al.: Terahertz wave enhances permeaility of the voltage-gated calcium channel. J. Am. Chem. Soc. **143**, 9 (2021)
6. Liu, X., et al.: Nonthermal and reversible control of neuronal signaling and behavior by midinfrared stimulation. Proc. Natl. Acad. Sci. **118**(10), e2015685118 (2021)
7. Zhang, J.X., et al.: Non-invasive, opsin-free mid-infrared modulation activates cortical neurons and accelerates associative learning. Nature Comm. **12**, 2730 (2021)
8. Xiang, Z., et al.: A primary model of THz and far-infrared signal generation and conduction in neuron systems based on the hypothesis of the ordered phase of water molecules on the neuron surface. Sci. Bull. **65**, 308 (2020)
9. Vicarelli, L., Vitiello, M.S., Coquillat, D., et al.: Graphene field-effect transistors as room-temperature terahertz detectors. Nat. Mater. **11**, 865–871 (2012)
10. Liu, C., Wang, L., Chen, X., et al.: Top-gated black phosphorus phototransistor for sensitive broadband detection. Nanoscale **10**, 5852–5858 (2018)
11. Tang, W., Liu, C., Wang, L., et al.: MoS2 nanosheet photodetectors with ultrafast response. Appl. Phys. Lett. **111**, 153502 (2017)
12. Kufer, D., Konstantatos, G.: Highly sensitive, encapsulated MoS2 photodetector with gate controllable gain and speed. Nano Lett. **15**, 7307–7213 (2015)
13. Qiu, Q., Huang, Z.: Photodetectors of 2D materials from ultraviolet to terahertz waves. Adv. Mater. **33**(15), 2008126-1-19 (2021)
14. Li, M., et al.: Ion-Bolometric Effect in Grain Boundaries Enabled High Photovoltage Response for NIR to Terahertz Photodetection. Adv. Funct. Mater. 2213970 (2023)
15. Burkov, A.A.: Topological semimetals. Nat. Mater. **15**(11), 1145–1148 (2016)

Design of an 89-183GHz Cold Optical and Quasi-Optical System

Jia-Hui Li[1,2(✉)], Hao-Tian Zhu[1,2], and Guang Liu[1]

[1] National Space Science Center, Chinese Academy of Sciences, Beijing 100190, China
lijiahui221@mails.ucas.ac.cn

[2] University of Chinese Academy of Sciences, Beijing 100049, China

Abstract. According to the basic principles of quasi-optical and gaussian beam, the quasi-optical mirror and lens are studied, and a set of millimeter-wave antenna quasi-optical feed system is designed, which can realize two channels of 89–183 GHz to receive electromagnetic radiation signals at the same time. The quasi-optical calculation and preliminary results are completed. The system works in low-temperature Dewar, in view of the actual needs, the locations and beam radius of the quasi-optical device are constrained and continuously optimized.

Keywords: Ellipsoidal reflector · Parabolic reflector · Quasi-optical · Terahertz

1 Introduction

Quasi-optical technology has become an important branch of millimeter-wave and terahertz science and technology, this technology has the characteristics of low-loss, high power, multi-polarization and wide frequency band [1]. The back-end of this design is the receiver of terahertz radiometer. The LNA of the back-end circuit works at a temperature of 20K, and a SIS mixer is used, which works at a temperature of 4K. The whole system works in a low temperature Dewar, and the whole quasi-optical system transits from 300 K to 4 K. According to the constraints of cold optical, it is necessary to reasonably arrange the different locations of the quasi-optical components in the Dewar and reduce the window size to reduce the heat leakage.

2 Quasi-Optical Design Principle

2.1 Gaussian Beam

Gaussian beam theory is the basis of quasi-optical system design. It is a transmission equation obtained by Helmholtz wave equation under the condition of paraxial approximation, which can be expressed as

$$E(r, z) = \sqrt{\frac{2}{\pi \omega_0}}\exp(-\frac{r^2}{\omega^2} - jkz - j\frac{\pi r^2}{\lambda R} + j\varphi_0) \tag{1}$$

C. Chang et al. (Eds.): YTHZ 2024, SPPHY 400, pp. 13–18, 2024.
https://doi.org/10.1007/978-981-97-3786-4_3

where r is the vertical distance from the propagation axis, ω is the beam radius of the Gaussian beam, R is the wavefront curvature radius of the Gaussian beam, φ_0 is the phase shift, and ω_0 is the beam waist of the Gaussian beam, that is, the beam radius at $z = 0$. Figure 1 shows the Gaussian beam propagating along the z-axis [2].

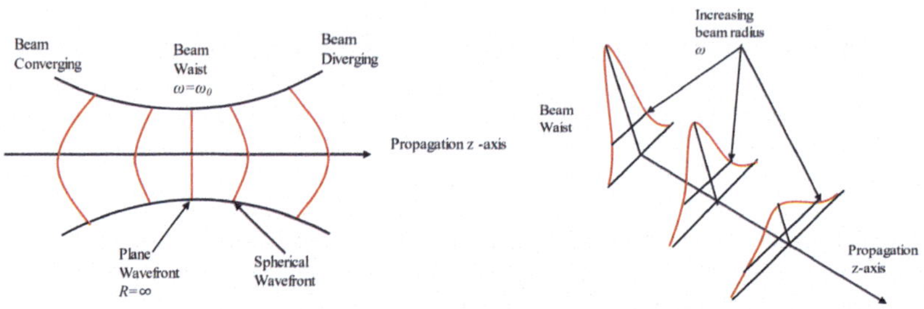

Fig. 1. Schematic of gaussian beam propagating.

2.2 Reflectors and Lenses

ABCD matrix is an important method to solve the beam transformation problem of quasi-optical system [3]. For a thin lens with refractive index n, its ABCD matrix is

$$M = \begin{bmatrix} A & B \\ C & D \end{bmatrix} = \begin{bmatrix} 1 & 0 \\ -\frac{1}{f} & 1 \end{bmatrix} = \begin{bmatrix} 1 & 0 \\ -(n-1)\left(\frac{1}{R_1} + \frac{1}{R_2}\right) & 1 \end{bmatrix} \tag{2}$$

where R_1 and R_2 are the radius of curvature of the left and right surfaces respectively. For ellipsoidal reflector and parabolic reflector, they can be equivalent to a lens system. The key to design is to find their equivalent focal length f. For ellipsoidal reflector, equivalent focal length is

$$f_e = \frac{R_1 R_2}{R_1 + R_2} \tag{3}$$

A common scheme is: when R_1 is equal to the distance from the ellipsoidal focus F_1 of the ellipsoidal reflector to the incident point and R_2 is equal to the distance from F_2 to the outgoing point, the incident beam and the outgoing beam match[4]–[6].

For parabolic reflector with an incident angle of 45°, equivalent focal length is

$$f_p = \frac{R}{2} \tag{4}$$

Figure 2 shows the corresponding figures of the parameters of the ellipsoidal mirror and the parabolic mirror. Note that d_1 and d_2 are the distance from the waist of the incident beam to the mirror and the distance from the waist of the outgoing beam to the mirror, respectively.

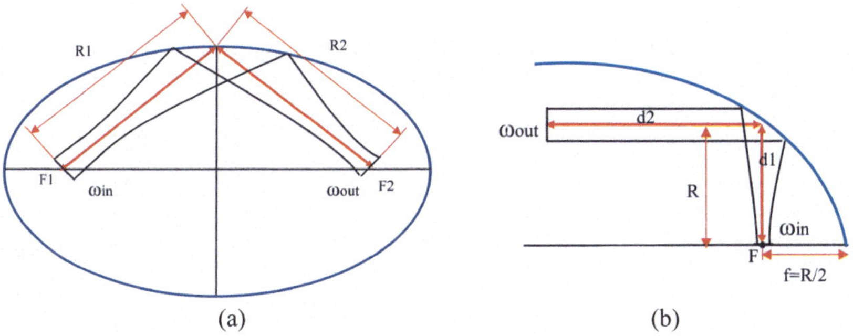

Fig. 2. Schematic of gaussian beam transformation (a) ellipsoidal mirror (b) parabolic mirror

3 Quasi-Optical Design

Firstly, considering the compactness of the optical path layout, two channels with relatively close frequencies can be multiplexed. The 89/110–115 GHz channel is received by a corrugated horn, and the 183 GHz channel is received by another corrugated horn. The system works in a low-temperature Dewar, and it is necessary to comprehensively consider the Dewar window size constraint and the mirror size constraint, and adjust the optical path space layout and the Gaussian beam radius to make the optical path structure reasonable and satisfy the constraints. In the far field, the relative field distribution of the Gaussian beam can be expressed as.

$$\frac{E(r)}{E(0)} = \exp[-(\frac{r}{\omega})^2] \tag{5}$$

where E(r) represents the electric field intensity at r from the Gaussian beam axis, E(0) represents the electric field intensity on the Gaussian beam axis at the same section, r is the intercept radius, and ω is the beam radius of the Gaussian beam. It should be noted that the beam radius corresponds to only 86.5% of the energy. We select 1.8 times the beam radius for design, about 99.85% [7].

The diagram of the Dewar structure is shown in Fig. 3(b). Layout is shown in Fig. 3(a). The 89/110–115 GHz frequency component is completely transmitted, and the 183 GHz frequency component is completely reflected. Finally, each frequency signal is received by the corresponding channel horn feed.

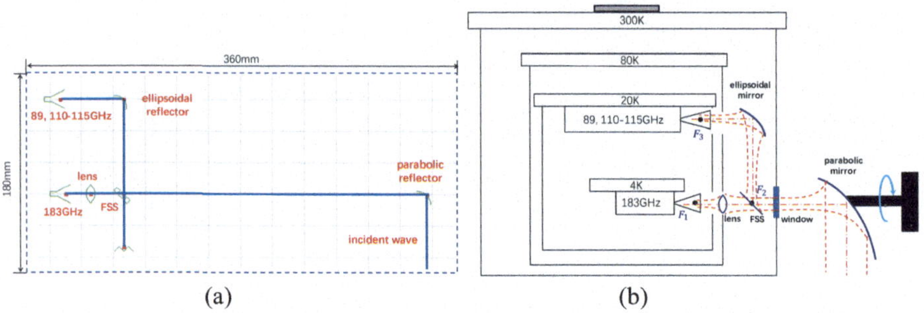

Fig. 3. (a) Quasi-optical network layout (b) Schematic of dewar structure

4 Quasi-Optical System Simulation

The design parameters of ellipsoidal mirror and parabolic mirror are shown in Table 1. The calculated results of the normalized pattern are shown in Fig. 4. Results of quasi-optical system are shown in Table 2.

Table 1. Parameters of the reflector

Reflector	Focal length	Distance of R_1	Distance of R_2
ellipsoidal	49.9 mm	65.9 mm	205.4 mm
parabolic	299.8 mm	299.8 mm	inf

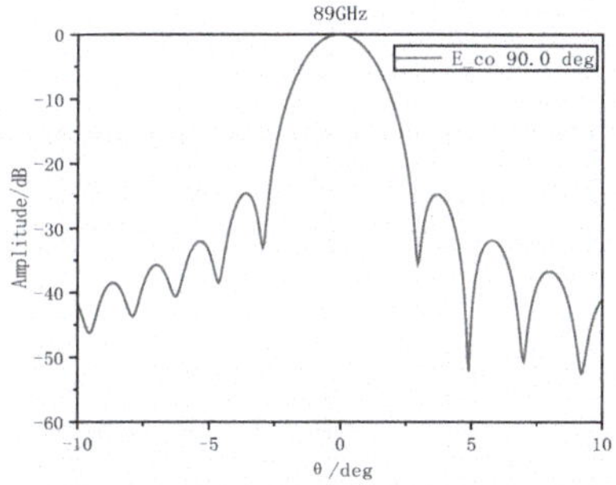

(a) 89 GHz beam radiation pattern(E-plane)

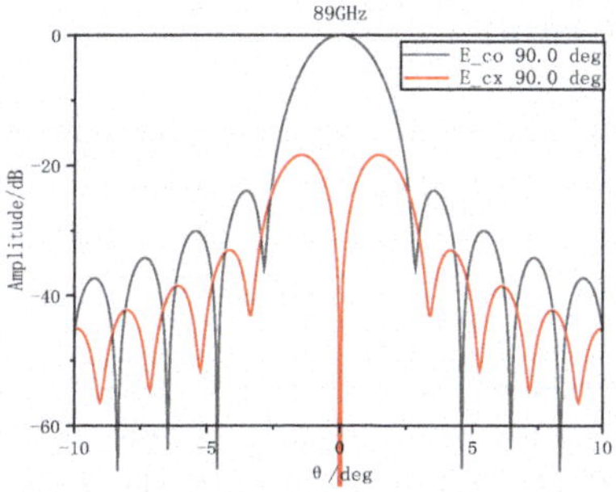

(b) 89 GHz beam radiation pattern(H-plane)

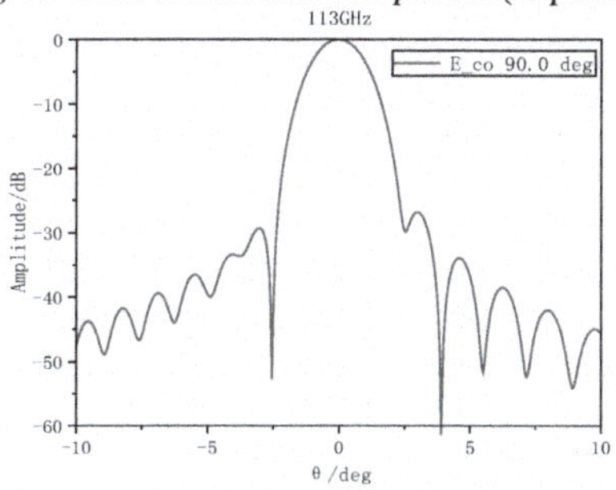

(c) 113 GHz beam radiation pattern(E-plane)

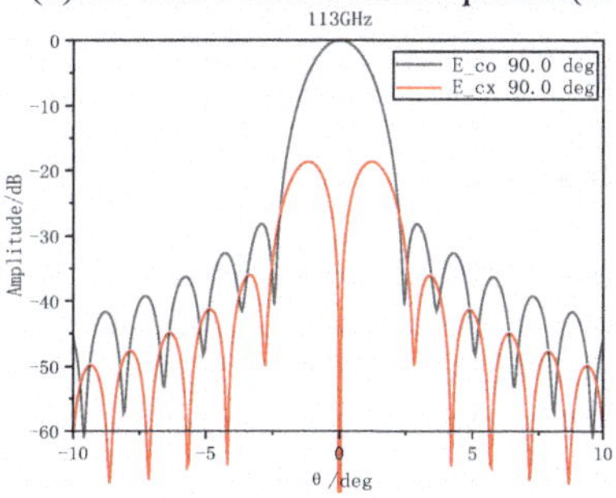

(d) 113 GHz beam radiation pattern(H-plane)

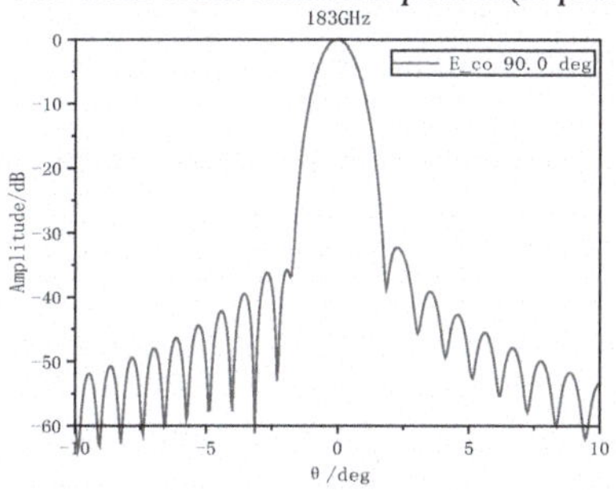

(e) 183 GHz beam radiation pattern(E-plane)

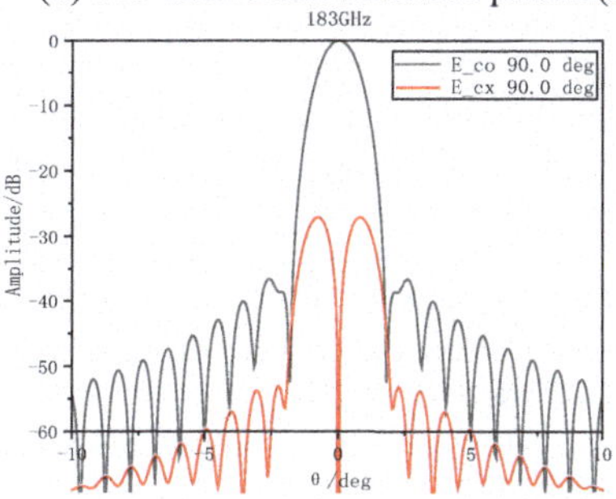

(f) 183 GHz beam radiation pattern(H-plane)

Fig. 4. Far-field normalized radiation pattern of each frequency channel

Table 2. Results of quasi-optical system

Frequency (GHz)	Directivity	Beam width(-3dB)	Main beam efficiency	Side lobe levels	Polarization purity
89	39.3dB	2.2°	99%	−24.6dB	18.3dB
113	41.1dB	1.8°	98.9%	−26.9dB	18.6dB
183	44.6dB	1.2°	99.1%	−32.3dB	27dB

5 Conclusion

According to the actual requirements and constraints, after many modifications and improvements, the quasi-optical path of 89–183 GHz system is designed theoretically and simulated. Follow-up work will continue to optimize and improve according to the requirements of the engineering.

Acknowledgments. The authors would like to thank Dr. Dong Liu, an Associate Researcher with Purple Mountain Observatory, Chinese Academy of Sciences, China, for providing useful suggestions of the quasi-optical design. This work was fully supported by the National Key R&D Program of China under Grant 2023YFB3905603.

References

1. Liu, X., et al.: Quasi-optical technology in the millimeter and terahertz wave ranges: theory, applications and development. J. Terahertz Sci. Elect. Inform. Tech. **22** 631–652 (2022)
2. Whale, M.: Optical Characterisation of Astronomical Submillimetre Receivers Including ALMA Bands 5 and 9. National University of Ireland Maynooth, Diss (2010)
3. Goldsmith, P.F.: Quasi-optical techniques. Proc. IEEE **80**(11), 1729–1747 (1992)
4. Liu, G., et al.: Myelin sheath as a dielectric waveguide for signal propagation in the mid-infrared to terahertz spectral range. Adv. Funct. Mater. **29**(7), 1807862 (2019)
5. Li, Y., et al.: Terahertz wave enhances permeability of the voltage-gated calcium channel. J. Am. Chem. Soc. **143**(11), 4311–4318 (2021)
6. Yao, C., et al.: A ten frequency bands quasi optical feed network. J. Terahertz Sci. Electron. Inf. Technol. **21**(2), 139–142 (2023)
7. Li, B., et al.: Research progress in terahertz quasi-optical feed network. J. Terahertz Sci. Electron. Inf. Technol. **17**(6), 939–943 (2020)
8. Liu, J., et al.: Design and analysis of the quasi-optical feed system. J. Terahertz Sci. Electron. Inf. Technol. **18**(2), 4 (2020)
9. Li, B., et al.: Terahertz quasi-optical feed network system optimization method. J. Detect. Control **44**(5), 112–115 (2022)

A Compact Graphene-Based Hot Electron Bolometer System for Terahertz Radiation Detection

Feiming Li[1,2], Wei Miao[1(✉)], Qingcheng Wang[1,2], Cui Yu[3], Zezhao He[3], Jiaqiang Zhong[1], Feng Wu[1], Zheng Wang[1], Kangmin Zhou[1], Yuan Ren[1], Wen Zhang[1], Jing Li[1], Shengcai Shi[1], Qingbin Liu[3], and Zhihong Feng[3]

[1] Purple Mountain Observatory, Chinese Academy of Sciences, Nanjing 210033, China
weimiao@pmo.ac.cn
[2] University of Science and Technology of China, Hefei 230026, China
[3] National Key Laboratory of Soild-State Microwave Devices and Circuits, Shijiazhuang 050051, China

Abstract. In this paper, we present the development of a compact hot electron bolometer (HEB) system for terahertz radiation detection. The bolometer system integrates a graphene-based HEB as its core detection unit and employs a compact infrared Dewar for efficient cooling of the graphene-based HEB. Utilizing 300 K/77 K loads as calibration sources, we quantified the sensitivity of the bolometer system at 4.2K. The measured results demonstrate a moderate sensitivity, with a noise equivalent power (NEP) of approximately 3.3×10^{-12} W/Hz$^{0.5}$. Furthermore, we evaluated the bolometer system's response time by applying a modulated terahertz signal, and the measured response time is found to be approximately 1 μs. These results underscore the promising potential of our graphene-based HEB bolometer system, especially in applications that require sensitive and high-speed detection capabilities.

Keywords: Terahertz · Compact · Graphene-based HEB · Noise equivalent power · Response time

1 Introduction

In recent years, there has been rapid development in the field of graphene, emerging as a prominent two-dimensional material. Due to its unique electronic structure, graphene exhibits distinct characteristics such as low electron heat capacity [1], weak electron-phonon interactions [2], and high electron mobility [3]. Moreover, its consistent absorption intensity across mid-infrared to ultraviolet wavelengths makes graphene well-suited for detector applications [4]. However, the relatively weak dependence of graphene resistance on temperature necessitates researchers to enhance this dependency through various methods for effective detection [5, 6]. An encouraging alternative, Johnson noise readout [7], stands out as it doesn't demand a strong resistance-temperature relationship

C. Chang et al. (Eds.): YTHZ 2024, SPPHY 400, pp. 19–23, 2024.
https://doi.org/10.1007/978-981-97-3786-4_4

or additional bias sources, allowing the detector to operate at the required temperature. This method facilitates the investigation of detector thermal transport properties, elucidation of their physical mechanisms, and provides guidance for detector development [8]. Although detectors employing Johnson noise readout have been demonstrated, the compactness of the detector systems is crucial for practical applications, especially for bio-information detection [9, 10]. To achieve system compactness, we integrated a graphene-based hot electron bolometer (HEB) and a low-temperature low-noise amplifier into a 4.2 K infrared Dewar. We determined the noise equivalent power of bolometer system by using 300 K/77 K loads as calibration sources, and characterized the response time by applying a modulated THz signal to the graphene-based HEB. Additionally, we compared our results with a commercial silicon-based bolometer.

2 Bolometer System

The bolometer system mainly consists of an HEB, a low-temperature low-noise amplifier, a thermometer, a room-temperature amplifier, a square-law detector, and a compact infrared Dewar. Figure 1a shows a SEM image of the graphene-based HEB, consisting primarily of a spiral antenna and a graphene microbridge. The frequency range of the spiral antenna is 0.3–1.6 THz, which is used to absorb radiation signals. The graphene microbridge has dimensions of 0.4 μm in length and 3 μm in width, and it is fabricated using epitaxial growth of bilayer graphene, as described in Refs. 11 and 12. Superconducting material Nb was used as electrodes connected to graphene to enhance detector performance by suppressing electron diffusion due to Andreev reflection. Subsequently, the graphene-based HEB was mounted on an elliptical high-resistance silicon lens used to collect THz radiation and placed in an oxygen-free copper block connected to the cold plate of the infrared Dewar. A low-temperature low-noise amplifier (1–12 GHz) and a thermometer were also placed in the infrared Dewar, as shown in Fig. 1d. Here, the low-temperature low-noise amplifiers are used to amplify detection signals, and the thermometer was used to monitor the temperature of the cold plate to ensure that the bolometer system operates at the correct temperature. The transmission between the HEB and low-temperature low-noise amplifier is achieved through CPWG structured transmission line, as shown in Fig. 1b. In order to ensure effective signal transmission, the design was carried out using the commercial software package HFSS based on finite element method to simulate the transmission line. The specific parameters are shown in Table 1. The simulation results are shown in Fig. 1c, and it can be seen that there is still good transmission characteristics at 12 GHz, which meets the requirements of the experiment. Figure 1e depicts the graphene-based HEB system with Johnson noise readout. After assembling the internal system of the Dewar, we further amplified the signal using room-temperature amplifiers (1–8 GHz) and detected the signal using a square-law detector. Additionally, impedance matching is a crucial issue to ensure the efficient operation of the system. According to the formula $\eta = (4R_O R_L)/(R_O + R_L)^2$, we evaluated the efficiency between the detector and the low-temperature low-noise amplifier to be 97%, here the resistance measured experimentally at 4.2 K was approximately 36 Ω, thus allowing us to ignore the effects of reflection.

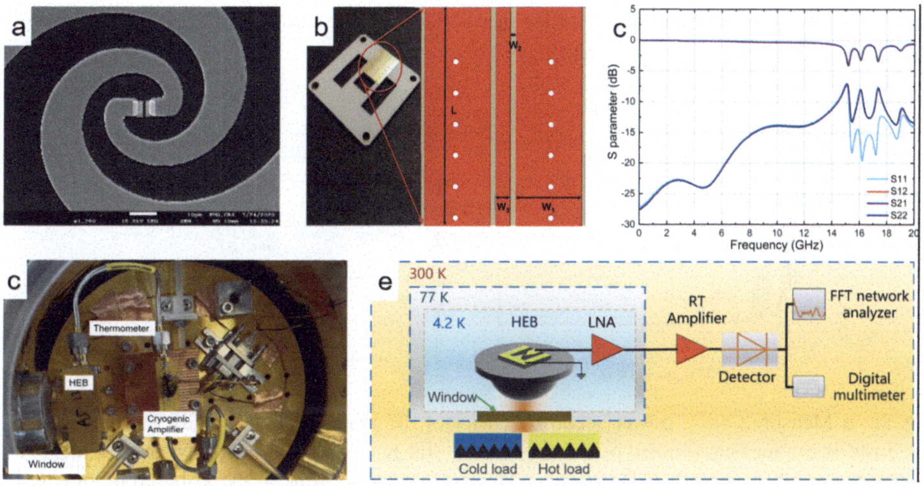

Fig. 1. **a** SEM image of graphene HEB detector. **b** Design and processing of transmission lines for CPWG structures. **c** simulation results of CPWG transmission line. **d** Internal image of graphene-based HEB system in infrared Dewar. **e** Schematic diagram of Johnson noise readout for graphene-based HEB system.

Table 1. Parameters of CPWG transmission line

Parameters	Symbol	Size (mm)
Thickness	T	0.813
Length	L	10.5
Width 1	W1	0.69
Width 2	W2	0.29
Width 3	W3	3.2

3 Results

In the experiment, we initially pre-cooled the system using liquid nitrogen and monitored the temperature to ensure the system reached 77 K. Subsequently, we further cooled the system using liquid helium to achieve a temperature of 4.2 K. Two blackbody loads of 77 K/300 K are used as calibration sources, then the radiation power is absorbed by HEB through HDPE cryostat window and a Zitex G104 infrared filter, and the variation of noise power of HEB is shown in the inset of Fig. 2a. It can be seen that the variation in response of the system, represented by ΔP_{out}, is approximately 28 μW, and the radiation power can be obtained by the formula $P_{Rad} = \int(\varepsilon(f, T_{300\,K})\text{-}\ \varepsilon(f, T_{77\,K}))df$, where $\varepsilon(f, T) = hf/(e^{hf/kT}\text{-}1)$, f is the frequency range of the spiral antenna, which we estimate to be about 3.97 nW. Figure 2a shows the noise power spectral density S_v obtained through FFT network analyzer. It can be observed that above 1.5 kHz, the noise power spectral density is nearly flat, approximately 23.3 nW/Hz$^{0.5}$. According to the formula NEP =

$S_v/(\Delta P_{out}/P_{Rad})$, we obtained the noise equivalent power of the graphene-based system to be approximately 3.3 x 10^{-12} W/Hz$^{0.5}$.

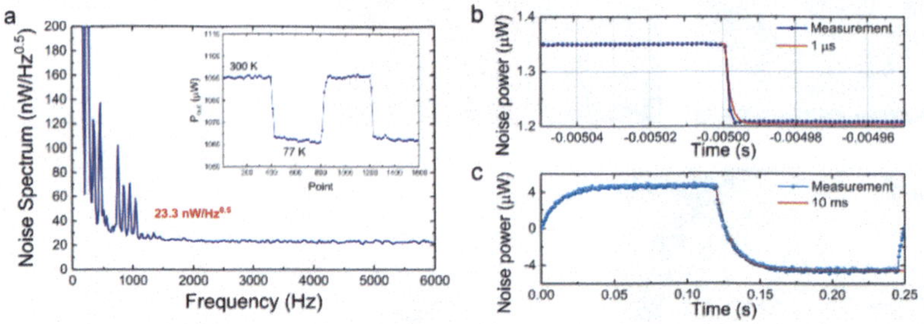

Fig. 2. **a** Measured noise power spectral density of the graphene-based HEB system at 4.2 K. The inset is the measured noise power for altered 300 K and 77 K loads. **b** Measured response time of the graphene-based HEB system. **c** Measured response time of the Si-based bolometer.

The response time, as an important indicator of the HEB system, reflects the speed of imaging. In our measurement, we applied a square wave excitation of 850 GHz from a signal source to the graphene-based HEB system and collected the corresponding noise power response through an oscilloscope. We extract the response time from the exponential decay of noise power over time using the formula $P_0 + (P_1\text{-}P_0)e^{-t/\tau}$. As shown in Fig. 2b, the response time of the graphene-based HEB system is approximately 1 μs, this is longer than the results evaluated in Ref. 13. The difference may be attributed to the distinct physical mechanisms of excitation via light and electricity, which require further investigation. For comparison, we also measured the response time of a commercial Si-based bolometer. As shown in Fig. 2c, the response time is approximately 10 ms, which is much slower than that of the graphene-based HEB system. Although the sensitivity of the graphene-based HEB system is not the highest, the fast response time provides the potential for rapid imaging applications.

4 Conclusion

We have successfully developed a compact graphene-based HEB system for THz radiation detection. This bolometer system offers several advantages, including its simplicity, portability, and rapid cooling capability. The noise equivalent power of the bolometer system has been measured to be approximately 3.3 × 10^{-12} W/Hz$^{0.5}$ at 4.2 K, and the response time of the bolometer system has been found to be around 1 μs. These characteristics demonstrate the potential applications of such a compact graphene-based HEB system for sensitive and high-speed imaging.

References

1. Du, X., et al.: Graphene-based bolometers. Graphene 2D Mater. **1**, 1–22 (2014)
2. Fong, K.C., Schwab, K.C.: Ultrasensitive and wide-bandwidth thermal measurements of graphene at low temperatures. Phys. Rev. X **2**, 031006 (2012)
3. Hwang, E.H., Das Sarma, S.: Acoustic phonon scattering limited carrier mobility in two-dimensional extrinsic graphene. Phys. Rev. B. **77**, 115449 (2008)
4. Yan, J., et al.: Dual-gated bilayer graphene hot-electron bolometer. Nat. Nanotech. **7**, 472–478 (2012)
5. Han, Q., et al.: Highly sensitive hot electron bolometer based on disordered graphene. Sci. Rep. **3**, 3533 (2013)
6. Sun, L., et al.: A pixelated frequency-agile metasurface for broadband terahertz molecular fingerprint sensing. Nanoscale **14**, 9681–9685 (2022)
7. Miao, W., et al.: Demonstration of a high-sensitivity and wide-dynamic-range terahertz graphene hot-electron bolometer with Johnson noise thermometry. Appl. Phys. Lett. **118**, 013104 (2021)
8. Zihlmann, S., et al.: Nonequilibrium properties of graphene probed by superconducting tunnel spectroscopy. Phys. Rev. B **99**, 075419 (2019)
9. Liu, G., et al.: Myelin sheath as a dielectric waveguide for signal propagation in the midinfrared to terahertz spectral range. Adv. Funct. Mater. **29**(7), 1807862 (2019)
10. Li, Y., et al.: Terahertz wave enhances permeability of the voltage-gated calcium channel. J. Am. Chem. Soc. **143**(11), 4311–4318 (2021)
11. He, Z.Z., et al.: High temperature RF performances of epitaxial bilayer graphene field-effect transistors on SiC substrate. Carbon **164**, 435–441 (2020)
12. Yu, C., et al.: Preparation and electrical transport properties of quasi free standing bilayer graphene on SiC substrate by H intercalation. Appl. Phys. Lett. **105**, 183105 (2014)
13. Miao, W., et al.: A terahertz detector based on superconductor-graphene-superconductor Josephson junction. Carbon **202**, 112–117 (2023)

A 340 GHz Waveguide-Coupled GaN HEMT Terahertz Detector

You Zhu[1,2], Qin Hua[2(✉)], Li Xinxing[2], Sun Jiandong[2], and Shang Guanyang[2]

[1] School of Nanotechnology and Nanobionics, University of Science and Technology of China, Hefei 230026, China

[2] Suzhou Institute of Nanotechnology and Nanobionics, Chinese Academy of Sciences, Suzhou 215123, China

hqin2007@sinano.ac

Abstract. Lens and waveguide integration are generally used to couple terahertz waves for terahertz detectors. Waveguide coupling provides a higher terahertz wave coupling efficiency compared to lens coupling which is beneficial to achieve a higher sensitivity. Previously, high-electron-mobility transistors (HEMT) terahertz detectors/mixers made of AlGaN/GaN heterostructure had to use lens coupling schemes due to their high impedance, which resulted in an impedance mismatch with the waveguides. This paper presents a waveguide coupling scheme including the design and simulation of a 340 GHz AlGaN/GaN HEMT terahertz detector. The detector design incorporates a waveguide-microstrip E-plane probe transition structure and a microstrip line impedance matching structure to reduce the impedance and improve the sensitivity. The simulation results show that the noise-equivalent power (NEP) of the terahertz detector can reach 29 $pW/Hz^{1/2}$ and 21 $pW/Hz^{1/2}$ at 340 GHz with the gate lengths of 1.0 μm and 0.8 μm for the HEMT mixer chip and the sensitivity can be further improved by shortening the gate length.

Keyword: Waveguide · Microstrip Line · Impedance Matching · HEMT

1 Introduction

The coupling between the terahertz electromagnetic waves in free space and the terahertz antenna/probe is realized by means of waveguides [1–3] or silicon lenses. Lens-based quasi-optical coupling schemes have wide bandwidth but a lower coupling efficiency, a smaller field of view or difficulties in precision integration in the high-frequency band. In comparison, waveguide-based coupling schemes have a narrower bandwidth but a higher coupling efficiency and, therefore, higher sensitivity [1].In 2020, a len-coupled AlGaN/GaN HEMT direct detector achieved an NEP of 3.7 $pW/Hz^{1/2}$ [4], which is comparable to the sensitivity level of the state-of-art Schottky-barrier diode (SBD) terahertz detector [5]. While silicon lenses are widely used for low-cost assembly, waveguide-coupled HEMT detectors are still in the experimental research stage, and few HEMT-based direct terahertz detectors integrated within waveguide structures have been reported so far. The main difficulties come from the facts that 3-terminal HEMT mixers

C. Chang et al. (Eds.): YTHZ 2024, SPPHY 400, pp. 24–29, 2024.
https://doi.org/10.1007/978-981-97-3786-4_5

require more space for integration in the waveguide cavity, the impedance of HEMT-based mixers is usually in kΩ range or above which causes a significant impedance mismatch between the mixer and the waveguide. It is however possible to tackle the above problems by using a more accurate HEMT mixer model for waveguide integration. The design of a highly sensitive waveguide-coupled HEMT direct detector is a problem of great research significance.

This paper presents the design and simulation of a 340 GHz waveguide-coupled terahertz direct detector based on the AlGaN/GaN HEMT. The detector utilizes a waveguide-microstrip E-plane probe transition structure and an impedance transformation structure between microstrip lines to reduce the mixer impedance. The impedance matching can be achieved to improve the coupling efficiency of the detector to the incident terahertz waves and hence the sensitivity. This method surpasses the previous integration technique in which the detector chip was directly attached at the waveguide port [6]. The simulation process of the detector includes the independent functional design of the passive circuit module first, followed by full-wave simulation and data extraction and post-processing of the detector response by using finite element simulation software in conjunction with MATLAB. The tools developed allow for design and optimization of waveguide-coupled HEMT detectors.

2 Structure Design of the Detector

2.1 General Structure

The overall and spatial structure of the 340 GHz terahertz direct detector designed in this paper is shown in Fig. 1. A waveguide-microstrip E-plane probe transition structure is used to form the device structure on a sapphire substrate. The microstrip line probe converts the input rectangular waveguide TE10 mode into microstrip line quasi-TEM mode and transmit it to the HEMT mixer for mixing. Afterwards, the connection to external circuits is realized through the low-pass filters and pads for device control and signal readout. The detector has a center frequency of 340 GHz and operates in the 320 - 350 GHz band. A standard rectangular waveguide WR-2.8 with dimensions of 0.711 × 0.356 mm is used. The detector chip has dimensions of 0.8 mm in the trench direction and 2 mm in the direction of the width of the trench, and the chip is on a sapphire substrate with a dielectric constant of 9.6. Due to the high frequency of terahertz signals, the thickness of the substrate was kept at 50 μm in order to minimize the transmission loss from the waveguide into the device substrate fixed in the trench.

The HEMT mixer is the core device of the detector assembly, as schematically shown in Fig. 1(a).The mixing mechanism consists of coupling terahertz waves through microstrip lines into a two-dimensional electron gas (2DEG) channel controlled by a field-effect gate. Wherein the horizontal terahertz electric field in the 2DEG channel modulates the drift velocity of the electrons, while the vertical electric field modulates the localized 2DEG concentration. As a result, the gate-controlled 2DEG channel can generate a mixing current or voltage of terahertz waves [7]. In this case, the HEMT source-to-gate spacing, L_{gs}, is not equal to the drain-to-gate spacing, L_{gd}, due to the asymmetric structure in order to generate a terahertz response [8].Wherein the HEMT mixer has a length $L = 100$ μm, a width $W = 150$ μm, a gate length $L_g = 1.0$ μm (0.8 μm), a gate-source pitch $L_{gs} = 1.3$ μm, and a gate-drain pitch $L_{gd} = 2.5$ μm.

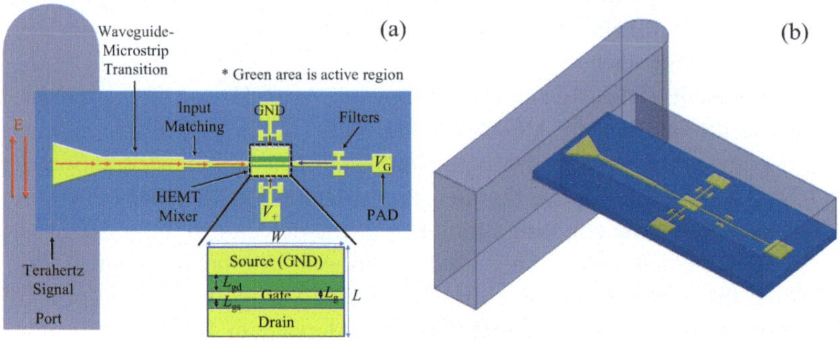

Fig. 1. **(a)** The overall structure of the detector; **(b)** The spatial structure of the detector

3 Detector Simulation and Optimization

3.1 Waveguide-Microstrip Transition Structure

The detector has been designed with a passive structure part first. The signal input structure, which is the waveguide-microstrip E-plane probe transition structure, is shown in Fig. 2(a). Through simulation, S-parameter results of the waveguide-microstrip transition structure have been obtained and presented in Fig. 2(b). The results show that within the range of 320–350 GHz, the waveguide-microstrip transition structure has a return loss of more than 10 dB and an insertion loss of less than 1 dB, indicating a good transmission of signals.

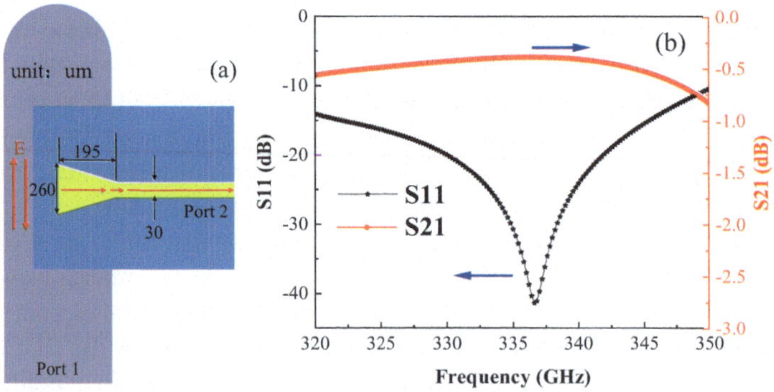

Fig. 2. **(a)** Waveguide-microstrip transition structure; **(b)** Transition structure S-parameters

The HEMT mixer employs a microstrip line as its gate, which has a gate length of only 1 μm (0.8 μm). This results in a significant impedance mismatch with the input microstrip line. Therefore, it is essential to perform an impedance transformation of the microstrip line to achieve impedance matching.

3.2 DC Differential Conductance

Once the design of each section of the detector is complete, the passive structure is assembled with the active module. The detector is then put through a complete full-wave simulation, followed by data extraction and post-processing calculations of the detector. The result of these calculations is the detector's performance index, known as the noise-equivalent power (NEP).

To calculate the differential conductance (G) of a detector, as well as the channel impedance (r_A) and the transconductance versus the gate voltage V_G (i.e., the first-order derivative of the differential conductance dG/dV_G), one can disregard the source-drain bias and only consider the gate voltage. This calculation is shown in Fig. 3.

Compared with the lens-coupled 300 nm gate-length detector designed by our group [4], the channel impedance r_A of the waveguide-coupled detector is theoretically reduced to at least 1/10 of that of the lens-coupled detector. The transconductance reaches the maximum value at the gate voltage $V_G = -3.6$ V, when the channel impedance of the waveguide-coupled detector is only 275 Ω (250 Ω). The low channel impedance provides the basis for the detector to achieve high-sensitivity detection.

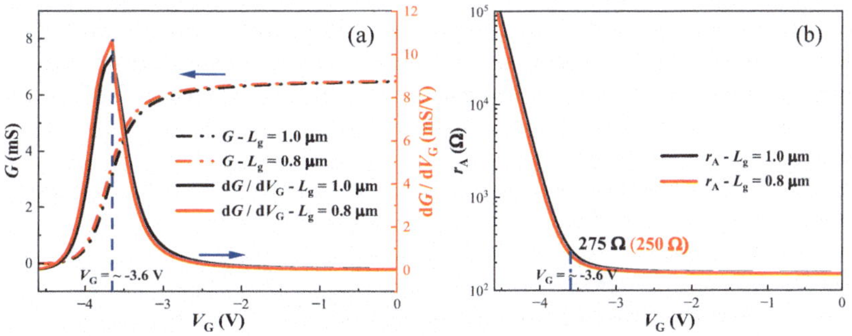

Fig. 3. **(a)** Differential conductance G and transconductance dG / dV_G of the detector versus V_G; **(b)** Channel impedance r_A of the detector versus V_G

3.3 Gate Control and Frequency Characteristics of the Detector Terahertz Response

At an incident terahertz power of 1 mW and an incident frequency of 340 GHz, the voltage responsivity R_v and the noise voltage v_n of the waveguide-coupled detector are shown in Fig. 4(a) with respect to the variation of the gate voltage V_G. Similarly, the variation of the noise-equivalent power (NEP) of the waveguide-coupled detector with the gate voltage V_G is shown in Fig. 4(b).

The R_v of the 1.0 μm gate-length detector reaches the maximum value of 34 V/W at the transconductance maximum point voltage of −3.6 V. Since source-drain bias is not

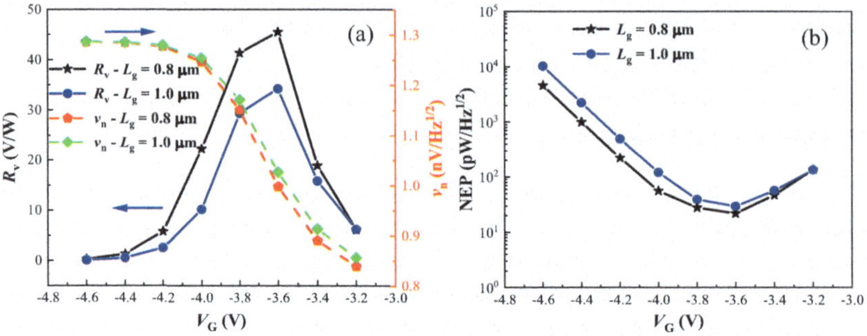

Fig. 4. Gate control characteristics of the detector **(a)** R_{V} and v_n; **(b)** NEP

considered in this paper, only the thermal noise voltage v_{n} of the detector itself needs to be considered. As the gate voltage is increased, the detector noise increases as well. The NEP of the 1.0 μm gate length detector reaches the minimum value of 29 pW/Hz$^{1/2}$ at the voltage of −3.6 V. It can be judged that the optimum operating voltage of the detector is −3.6 V. Compared with the 1.0 μm gate length design, the R_{V} of the 0.8 μm gate length detector at the optimum gate voltage operating point increases to 45 V/W, and the NEP decreases to 21 pW/Hz$^{1/2}$. The enhancement of the detection sensitivity is associated with the improvement of the field effect factor and antenna factor due to the reduction of the gate length [4].

The next step is to obtain the response spectrum of the detector. At the optimal operating voltage of −3.6 V, the R_{V} and v_n of the detector with terahertz frequency are shown in Fig. 5(a), and the NEP with terahertz frequency is shown in Fig. 5(b).

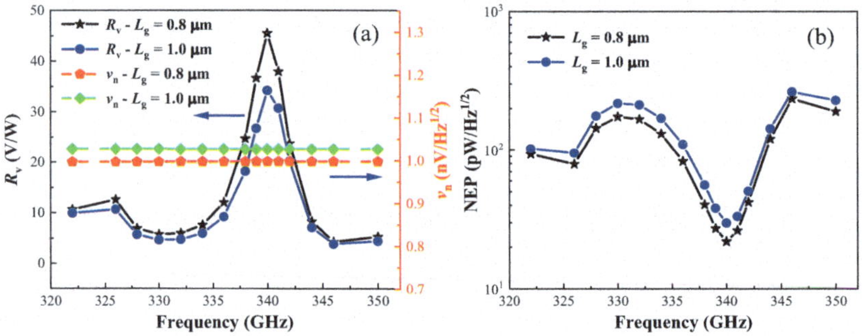

Fig. 5. Frequency characteristics of the detector **(a)** R_{V} and v_n; **(b)** NEP

At the center frequency of 340 GHz, the R_{V} of the 1.0 μm and 0.8 μm gate-length detector reaches the maximum values of 34 V/W and 45 V/W, respectively. The NEP of the 1.0 μm and 0.8 μm gate length detector at 340 GHz reaches the minimum values of 29 pW/Hz$^{1/2}$ and 21 pW/Hz$^{1/2}$, respectively. The frequency characteristics of this structure combined with waveguide, probe, and HEMT mixer suffer from the problem of a narrow bandwidth, which can be further optimized to increase the bandwidth, which is useful for THz detection of bio-information [9–12].

4 Conclusion

In this research paper, a 340 GHz waveguide-coupled AlGaN/GaN HEMT terahertz direct detector is presented. A waveguide-microstrip transition structure is used to reduce the impedance of the detector, resulting in 275 Ω (250 Ω). This also improves the sensitivity of the detector. At 340 GHz, the noise-equivalent power (NEP) of the detector designed with 1.0 μm and 0.8 μm gate-length is 29 $pW/Hz^{1/2}$ and 21 $pW/Hz^{1/2}$, respectively. By further shortening the gate length, the NEP can continue to be decreased, and the detection sensitivity can be improved even further.

References

1. Bowers, J.E., Burrus, C.A., Mitschke, F.: Millimeter-waveguide-mounted InGaAs photodetectors. Electron. Lett. **22**(12), 633–635 (1986)
2. Liu, G., et al.: Myelin sheath as a dielectric waveguide for signal propagation in the mid-infrared to terahertz spectral range. Adv. Funct. Mater. **29**(7), 1807862 (2019)
3. Zhang, Z., et al.: Advanced terahertz refractive sensing and fingerprint recognition through metasurface-excited surface waves. Adv. Mater **36**(14), 2308453 (2024)
4. Sun, J.D., Feng, W., Ding, Q.F., et al.: Smaller antenna-gate gap for higher sensitivity of GaN/AlGaN HEMT terahertz detectors. Appl. Phys. Lett. **116**(16), 161109 (2020)
5. VDI. The specifications of different SBD detectors coupling with waveguide or silicon lens. https://vadiodes.com/en/products/detectors
6. Li, X., Sun, J.D., Zhang, Z.P., et al.: Integration of a field-effect-transistor terahertz detector with a diagonal horn antenna. Chin. Phys. B **27**(6), 068506 (2018)
7. Qin, H., Huang, Y.D., Sun, J.D., et al.: Terahertz-wave devices based on plasmons in two-dimensional electron gas. Chinese Optics **10**(1), 51–67 (2017)
8. Sun, J.D., et al.: High-responsivity, low-noise, room-temperature, self-mixing terahertz detector realized using floating antennas on a GaN-based field-effect transistor. Appl. Phys. Lett. **100**(1), 013506 (2012)
9. Xiang, Z., et al.: A primary model of THz and far-infrared signal generation and conduction in neuron systems based on the hypothesis of the ordered phase of water molecules on the neuron surface. Sci. Bull. **65**, 308 (2020)
10. Li, Y., et al.: Terahertz wave enhances permeability of the voltage-gated calcium channel. J. Am. Chem. Soc. **143**(11), 4311–4318 (2021)
11. Wu, K., et al.: Terahertz wave accelerates DNA unwinding: a molecular dynamics simulation study. J. Phys. Chem. Lett. **11**(17), 7002–7008 (2020)
12. Zhang, J., et al.: Non-invasive, opsin-free mid-infrared modulation activates cortical neurons and accelerates associative learning. Nat. Commun. **12**, 2730 (2021)

High-Performance $MAPbI_3$ Infrared-Terahertz Wave Detector Based on Sulfonated Graphene Oxide Modification

Huanyu Ren[1], Mengyao Li[1], Guanchu Ding[1], Aosheng Zheng[1], Fan Yang[1], Zhengyi Zhao[1], Shuai Zhong[1], Yanyan Liu[2(✉)], Yating Zhang[1(✉)], and Jianquan Yao[1]

[1] Key Laboratory of Opto-Electronics Information Technology (Tianjin University), Ministry of Education, School of Precision Instruments and Opto-Electronics Engineering, Tianjin University, Tianjin 300072, China
yating@tju.edu.cn

[2] National Key Laboratory of Electromagnetic Space Security, Tianjin 300308, China
lyy13@126.com

Abstract. Infrared and terahertz waves contain a wealth of information that can be used effectively in medicine, agronomy and environmental monitoring. Terahertz wave detectors can also be used to detect small changes in semiconductor chips to improve their functional performance. Among many advanced materials such as room-temperature infrared-terahertz wave detectors, perovskites are favored due to their long carrier diffusion distance, low processing cost and large-scale preparation. In this paper, we used sulfonated graphene oxide (s-GO) to modify the grain boundaries of methylaminolead iodide perovskites ($MAPbI_3$) to reduce the obstruction of grain boundary defects to the generation and transport of $MAPbI_3$ photothermal carriers. The s-GO was used to stabilize the grain boundary defects and promote the carrier transport of $MAPbI_3$, so as to improve the responsivity of the s-GO/$MAPbI_3$ bolometer detector to infrared-terahertz waves, so that the current responsivity to 1064 nm infrared light was 47.5 μA/W and the current responsivity to 1.36 mm terahertz light was 0.17 μA/W.

Keywords: Terahertz · Sulfonated graphene oxide · Perovskite Detector

1 Introduction

Terahertz (THz) waves hold the frequency of 0.1 THz~10 THz (1 THz = 10^{12} Hz, corresponding wavelength range is 3 μm–3 mm) have shown great potentials in various applications in medicine, agronomy and environmental monitoring. Terahertz wave detectors, which convert optical signals into electrical information, have been widely used in terahertz imaging, communications, and military fields [1–5]. Therefore, developing high-performance THz detectors is of great significance in THz application development.

Huanyu Ren and Mengyao Li—These authors contribute equally in this work

C. Chang et al. (Eds.): YTHZ 2024, SPPHY 400, pp. 30–36, 2024.
https://doi.org/10.1007/978-981-97-3786-4_6

In recent years, perovskite detectors have been increasingly used in the terahertz band, and have been favored by the high carrier mobility, large optical absorption coefficient and wide spectral range of the materials themselves [6, 7]. In addition, the low thermal conductivity and high Seebeck coefficient make it a promising thermoelectric material [8]. However, it is difficult to precisely control the specific defects of perovskite materials, i.e. the possibility of doping with organic cations with structural defects which harms the film quality, and further become a huge obstacle to the development of perovskite THz detectors [9].

We adopted the method reported by Song et al. [10] and designed a high-performance room-temperature $MAPbI_3$ infrared-terahertz detector. The device structure is to prepare a perovskite thin film on a conventional Si/SiO_2 substrate following by the evaporate-deposit gold electrodes. We introduced sulfonated graphene oxide (s-GO) onto the perovskite layer [11–13], and adopted the sulfonic acid group in s-GO to interact with a large number of Pb atoms in $MAPbI_3$ to form new coordination groups, which effectively filled the iodine vacancy and improved the carrier transfer efficiency, further improved the crystallization quality of the perovskite thin film and thus improved the device performance.

2 Results and Discussions

The s-GO/$MAPbI_3$ device was prepared using the antisolvent method. Figure 1 illustrates the specific process of s-GO/$MAPbI_3$ film preparation. The treated substrate was spun coated with a mixed solution of s-GO/perovskite precursor, and then chlorobenzene solution was added to turn the surface brownish-black and annealed to form an s-GO/$MAPbI_3$ film.

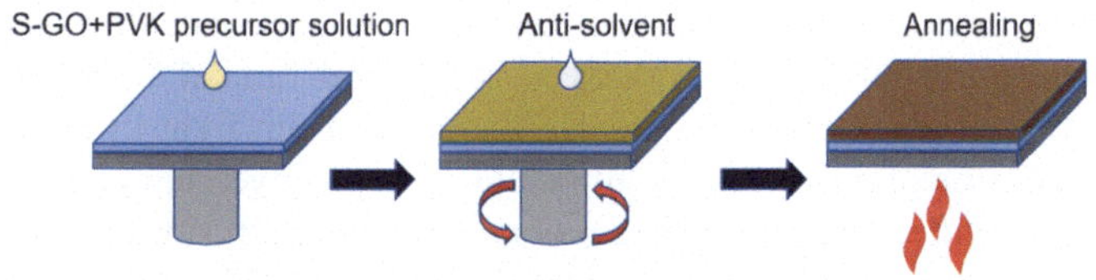

Fig. 1. Preparation of s-GO/$MAPbI_3$ film

We have compared the Fourier Transform Infrared Spectrometer(FTIR)spectrums of the s-GO/$MAPbI_3$ film and the $MAPbI_3$ film. As depicted in Fig. 2(a), the FTIR absorption peaks of the two films can be observed at 3179 cm^{-1}. Besides, the two absorption peaks corresponding to 1124 cm^{-1} and 1005 cm^{-1} of the s-GO/$MAPbI_3$ film indicate the presence of sulfonic acid groups (-SO_3H), and the convergence of the stretching peaks towards the low wavenumber also proves the interaction of -SO_3H with lead. To analyze the crystalline composition of the s-GO/$MAPbI_3$ film, we performed X-Ray diffraction (XRD) on the s-GO/$MAPbI_3$ film as illustrated in Fig. 2(b), the XRD pattern indicates the $MAPbI_3$ doped with s-GO possesses the same position as the diffraction peak of the typical $MAPbI_3$ crystal plane, which indicates that s-GO can be effectively doped in $MAPbI_3$ without changing the characteristics of $MAPbI_3$. As depicted in Fig. 2(c), the Scanning Electron Microscope (SEM) image of s-GO/$MAPbI_3$ film indicates a good crystallization, the morphology exhibit dense surface and the boundaries are clear, and the grain morphology is still clearly visible.

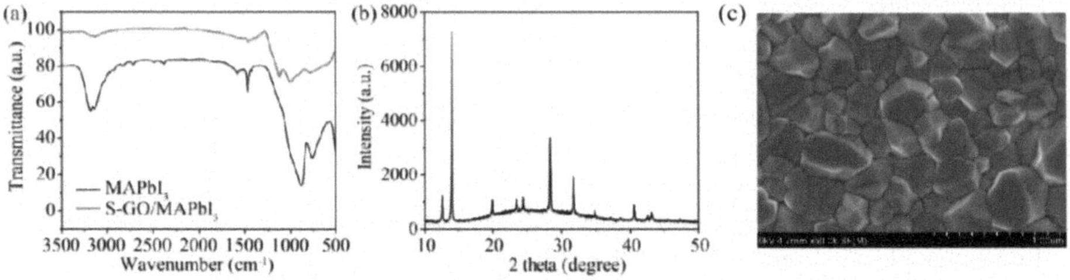

Fig. 2. s-GO/$MAPbI_3$ film (a) Fourier transform infrared spectra; (b) XRD images; (3) SEM images

To further explore the device working mechanism of the detector, the *I-V* curves of the s-GO/$MAPbI_3$ detector under infrared light irradiation and dark state were collected. As the voltage increases, the current increases linearly, indicating a ohmic contact between the film and the electrodes. In addition, 1064 nm light irradiation causes a significant increase in current and a decrease in resistivity. To shed more light on the device performance, the transfer characteristic curves and gate leakage current (IGSS) curves of the device from −60 V–60 V gate voltage with or without 1064 nm light irradiation were collected, as shown in Fig. 3(c), within a certain range, with the increase of gate voltage, the current increases greatly surpasses a threshold, exhibiting the "V" shaped characteristics, revealing the transition of the main carriers from holes to electrons, and proving that there are a large number of carriers with similar concentrations in the holes and electron transport channels of the device. At the same time, the current changes are observed at different positions of the device under 1064 nm light irradiation, as illustrated in Fig. 3(d), the photocurrent at all positions is positive, and is greater the more irradiated to the middle. The photocurrent obtained from both sides of the electrode is the smallest, indicating that the device is mainly worked on the radiant heat effect.

To further analyze the device's performance, the photocurrent of the device was measured and recorded using three wavelengths of light, 1064 nm (near-infrared), 10.6 μm (long-wave infrared), and 1.36 mm (terahertz) under a modulated square wave signal of 1 Hz. As can be inferred from Fig. 4(a) that the photocurrent of the device exhibits a stable square waveform with the periodic change of incident light signal. Due to the low optical power density of terahertz waves, we amplify their response curves by a factor of 10.

We further recorded the photocurrent as a function of incident light power, as shown in Fig. 4(b), and found linearly proportional relationship. This is because as the optical power gradually increases, more carriers are excited, causing the photocurrent of our device to increase linearly. To quantitively evaluate the device performance, responsivity (R) is introduced, which is defined as the strength of the output electrical signal relative

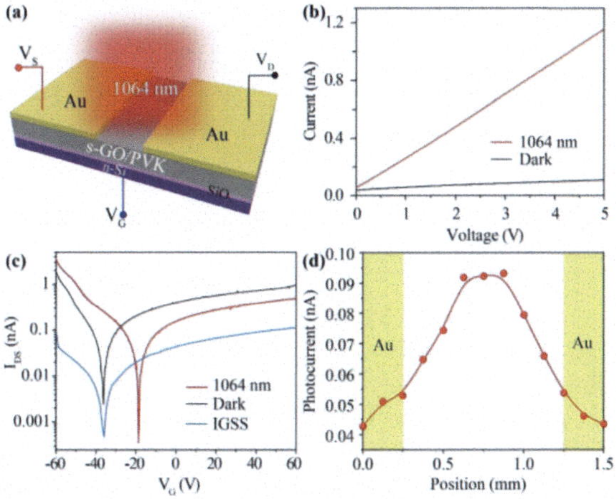

Fig. 3. (a) Schematic diagram of photoelectric test of s-GO/$MAPbI_3$ device; (b) The s-GO/$MAPbI_3$ detector at 1064 nm light I-V curves with and without illumination; (c) Transfer of the s-GO/$MAPbI_3$ detector at 1064 nm with and without light characteristic curves and IGSS curves; (d) Relationship between the photocurrent of the s-GO/$MAPbI_3$ detector and the irradiation position of infrared light

to the strength of the input optical signal, and can be expressed as:

$$R = I_{ph}/P_{in} \tag{1}$$

where I_{ph} is the photocurrent generated by the optical signal, and P_{in} is the incident optical power. We also collected the relationship between responsivity and optical power, and as shown in Fig. 4(c), we find the inversely linearly proportional relationship, which can be attributed to the excessive optical power leads to a large number of non-radiative recombination in the crystal lattice, which significantly reduces the carrier lifetime of the material itself, and thus leads to the reduction of current responsivity.We also plotted R in various wavelength as illustrated in Fig. 4(d), the R of the device is 47.5 μA/W, 6.6 μA/W, and 0.17 μA/W in 1064 nm, 10.6 μm, 1.36 mm, respectively. Despite the large difference in wavelengths, the device still exhibits good responsivity in the terahertz band, proving that the device has good performance in the high frequency band.

In addition, *NEP* and $D*$ are also important characterization parameters for detectors. Noise equivalent power (NEP) is used to define the optimum of the noise level and is defined by the following formula:

$$NEP = i_n/R \tag{2}$$

where i_n is the noise current density. The *NEP* of s-GO/$MAPbI_3$ is 518 nW/$Hz^{1/2}$. The detectiviety ($D*$) measures the sensitivity of the detector and is formulated as:

$$D^* = (A \times \Delta f)^{1/2}/NEP \tag{3}$$

where A is the device effective area and Δf is the device bandwidth, and the D* of s-GO/$MAPbI_3$ is 6.1 × 104 Jones.

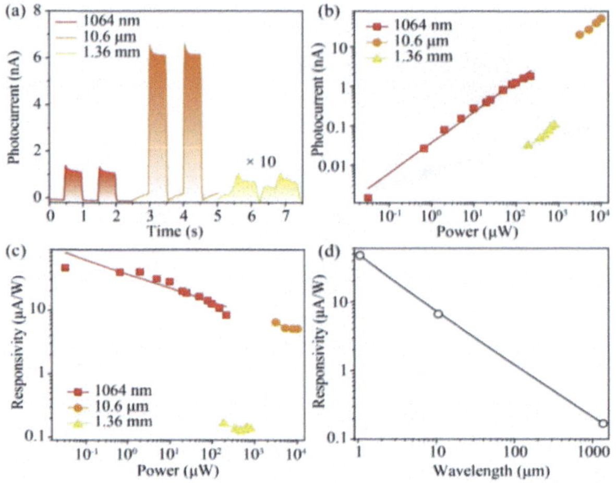

Fig. 4. (a) Infrared-terahertz modulated at 1064 nm, 10.6 μm, and 1.36 mm wavelengths by the s-GO/$MAPbI_3$ device the current response curve of the light; (b) photocurrent and (c) current responsivity curves with optical power; (d) The current responsivity of the s-GO/$MAPbI_3$ device as a function of optical wavelength.

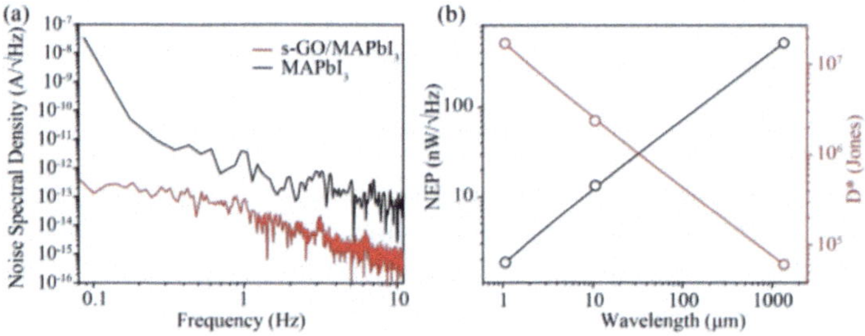

Fig. 5. In the terahertz band (a) Noise spectral density of s-GO/$MAPbI_3$ devices and $MAPbI_3$ devices; (b) s-GO/$MAPbI_3$ devices NEP and D* wavelength curves

As another important parameter of the detector, response time is calculated as the amount of time required to increase an electrical signal from 10% to 90% (or from 90% to 10%) time, which is the rising (or falling) time. To shed light the response time of our device, we collected the corresponding I-t curve of our device under 0.22 THz irradiation. The influence of s-GO on the response speed of the detector can be inferred from Fig. 5, and the rise and fall time of the s-GO/$MAPbI_3$ device is 17 and 38 ms, respectively. Meanwhile, under 0.22 THz irradiation, the response time of the MAPbI3 device is 31.2 ms and 89.3 ms, respectively. The results indicate that the higher carrier mobility and the faster carrier migration speed of s-GO/$MAPbI_3$ device compared to $MAPbI_3$ device, resulting in a shorter response time (Fig. 6).

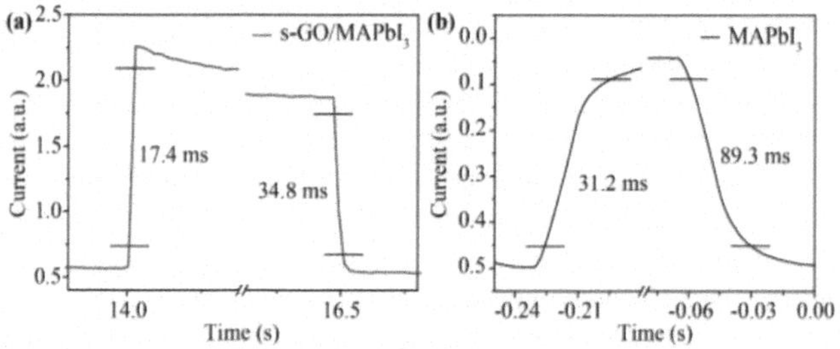

Fig. 6. In the terahertz band (a) Response of the s-GO/$MAPbI_3$ bolometer detector and (b) $MAPbI_3$ bolometer detector

3 Conclusion

We designed and prepared $MAPbI_3$ thin films modified by s-GO using the anti-solvent method to improve the carrier mobility and thus improve the device performance. This method results in a significant improvement in the overall performance of the s-GO/$MAPbI_3$ bolotherm detector. Especially in the terahertz band, the current responsivity of the s-GO/$MAPbI_3$ detector at 1.36 mm is 0.17 μA/W, NEP is 518 nW/Hz1/2, and D* is 6.1×10^4 Jones. The s-GO/$MAPbI_3$ device exhibit a rise time of 17 ms and a fall time of 38 ms. Our work clearly prove that the modification of the $MAPbI_3$ film by s-GO can effectively improve the performance of the detector.And it has a good response in the terahertz band.

Acknowledgement. This work was supported by the National Natural Science Foundation of China (U22A2008, 62175181), the National Key Research and Development Program of China (2021YFB2800703), the Opening Funding of National Key Laboratory of Electromagnetic Space Security, and Laoshan Laboratory Science and Technology Innovation Project (No.LSKJ202200801).

References

1. Zhong, S.: Progress in terahertz nondestructive testing **14**(3), 273–281 (2019)
2. Delrue, C., De Bruyne, S., Speeckaert, M.M.: The potential use of near- and midinfrared spectroscopy in kidney diseases **24**(7), 6740 (2023)
3. Liu, G., et al.: Myelin sheath as a dielectric waveguide for signal propagation in the mid-infrared to terahertz spectral range. Adv. Funct. Mater. **29**(7), 1807862 (2019)
4. Li, Y., et al.: Terahertz wave enhances permeaility of the voltage-gated calcium channel. J. Am. Chem. Soc. **143**, 9 (2021)
5. Zhang, J.X., et al.: Non-invasive, opsin-free mid-infrared modulation activates cortical neurons and accelerates associative learning. Nature Comm. **12**, 2730 (2021)
6. Kojima, A., Teshima, K., Shirai, Y., et al.: Organometal halide perovskites as visiblelight sensitizers for photovoltaic cells **131**(17), 6050–6051 (2019)
7. Yin, W.-J., Shi, T., Yan, Y.: Unique properties of halide perovskites as possible origins of the superior solar cell performance **26**(27), 4653–4658 (2014)

8. Zhao, Y., Zhu, K.: Organic–inorganic hybrid lead halide perovskites for optoelectronic and electronic applications **45**(3), 655–689 (2016)
9. Jena, A.K., Kulkarni, A., Miyasaka, T.: halide perovskite photovoltaics: background, status, and future prospects **119**(5), 3036–3103 (2019)
10. Hu, X., et al.: Cementitious grain-boundary passivation for flexible perovskite solar cells with superior environmental stability and mechanical robustness **66**(6), 527–535 (2021)
11. Chung, C.-C., Narra, S., Jokar, E., et al.: Inverted planar solar cells based on perovskite/graphene oxide hybrid composites **5**(27), 13957–13965 (2017)
12. Zong, Y., Zhou, Y., Zhang, Y., et al.: Continuous grain-boundary functionalization for high-efficiency perovskite solar cells with exceptional stability **4**(6), 1404–1415 (2018)
13. Gu L, Wang S, Chen Y, et al.: Stable high-performance perovskite solar cells via passivation of the grain boundary and interface **4**(7), 6883–6891 (2021)

A 195–245 GHz Direct Up-Conversion Mixer with Wideband Active Balun

Yuhan Ding, Kailei Wang, Qian Xie, and Zheng Wang(✉)

School of Integrated Circuit Science and Engineering, University of Electronic Science and Technology of China, Chengdu, China
wangzheng@uestc.edu.cn

Abstract. This paper presents a sub-Terahertz modified double-balanced Gilbert mixer in 130 nm SiGe BiCMOS process. An active transconductance balun is designed for supporting single ended large bandwidth baseband signal input. The active balun consists of a common-emitter (CE) and a common-base (CB) structure, the negative capacitor at input node of CB stage cancels the parasitic capacitor of CE stage and bias circuit. Therefore, the active balun achieves a large bandwidth from DC. A transformer-based balun is used for LO port matching and single to differential conversion. At RF port, The impedance transformation using transformer and marchand balun to achieve wide RF bandwidth. The proposed up-conversion mixer achieves 3-dB bandwidth form 195 GHz to 245 GHz. The peak conversion gain (CG) is −6 dB, with output referred 1 dB compression point (OP1dB) of −16 dBm. The Mixer is driven by a −4 dBm LO signal. The LO-to RF leakage rejection is 30 dB. The power comsumption is 16 mW.

Keywords: BiCMOS · Sub-terahetz · Mixer · Active Balun · Diretct Conversion

1 Introduction

The rapid increasing in demand of high data rate expands the operating frequency of transceiver chips into the sub-terahertz band, in which a large RF bandwidth of tens of gigahertz is available.

In recent years, several transmitters with operating frequency above 200 GHz have been proposed in the literature [1–6]. They employed direct-conversion architecture to avoid the challenge of IF circuit design caused by large bandwidth baseband signal. Meanwhile, the system structure is simpler and has more advantages in cost and power consumption. An important building block of transmitter is up-conversion mixer, with high demands on its bandwidth, linearity and power consumption, usually. In terahertz direct-conversion application, particularly, we are concerned with the LO power requirements of the mixer and LO-to-RF isolation.

In this paper, a modified micromixer [7] is explored with focus on large bandwidth to enable high data rate. It consists of double balanced mixing-quad and a constant input impedance active balun. Transformers are used for LO and RF matching to achieve wide RF bandwidth.

C. Chang et al. (Eds.): YTHZ 2024, SPPHY 400, pp. 37–41, 2024.
https://doi.org/10.1007/978-981-97-3786-4_7

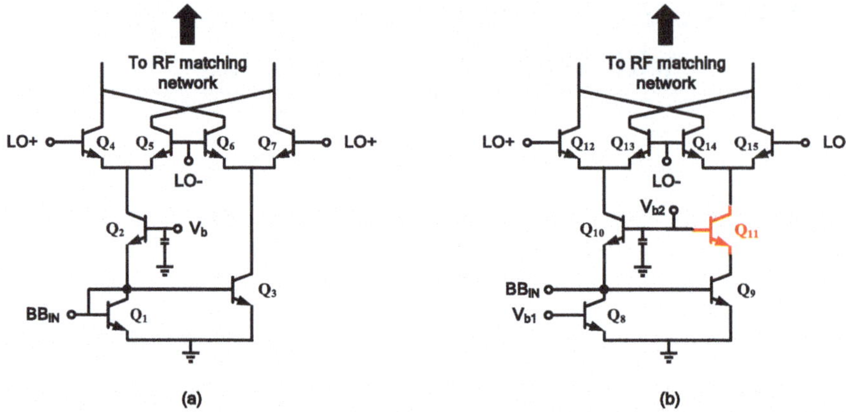

Fig. 1. (a) Schematic of Gilbert's micromixer [7], (b) modified micromixer in this paper

This paper is organized as follows. Section 2 describes the detailed consideration of circuits design, including matching network and active balun design. Section 3 presents the results of simulation. Section 4 concludes the paper.

2 Circuits Implementation

In the sub-terahertz direct-conversion application. We are concerned with LO-to-RF leakage rejection because LO is located at the spectrum of RF signal, and parasitic capacitance is very sensitive for such high frequency signal. Therefore, we use a double balanced mixing-quad ($Q_{4\text{–}7}$, as shown in Fig. 1 (a)) as mixing core. It has inherent good LO-to-RF isolation. To some extent, it cancels part of parasitic capacitance, benefits from its cross-coupled structure, so the double balance structure is suitable for wideband applications.

2.1 Active Balun Design

A double balanced up-conversion mixing-quad requires a differential baseband input. An extra demand created by tens of gigahertz baseband is that the interface need to be simpler. A single-ended form baseband input could avoid imbalance between signal path (coaxial cable, adapter and bonding wire). So a DC coupled single to differential conversion (SDC) active balun is employed as baseband transconductance stage. Moreover, passive baluns are not of interest since they are by nature ac-coupled [8].

SDC Active balun has several structures [9, 10]. One of them in combination with mixers is the micromixer, which was originally designed for better linearity [7]. Recently, one research about this combination of CB and CE structure used in wideband amplifier, offers new insight of its advantage [11].

As shown in Fig. 2 (a), due to the delay τ_1 of transconductance of CB stage (Q1), g_{m1} is a complex number rather than a real number, the imaginary part of gm_1 is equivalent to a negative shunt capacitor $C_n = -C_{gm1}$, derived as following expression [11]:

$$g_{m1} = |g_{m1}|e^{-j\omega\tau_1} = G_{gm1} + j\omega(-C_{gm1}) \quad (1)$$

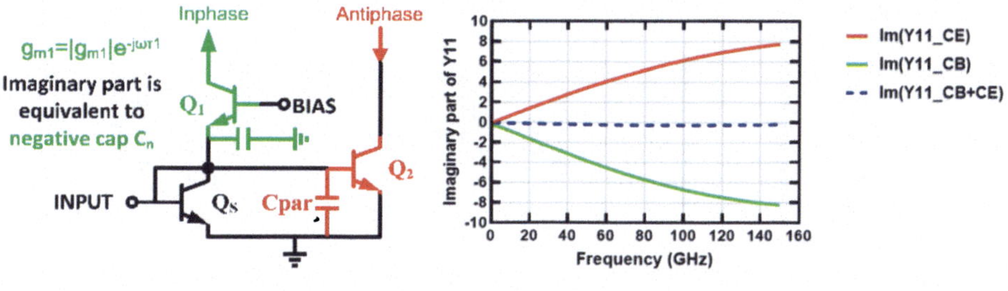

(a) (b)

Fig. 2. (a) Capacitance cancellation in CB+CE structure, (b) imaginary part of Y11 of CB, CE, CE+CB structure, results from EM and transistor simulation.

This C_n could cancel out the parasitic capacitance at input node, which means a constant input impedance. The shunt capacitance could be represented by imaginary part of Y11. The results of simulation proves the cancellation, as shown in Fig. 2(b), and routing metal has been taken into account in the electromagnetic (EM) simulation. The CE stage is natural antiphase transconductance, so we get a wideband SDC active balun. In additional, it is very compact because it is inductor-less[12].

The active balun employed in this paper is shown in Fig. 1 (b). The diode-connected Q_s is replaced by a current source Q_8, so that the input impedance increases form $1/2g_m$ to $1/g_m$. For a given impedance, we can use larger g_m to improve the CG of mixer. Q_{11} improves amplitude balance. Moreover, it consumes some voltage headroom, which avoid V_{CE9} exceeding breakdown voltage.

2.2 RF and LO Matching Network

The matching networks of RF and LO port are shown in Fig. 3(a) and Fig. 3 (b), respectively. The top two layers, TM2 and TM1, are thick metals with the thicknesses of 3 and 2 μm, respectively. They are suitable for transformer and Marchand balun design. The network of RF ports uses one more Marchand balun than that of RF port. At the windings on the side of circuits, we want to use shunt capacitance so that we can easily provide a bias voltage or V_{DD}. A series transmission line transforms the impedance to the value that we can matching with this topology.

3 Simulation Results

All the passive elements including connection metal wires have been fully EM simulated. The results of CG and return loss are shown in Fig. 4. (a). The −6 dB peaking conversion gain is obtained by sweeping the baseband frequency from 0 to 30 GHz with −4 dBm LO power while keeping the LO frequency 220 GHz. The simulated OP1dB is −6 dBm. The LO-to-RF rejection is 30 dB simulated by HB simulation. Total power is 16 mW under −4 dBm LO signal. The layout of our design is shown in Fig. 4 (b). Table 1 summarizes the performance and makes comparison to other research.

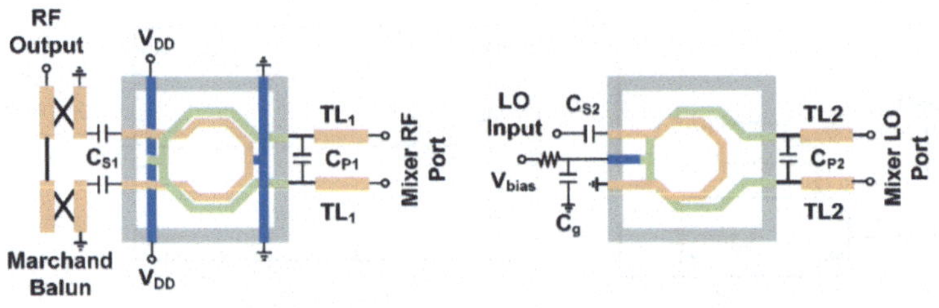

(a) (b)

Fig. 3. (a) RF matching network, (b) LO matching network.

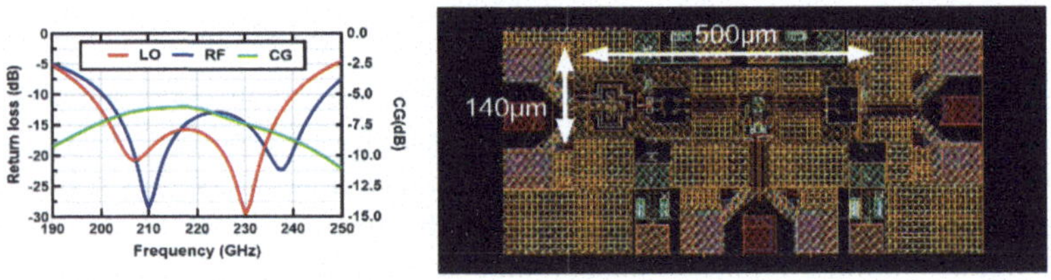

(a) (b)

Fig. 4. (a) Results of return loss and CG, (b) layout of the Mixer, core area: 500 μm * 140 μm

Table 1. Performance summary and comparison.

	This work	Ref [1][a]	Ref [13]	Ref [14]
Frequency (GHz)	195–245	215–250	270–314	193–297
process	130 nm SiGe BiCMOS	130 nm SiGe BiCMOS	130 nm SiGe BiCMOS	130 nm SiGe BiCMOS
conversion gain (dB)	−6	0	−8.4	−13.2
P1dB (dBm)	−16	−12	−3.6	−20.3
Isolation (dB)	30	23	–[b]	20.7
LO power (dBm)	−4	−5	8	0.6
Power (mW)	16	25	42.5	27.2

[a]Performance of this design is derived from TX and PA. [b]This design has IF = 30 GHz.

4 Conclusion

This paper presents a modified direct up-conversion double balanced Gilbert mixer operating from 195 GHz to 245 GHz, driven by −4 dBm LO signal.

It has a peak CG of −6 dB, peak OP1dB of −16 dBm. It offers 30 dB isolation between LO and RF. The realized mixer has relatively low power consumption of 16 mW.

Acknowledgment. This work was supported by the National Natural Science Foundation of China under Grant 62034002.

References

1. Eissa, M.H.: Wideband 240-GHz transmitter and receiver in BiCMOS technology with 25-Gbit/s data rate. IEEE J. Solid-State Circuits **53**(9), 2532–2542 (2018)
2. Holloway, J.W., et al.: 11.9 A 105Gb/s dielectric-waveguide link in 130nm BiCMOS using channelized 220-to-335GHz signal and integrated waveguide coupler. In: IEEE International Solid-State Circuits Conference (ISSCC), San Francisco, CA, USA, pp. 196–198 (2021)
3. Rodriguez, P.: A QPSK 110-Gb/s polarization-diversity MIMO wireless link with a 220–255 GHz tunable LO in a SiGe HBT technology. IEEE Trans. Microw. Theory Tech. **68**(9), 3834–3851 (2020)
4. Zhang, Z., et al.: Advanced terahertz refractive sensing and fingerprint recognition through metasurface-excited surface waves. Adv. Mater., 2308453 (2024)
5. Liu, G., et al.: Myelin sheath as a dielectric waveguide for signal propagation in the mid-infrared to terahertz spectral range. Adv. Funct. Mater. **29**(7), 1807862 (2019)
6. Zhang, J., et al.: Non-invasive, opsin-free mid-infrared modulation activates cortical neurons and accelerates associative learning. Nat. Commun. **12**, 2730 (2021)
7. Gilbert, B., et al.: A highly linear variant of the Gilbert mixer using a bisymmetric class-AB input stage. IEEE J. Solid-State Circuits **32**(9), 1412–1423 (1997)
8. Iseini, F.: DC-coupled ultra broadband differential to single-ended active balun in 130-nm SiGe BiCMOS technology. IEEE Microw. Wirel. Technol. Lett. **33**(3), 307 (2023)
9. Huang, B.-J.: A 2–40 GHz active balun using 0.13 μm CMOS process. IEEE Microw. Wirel. Compon. Lett. **19**(3), 164–166 (2009)
10. Franzese, A.: Ultra broadband low-power 70 GHz active balun in 130-nm SiGe BiCMOS. In: IEEE BiCMOS and Compound Semiconductor Integrated Circuits and Technology Symposium, CA, USA, pp. 1–4 (2020)
11. Tang, D.: An ultra-wideband amplifier with a novel non- distributed butterfly topology achieving 2-250 GHz bandwidth and 4.67 THz GBW in 130nm SiGe BiCMOS. In: 2023 IEEE Custom Integrated Circuits Conference (CICC), TX, USA, pp. 1–2 (2023)
12. Tsai, M.-D., et al.: A 0.3–25-GHz ultra-wideband mixer using commercial 0.18-μm CMOS technology. IEEE Microw. Wirel. Compon. Lett. **14**(11), 522 (2004)
13. Zheng, S.: A 300-GHz upconverter with 0-dBm psat and 44-GHz bandwidth in 130-nm SiGe BiCMOS. IEEE Microw. Wirel. Technol. Lett. **33**(12), 1646 (2023)
14. Steinweg, L., et al.: A 100-GHz-RF-bandwidth up-conversion mixer in 130 nm SiGe BiCMOS. In: 29th IEEE International Conference on Electronics, Circuits and Systems, Glasgow, United Kingdom, pp. 1–4 (2022)

Terahertz Narrowband Metasurface Filter Based on Electromagnetically Induced Transparency

Mingming Chen[1(✉)] and Xue-Xia Yang[1,2]

[1] School of Communication and Information Engineering, Shanghai University, Shanghai 200444, China
mmchen@shu.edu.cn

[2] Key Laboratory of Specialty Fiber Optics and Optical Access Networks, Shanghai University, Shanghai 200444, China

Abstract. A terahertz narrowband metasurface filter based on electromagnetically induced transparency (EIT) effect is proposed in this paper. The unit cell of the proposed terahertz metasurface filter consists of two split ring resonators (SRRs) and two cut-wires (CWs). A transparent window with narrowband at 1.22 THz can obtained and the maximum transmission amplitude is 0.92. The narrow passband with high transmission amplitude is created by the destructive interference between two SRRs and two CWs. Two SRR and two CWs can be viewed as two different bright modes of EIT effect. In addition, the magnetic field distributions and surface current distributions at 1.22 THz can also certify the high transmission amplitude. Moreover, a comparison of different terahertz metasurface filters is shown in this work. The comparison shows that our terahertz metasurface shows high transmission amplitude in the terahertz band. Therefore, our proposed terahertz metasurface filter exhibits excellent performance and shows great application prospect in 6G communication.

Keywords: Terahertz Filter · Electromagnetically Induced Transparency · Metasurface

1 Introduction

Electromagnetically induced transparency (EIT) is firstly found in a quantum system [1]. The destructive interference between two excitation pathways creates the EIT effect. EIT effect can show a narrow window in the opaque medium with the strong dispersion. In general, EIT effect has wide applications in slow light and nonlinear effects [2–5]. However, strict experimental conditions render the developments in atomic system. Metasurface can overcome the difficulties in the atomic system. EIT effect based on metasurface has been applied in many aspects, such as biosensors, modulators and filter.

Due to the high loss in the terahertz band, terahertz passband filter cannot obtain high transmission amplitude. EIT effect can excites a low-loss transparent window at a

C. Chang et al. (Eds.): YTHZ 2024, SPPHY 400, pp. 42–46, 2024.
https://doi.org/10.1007/978-981-97-3786-4_8

broad spectrum. Therefore, the terahertz filter based on EIT effect can reduce the loss and obtain the high transmission amplitude.

In this paper, we propose a terahertz metasurface filter based on EIT effect. The unit cell of the proposed terahertz metasurface filter consists of two split ring resonators (SRRs) and two cut-wires (CWs). A narrow passband can be obtained and the maximum transmission amplitude at 1.22 THz is 0.92. The high transmission amplitude is created by the destructive interference between two SRRs and two CWs. Therefore, our proposed terahertz metasurface filter exhibits excellent performance and shows great application prospect in 6G communication.

2 Structural Design

Figure 1 (a) illustrates the schematic illustration of the proposed terahertz metasurface filter under the illumination of *y*-polarized electromagnetic wave. The unit cell of the proposed terahertz metasurface filter consists of two SRRs and two CWs, as shown in Fig. 1 (b). The substrate of unit cell is made of silicon dioxide (SiO_2) with the thickness of 3 μm. The metal of two SRRs is gold, which is described by the Drude model [6]:

$$\varepsilon(\omega)= 1 - \omega_p/\left(\omega^2 + i\omega\omega_c\right) \tag{1}$$

where ω is the angle frequency of the incident wave. The plasma frequency and collision frequency of gold are $\omega_p = 2\pi \times 2.175 \times 10^{15}$ rad/s and $\omega_c = 2\pi \times 6.5 \times 10^{12}$ rad/s.

The unit cells of the proposed terahertz metasuface filter are arranged in a periodical array with the length of $P = 150$ μm. The length of two CWs is $l_1 = 130$ μm. Two lengths of two SRRs are $l_2 = 55$ μm and $l_3 = 60$ μm, respectively. The distance between two SRRs is $g = 10$ μm. The distance between two SRRs and two CWs is $s = 10$ μm. In addition, the thicknesses of SRRs and CWs are 0.2 μm.

A simulator (CST Microwave Studio) based on finite element method (FEM) carries out all simulations of the proposed terahertz metasurface filter. All simulations use adaptive fine mesh settings. The unit cell is simulated using a periodic boundary condition.

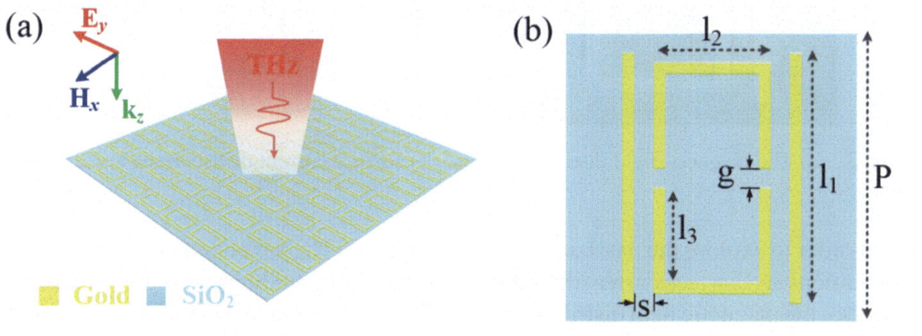

Fig. 1. (a) Schematic illustration of the proposed terahertz metasurface filter under the illumination of *y*-polarized electromagnetic wave. (b) The geometrical parameters of unit cell of the proposed terahertz metasurface filter.

3 Results and Discussion

Figure 2 shows the transmission spectrum of the proposed terahertz metasurface filter. A transmission window can be observed at 1.22 THz. A high transmission amplitude of 0.92 can be obtained.

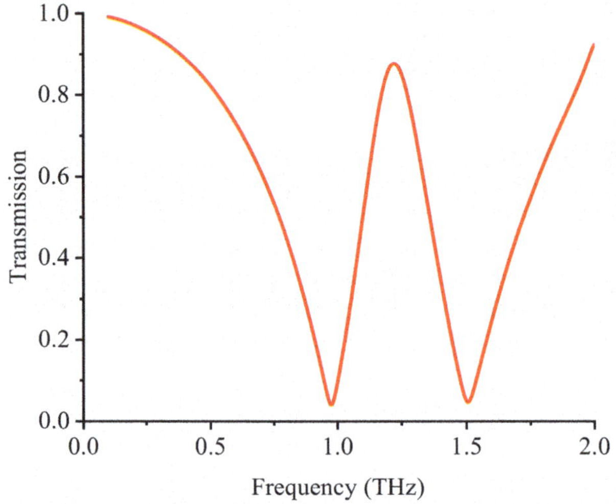

Fig. 2. The transmission spectrum of the proposed terahertz metasurface filter.

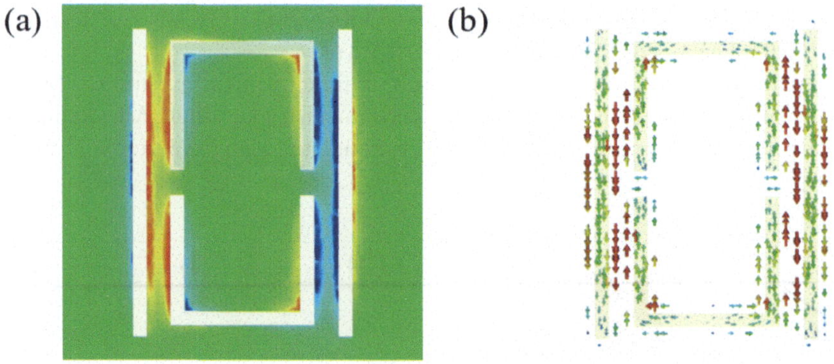

Fig. 3. (a) The magnetic field distributions and (b) surface current distributions at 1.22 THz.

In order to explore the passband generated at 1.22 THz, we obtain the magnetic field distributions and surface current distributions at 1.22 THz, as shown in Fig. 3. Figure 3 (a) describes the magnetic field distributions at 1.22 THz. We notice that the strong magnetic field can be found between two SRRs and two CWs. The weak magnetic field can be observed on the horizontal axis of two SRRs. In order to observe this phenomenon more clearly, we obtain the surface current distributions at 1.22 THz, as shown in Fig. 3 (b).

We notice that the opposite surface current distributions can be observed between two SRRs and two CWs. The similar phenomenon can be observed. Therefore, the narrow passband with high transmission amplitude is created by the near-field coupling between two SRRs and two CWs.

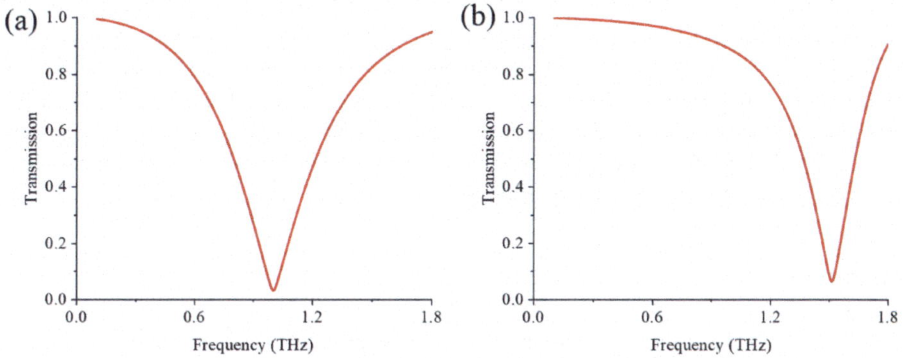

Fig. 4. (a) The transmission spectra of isolated CWs and (b) SRRs.

In order to explain the physical mechanism, we obtain the transmission spectra of isolated CWs and SRRs, as shown in Fig. 4. Figure 4 (a) shows the transmission spectrum of two CWs. A transmission dip at 1.00 THz can be observed. In addition, the transmission spectrum of isolated SRRs can be observed at 1.52 THz, as shown in Fig. 4 (b). Two CWs and two SRRs can be directly excited by the incident electromagnetic wave. Therefore, two SRR and two CWs can be viewed as two different bright modes [7]. Therefore, the destructive interference between two SRR and two CWs creates the passband of the proposed terahertz metasurface filter.

A comparison of different terahertz metasurface filter is shown in Table 1. We notice that the proposed terahertz metasurface filter exhibits high transmission amplitude at terahertz band.

Table 1. Comparison with other terahertz metasurface filter.

References	Frequency range	Transmission amplitudes
[8]	0.4–1.0 THz	0.70
[9]	2–5 THz	0.89
[10]	0–1.2 THz	0.85
Our work	0–2 THz	0.92

4 Conclusion

We propose a terahertz narrowband metasurface filter based on EIT effect in this paper. The unit cell of the proposed terahertz metasurface filter is composed of two SRRs and two CWs. A transparent window with narrowband at 1.22 THz can obtained and the maximum transmission amplitude is 0.92. The narrow passband with high transmission amplitude is created by the destructive interference between two SRRs and two CWs. The magnetic field distributions and surface current distributions at 1.22 THz can certify the high transmission amplitude. Moreover, a comparison of different terahertz metasurface filters shows that our terahertz metasurface shows high transmission amplitude in the terahertz band. Therefore, our proposed terahertz metasurface filter exhibits excellent performance and shows great application prospect in 6G communication.

References

1. Boller, K.J., Imamoğlu, A., Harris, S.E.: Observation of electromagnetically induced transparency. Phys. Rev. Lett. **66**(20), 2593 (1991)
2. Novikova, I., Walsworth, R.L., Xiao, Y.: Electromagnetically induced transparency-based slow and stored light in warm atoms. Laser Photonics Rev. **6**(3), 333–353 (2012)
3. Zhang, Y., Brown, A.W., Xiao, M.: Opening four-wave mixing and six-wave mixing channels via dual electromagnetically induced transparency windows. Phys. Rev. Lett. **99**(12), 123603 (2007)
4. Lou, J., et al.: Calibration-free, high-precision, and robust terahertz ultrafast metasurfaces for monitoring gastric cancers. PNAS **119**(43), e2209218119 (2022)
5. Jiao, Y., et al.: Photoactive terahertz metasurfaces for ultrafast switchable sensing of colorectal cells. Mater. Horiz. **9**, 2984–2992 (2022)
6. Zhu, L., Meng, F.Y., Fu, J.H., Wu, Q., Hua, J.: An approach to configure low-loss and full transmission metamaterial based on electromagnetically induced transparency. IEEE Trans. Magn. **48**(11), 4285–4288 (2012)
7. Niakan, N., Askari, M., Zakery, A.: High Q-factor and large group delay at microwave wavelengths via electromagnetically induced transparency in metamaterials. J. Opt. Soc. Am. B **29**(9), 2329–2333 (2012)
8. Zhang, Y., et al.: Dual-frequency switchable bandpass filter in the terahertz range based on enhanced trapped-mode resonances. Chem. Phys. Lett. **826**, 140637 (2023)
9. Zhong, M.: Realization of terahertz single-band metamaterial filter based on the bright-bright coupling. Opt. Mater. **105**, 109914 (2020)
10. Hu, F., et al.: Electrically triggered tunable terahertz band-pass filter based on VO 2 hybrid metamaterial. IEEE J. Sel. Top. Quantum Electron. **25**(3), 1–7 (2018)

A THz Detection Method for GFRP Delamination Based on THz Time-Domain Signal Model and Transfer Learning

Guanghui Lian, Yafei Xu, and Liuyang Zhang(✉)

Xi'an Jiaotong University, Xi'an 710049, Shaanxi, People's Republic of China
liuyangzhang@xjtu.edu.cn

Abstract. Composite materials, with their superior properties, have found extensive applications in aerospace and other fields. Compared to metallic materials, composite materials pose more stringent and specific requirements for non-destructive testing technologies. Data-driven terahertz (THz) non-destructive testing techniques have shown great potential in identifying the composite material damage. However, most previous data-driven methods rely on a sufficient amount of labeled data and assume that the training data (source domain) and test data (target domain) follow similar distributions. Due to variations in the testing environments, materials, and other conditions, THz signals can exhibit significant differences, leading to decreased performance of the models in the target domain. We propose a THz detection technology for composite materials using a THz time-domain signal model and deep domain adaptation, capable of imaging delamination defects without labeled data training. A THz echo model was developed to generate a simulation dataset, addressing dispersion in multi-layered composites. Transfer learning was applied to bridge the gap between THz signals and simulations. Our method shows superior generalization and accuracy in defect recognition compared to traditional deep learning models, as evidenced by experimental results.

Keywords: Terahertz NDT · Transfer Learning · GFRP Delamination

1 Introduction

Composite materials are crucial in engineering for their strength and resistance properties but are prone to inter-laminar delamination, leading to performance degradation. The Terahertz (THz) technique has achieved breakthroughs in the field of biological testing due to its excellent penetrability, accuracy, and safety [1–9], and these advantages are equally significant in the composite material delamination detection. THz waves can detect minor defects due to their unique properties, offering superior resolution and safety. However, THz signals are sensitive to noise, interlayer reflections, and dispersion, which can impede feature extraction and reduce inspection accuracy [10]. Advanced THz signal processing is crucial to ensure reliable and precise NDE for composite materials.

C. Chang et al. (Eds.): YTHZ 2024, SPPHY 400, pp. 47–51, 2024.
https://doi.org/10.1007/978-981-97-3786-4_9

THz NDT signal processing techniques are categorized into prior-based manual methods and data-driven methods. Prior-based methods develop algorithms using physical THz signal knowledge to extract damage features accurately. Despite their effectiveness in complex THz signal analysis, these methods are limited by complexity, specialization, and limited applicability, constraining their use in practical THz NDT.

Data-driven THz NDT methods focus on rapid, automatic damage assessment in materials, leveraging large datasets to identify discriminative features without relying on physical signal priors. However, most data-driven THz NDT methods rely on supervised learning, requiring carefully labeled training and testing datasets. While collecting a comprehensive dataset covering all scenarios could train a robust, generalized deep learning model, doing so is impractical due to the immense effort and limited accessibility in real-world THz NDT applications.

Here, we propose a THz detection technology for composite materials based on a THz time-domain signal model and transfer learning. This technology can locate delamination defects in composite materials of different materials or structure without training the classifier using labeled data. By studying the dispersion of THz signals in multi-layered composites, we built a THz echo model and used it to create a simulation dataset. The classifier, trained on the simulation dataset, is then applied to classify real-world measured data. Our approach outperforms traditional deep learning models in generalization, defect recognition accuracy, and imaging resolution across various THz datasets, as demonstrated by experiments.

2 Proposed Methods

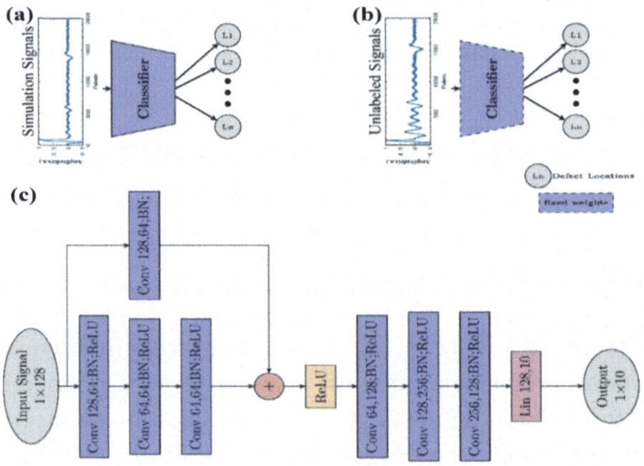

Fig. 1. The overall framework of the proposed THz detection technology. (a) Use the simulation dataset generated based on the terahertz echo signal model to train a feature classifier. (b) Directly detect the actual terahertz signals with the model pretrained in (a). (c) The detailed structure of the Classifier.

The overall framework of the proposed THz detection technology is depicted in Fig. 1. Initially, a simulation dataset generated based on a terahertz echo signal model is utilized to train a feature classifier. The role of this classifier is to provide the depth location of the delamination damage in the composite material corresponding to the input signal. Once the parameters of the feature classifier are fixed, it can directly detect the actual terahertz signals, indicating the layer in which the delamination damage is located. The feature classifier is a 6-layer convolutional neural network with residual connection. In the experiment, the output dimension of the network is 10, corresponding to 9 interfaces at different depths where delamination damage might occur, as well as specimens without delamination damage. The specific network architecture is shown in Fig. 1(c).

2.1 Terahertz Echo Signal Model

To establish a terahertz echo signal model and eliminate the dependence of traditional deep learning models on a large amount of labeled data, a detailed analysis of the terahertz detection signals for composite materials was conducted. The reflection pulses within the signals contain structural information of the composite material and can be decomposed into a linear combination of a series of Gaussian pulses [11]. Therefore, a pulse echo of the terahertz echo signal can be described in the following form:

$$g(t) = \beta_1 * \exp(-(t - t_1)^2/\alpha_1^2) + \beta_2 * \exp(-(t - t_2)^2/\alpha_2^2) \quad (1)$$

Where $g(t)$ is a THz pulse echo; β_1, β_2 represent the amplitude of the Gaussian pulses; t_1, t_2 denote the time delay of the Gaussian pulses; and α_1, α_2 are width factors that control the duration of the pulses. Each simulated signal contains n such terahertz echo pulses, where n represents the number of layers in the simulated composite material.

2.2 Deep Learning Model for Delamination Detection

We propose a hybrid THz detection model that combines the strengths of ResNet [13] and CNN for composite material analysis [14]. This approach utilizes residual learning blocks with shortcut connections to capture complex patterns without overfitting. We omit pooling layers to preserve feature location precision. Batch normalization is applied before activation to improve initialization sensitivity and learning efficiency.

As shown in Fig. 1, after training using simulation dataset, the classifier can be used for classification by fixing the model parameters without further training using the target data. The key advantage of this method is that it avoids the process of training with target data, thereby preventing the adverse effects of outliers in the target data on model parameters. Compared to traditional data-based transfer learning methods, this approach is more efficient and controllable, and allows for the use of a pre-trained classifier for classification tasks without accessing the target data.

3 Experiments

3.1 Experiment Settings

To validate the effectiveness of the method, five multi-layer composite material samples (A–E) were fabricated using 3D printing technology. As shown in Fig. 2, an air gap was introduced between the layers of the samples to simulate delamination damage in multi-layer composites. The samples employed different materials or thicknesses to simulate the requirements under various detection scenarios. The materials and layer thicknesses for each sample are listed in Table 1.

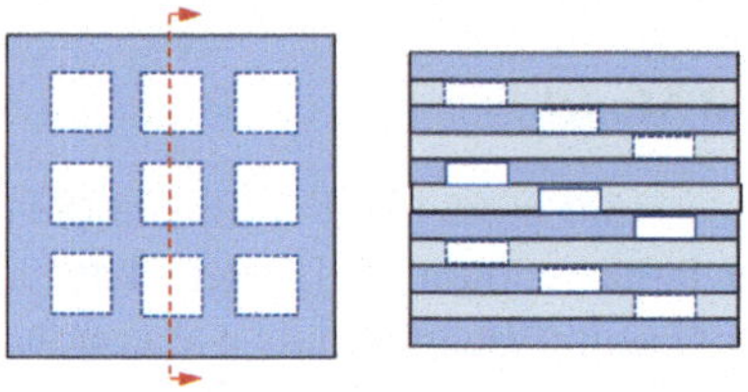

Fig. 2. Air gaps were introduced between the layers to simulate delamination

Table 1. Materials and layer thicknesses for each sample

Samples	A	B	C	D	E
Material	PLA	PLA+GF	ABS+GF	PLA+GF	ABS+GF
Layer Thickness (μm)	150	150	150	200	200

Table 2. Average classification accuracy of different methods on five target datasets.

Methods	Dataset A	Dataset B	Dataset C	Dataset D	Dataset E
CNN-DADA	83.61%	78.65%	70.4%	25.3%	22.84%
Proposed	96.80%	**89.54%**	85.85%	81.46%	96.64%
Proposed+CNN-DADA	**97.77%**	89.46%	**94.23%**	**95.58%**	**98.30%**

3.2 Experiment Results

The THz Time-Domain Spectrometer was used to test the five samples, and the test signals were classified using the transfer learning method (CNN-DADA [12]), the proposed method, and the method with the additional introduction of CNN-DADA. The accuracy obtained is shown in Table 2.

4 Conclusion

We propose a novel THz detection technology for composite materials that leverages a THz time-domain signal model and deep domain adaptation. This approach can accurately locate and image delamination defects in composite materials across various testing conditions, eliminating the need for labeled data to train the feature extractor. Specifically, we first studied the dispersion process of THz signals in multi-layered composite materials and established a THz echo signal model to generate a simulation dataset. Then, we preprocessed the collected THz signals to initially reduce the distribution discrepancy between the signals and the simulation dataset, and further reduced the distribution distance through deep adversarial domain adaptation. Experiments confirm that the proposed method, without using measured data to train the feature extractor, exhibits better generalization performance, higher defect recognition accuracy compared to traditional deep learning models on different THz datasets.

References

1. Wang, P., et al.: Progress on cutting-edge infrared-terahertz biophysics. IEEE Trans. Microw. Theory Tech. **70**(11), 5117–5140 (2022)
2. Lou, J., et al.: Calibration-free, high-precision, and robust terahertz ultrafast metasurfaces for monitoring gastric cancers. Proc. Natl. Acad. Sci. **119**(43), e2209218119 (2022)
3. Lang, S., et al.: A pixelated frequency-agile metasurface for broadband terahertz molecular fingerprint sensing. Nanoscale **14**(27), 9681–9685 (2022)
4. Zhang, Z., et al.: Advanced terahertz refractive sensing and fingerprint recognition through metasurface-excited surface waves. Adv. Mater., 2308453 (2024)
5. Liu, G., et al.: Myelin sheath as a dielectric waveguide for signal propagation in the mid-infrared to terahertz spectral range. Adv. Funct. Mater. **29**(7), 1807862 (2019)
6. Li, Y., et al.: Terahertz wave enhances permeability of the voltage-gated calcium channel. J. Am. Chem. Soc. **143**(11), 4311–4318 (2021)
7. Zhang, J., et al.: Non-invasive, opsin-free mid-infrared modulation activates cortical neurons and accelerates associative learning. Nat. Commun. **12**, 2730 (2021)
8. Xiang, Z., et al.: A primary model of THz and far-infrared signal generation and conduction in neuron systems based on the hypothesis of the ordered phase of water molecules on the neuron surface. Sci. Bull. **65**, 308 (2020)
9. Wu, K., et al.: Terahertz wave accelerates DNA unwinding: a molecular dynamics simulation study. J. Phys. Chem. Lett. **11**(17), 7002–7008 (2020)
10. Yafei, X.: Three-dimensional nondestructive characterization of delamination in GFRP by terahertz time-of-flight tomography with sparse Bayesian learning-based spectrum-graph integration. Compos. Part B Eng. **225** (2021)
11. Yafei X.: Double Gaussian mixture model-based terahertz wave dispersion compensation method using convex optimization technique. Mech. Syst. Signal Process. **164** (2022)
12. Yafei, X.: Terahertz transfer characterization for composite delamination under variable conditions based on deep adversarial domain adaptation. Comput. Sci. Technol. **232** (2023)
13. Kaiming, H.: Deep residual learning for image recognition. arXiv:1512.03385 (2015)
14. Xiaowu, Z.: Integration of residual network and convolutional neural network along with various activation functions and global pooling for time series classification. Neurocomputing **367** (2019)

All-Metal Nested Square Rings THz Metamaterial for Sensitive Biosensor in the Detection of Glucose

Chunyue Zhao[1], Wei Wang[1,2(✉)], Shengqin Yang[1], and Mei Yang[1]

[1] School of Physics and Electrical Information, Yunnan Normal University, Kunming, China
wangwvin@163.com

[2] Yunnan Key Laboratory of Opto-Electronic Information Technology, Kunming, China

Abstract. This article presents an all-metal THz metamaterial biosensor, which adopts two nested square rings interconnected by metal bars as a unit cell. The metamaterial is fabricated on a copper foil using laser micro-machining. The copper foil is periodically patterned with a continuous array of nested rings. The simulation results show that the refractive index sensitivity of the metamaterial sensor is 210 GHz/RIU. To evaluate its sensing capabilities, glucose solution is selected as the detection substance. The experimental results demonstrate that the sensor's minimum detection concentration is 0.05 mol/L. The metamaterial biosensor exhibits simple structure, high repeatability, and easy fabrication, rendering it a promising candidate for biological detection and disease diagnosis applications.

Keywords: All-metal metamaterial · nested square rings · terahertz biosensor · glucose solution

1 Introduction

Terahertz waves, which bridge the transition from electronics to photonics in the electromagnetic spectrum between microwaves and infrared, have low energy compared to optical waves and do not cause radiation ionization damage to biomolecules. Furthermore, many biomolecules exhibit unique fingerprint characteristics in the terahertz range [1, 2], which makes terahertz technology widely applicable in areas such as chemistry [3], medicine [4, 5], food [6], and biosciences [7–12]. Currently, terahertz metamaterial sensors have achieved high sensitivity performance. However, traditional fabrication methods for terahertz metamaterials, such as photolithography [13] and electron beam exposure [14], suffer from complex processes, long fabrication times, and high costs, which restrict their application in nondestructive testing. Therefore, the development of a rapid, low cost, convenient, and high precision terahertz metamaterials is of significant importance. In addition, laser etching technology, as a new metamaterial processing method [15], has become an effective solution due to its good process flexibility, convenience, and low cost.

C. Chang et al. (Eds.): YTHZ 2024, SPPHY 400, pp. 52–56, 2024.
https://doi.org/10.1007/978-981-97-3786-4_10

In this paper, we report a terahertz metamaterial biosensor based on all-metallic medium fabricated by laser etching, which can detect glucose concentration by analyzing terahertz spectra. The refractive index sensitivity of the sensor is as high as 210 GHz/RIU through three-dimensional electromagnetic simulation software. The experimental results demonstrate that the concentration and frequency of diverse glucose solutions demonstrate systematic variations on the sensor, which aligns with the simulation findings. The experiment enables the detection of trace glucose by analyzing the frequency shift of the transmission peak in the terahertz spectrum caused by different concentrations of glucose solution, and the lowest detection concentration is 0.05 mol/L.

2 Structure Design

Figure 1 (a) depicts the 3D array diagram of the designed metamaterial sensor, where terahertz waves are vertically incident on its surface to generate resonance. The unit structure diagram and parameter settings are shown in Fig. 1 (b). The lattice constant is set to Px = 300 μm and Py = 320 μm. The structure consists of nested rings and metal rods connecting the nested rings. The inner width L1 of the outer ring is 240 μm. The outer width L2 and inner width L3 of the inner ring are 160 μm and 70 μm, respectively. The width w of the metal rod connecting the nested ring is 30 μm. The thickness h of the sensor is 80 μm.

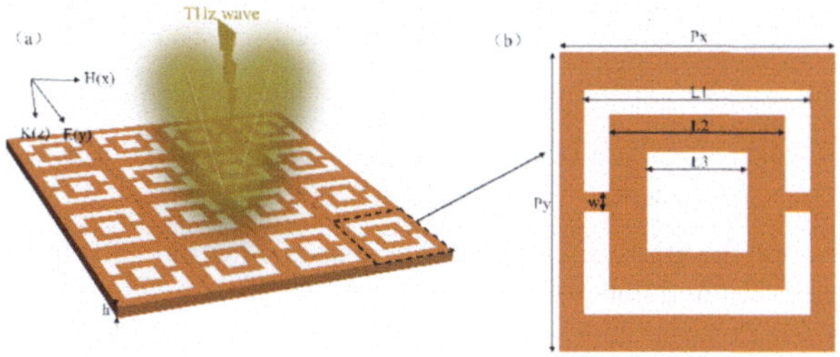

Fig. 1. Metamaterial sensor. (a) 3D array diagram. (b) Cell structure diagram.

The resonance performance is simulated by three-dimensional electromagnetic simulation software. The metal material is set to copper. The terahertz wave is vertically incident to the surface of the structure along the z axis. Periodic boundary conditions are set in the x and y directions, and open boundary conditions are set in the z direction. Figure 2 presents the transmission and reflection spectra of the sensor, from which it can be observed that the sensor exhibits a peak transmission at 0.456 THz with a transmission rate as high as 99%.

To evaluate the sensing performance of the sensor, a 15 μm-thick specimen was added to the sensor, and the different transmission spectra of the metamaterial biosensor were simulated as depicted in Fig. 3(a). With an increase in refractive index of the specimen, there was a redshift observed in the resonance peak frequency towards longer wavelengths. The sensitivity of the sensor was calculated according to $S = \Delta f/\Delta n$, where Δf represents the magnitude of the frequency shift and Δn denotes the change in

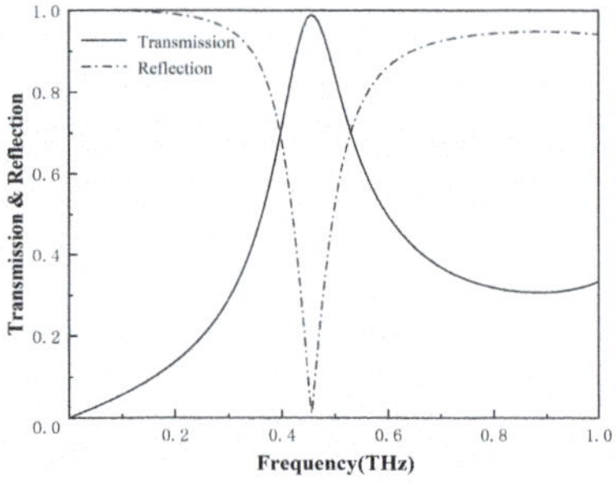

Fig. 2. Simulated transmission spectrum T and reflection spectrum R of the sensor.

refractive index of the specimen. The sensitivity unit is expressed as GHz/RIU (Refractive Index Unit) [15]. According to the resonance peak frequency shift corresponding to the refractive index of different test substances, the fitting result in Fig. 3(b) shows that the refractive index sensitivity of this sensor is 210 GHz/RIU.

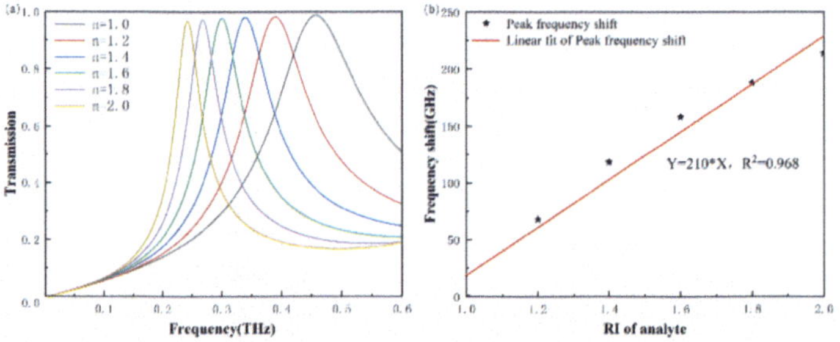

Fig. 3. (a) The curve of the transmission coefficient as a function of refractive index for the analyte. (b) Frequency shift and linear fitting.

3 Experimental Results and Discussion

The sensor was fabricated on an 80 μm thick copper foil using laser etching technology, based on the design illustrated in Fig. 1(a). The sensor sample has overall dimensions of 1.5 cm × 1.5 cm, as depicted in Fig. 4(a). Figure 4(b) shows the photomicrograph of the sample taken under an optical microscope. Compared with the simulated structure, the process errors are relatively small and negligible. To minimize experimental errors during terahertz time-domain spectroscopy analysis, the indoor air moisture level of the sample was reduced to below 3% using an air compressor, thereby eliminating any potential interference from moisture on the terahertz signal.

The transmission spectra of the sensor is depicted in Fig. 5 (a), presenting both simulation and experimental results. It can be observed from the figure that there is agreement between the theoretical simulation and experimental findings. The slight deviation in the transmission curve is due to the limited time delay window in the THz-TDS experiment

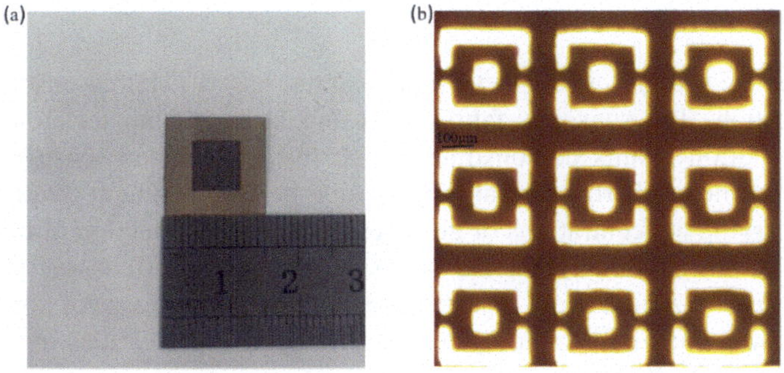

Fig. 4. Sensor sample. (a) Sample photo. (b) Optical microscope photograph.

and possible system parameter errors between simulation and experimental environments. Furthermore, apart from inherent experimental errors, scattering losses on the rough surface of the sensor may also contribute to this slight deviation [16].

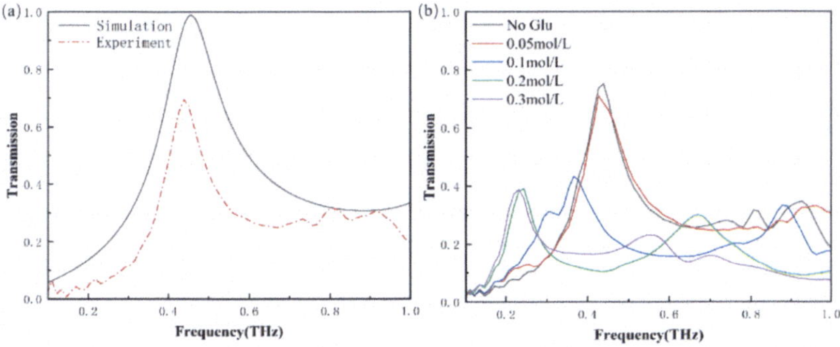

Fig. 5. (a) Simulated and experimental transmittance spectra of the sensor. (b) Spectra measurements for different concentrations of glucose solutions.

To demonstrate the biopotential of the proposed terahertz metamaterial sensor, glucose (Glu) solutions were used for detection in the experiments. Glucose was purchased from Aladdin. Glucose solutions were prepared with concentrations of 0.05 mol/L, 0.1 mol/L, 0.2 mol/L, and 0.3 mol/L respectively. Each time, a volume of 80 μL of glucose solution was dispensed onto the sensor surface using a pipette and then dried for 30 min at a constant temperature of 40 °C in a thermostatic chamber.

The tests were conducted using the THzTDS, and the corresponding test results are shown in Fig. 5(b). With the addition of concentrations of glucose solution, the resonant frequency of the sensor exhibited varying degrees of deviation. A higher concentration of glucose led to an increase in refractive index, and consequently resulted in a greater redshift in the resonant frequency. These experimental findings align with the simulation results.

4 Conclusion

In conclusion, we propose an all-metal metamaterial-based THz biosensor for glucose solution detection. The sensor is fabricated using laser etching technology on copper foil. Simulation results demonstrate a high refractive index sensitivity of up to 210 GHz/RIU. The terahertz spectral transmission peak experiment frequency shift in response to varying concentrations of glucose, enabling the detection of trace amounts down to 0.05 mol/L in our experimental setup. This biosensor offers several advantages including compact size, excellent repeatability, high sensitivity, ease of fabrication, and cost-effectiveness, making it highly promising for future applications in bio-molecule detection and disease diagnosis.

Acknowledgements. This work was supported by the project fund for Xingdian Talents Support Plan and the Graduate Research and Innovation Fund of Yunnan Normal University.

References

1. Xiong, W., Shen, J.: Fingerprint extraction from interference destruction terahertz spectrum. Opt. Express **18**(21), 21798–21803 (2010)
2. Pickwell, E., Wallace, V.P.: Biomedical applications of terahertz technology. J. Phys. D Appl. Phys. **39**(17), R301 (2006)
3. Shuai-nan, et al.: Study on terahertz spectroscopy detection of low concentration imidaclopridon metamaterials. Spectrosc. Spectr. Anal. **41**(4), 1044–1049 (2021)
4. Xie, L., et al.: Extraordinary sensitivity enhancement by metasurfaces in terahertz detection of antibiotics. Sci. Rep. **5**(1), 8671 (2015)
5. Zhang, C., et al.: Terahertz toroidal metasurface biosensor for sensitive distinction of lung cancer cells. Nanophotonics **11**(1), 101–109 (2021)
6. Redo-Sanchez, A., et al.: Assessment of terahertz spectroscopy to detect antibiotic residues in food and feed matrices. Analyst **136**(8), 1733–1738 (2011)
7. Wang, G., Zhu, F., Lang, T., et al.: All-metal terahertz metamaterial biosensor for protein detection. Nanoscale Res. Lett. **16**, 1–10 (2021)
8. Zhang, J., et al.: Non-invasive, opsin-free mid-infrared modulation activates cortical neurons and accelerates associative learning. Nat. Commun. **12**, 2730 (2021)
9. Liu, G., et al.: Myelin sheath as a dielectric waveguide for signal propagation in the mid-infrared to terahertz spectral range. Adv. Funct. Mater. **29**(7), 1807862 (2019)
10. Xiang, Z., et al.: A primary model of THz and far-infrared signal generation and conduction in neuron systems based on the hypothesis of the ordered phase of water molecules on the neuron surface. Sci. Bull. **65**, 308 (2020)
11. Wu, K., et al.: Terahertz wave accelerates DNA unwinding: a molecular dynamics simulation study. J. Phys. Chem. Lett. **11**(17), 7002–7008 (2020)
12. Li, Y., et al.: Terahertz wave enhances permeability of the voltage-gated calcium channel. J. Am. Chem. Soc. **143**(11), 4311–4318 (2021)
13. Won, S., et al.: Graphene-based crack lithography for high-throughput fabrication of terahertz metamaterials. Carbon **158**, 505–512 (2020)
14. Liu, M., et al.: Deeply subwavelength metasurface resonators for terahertz wavefront manipulation. Adv. Opt. Mater. **7**(21), 1900736 (2019)
15. Serita, K., et al.: A terahertz-microfluidic chip with a few arrays of asymmetric meta-atoms for the ultra-trace sensing of solutions. Photonics **6**(1), 12 (2019). MDPI
16. Lin, Y., et al.: Free-standing double-layer terahertz band-pass filters fabricated by femtosecond laser micro-machining. Opt. Express **25**(21), 25125–25134 (2017)

High Circular Dichroism Terahertz Chiral Metasurfaces Enabled by Quasi-Bound States in the Continuum

Weizhe Yuan, Zhonglei Shen, Chenglin Yang, and Liuyang Zhang(✉)

School of Mechanical Engineering, Xi'an Jiaotong University, Xi'an 710049, Shanxi, China
liuyangzhang@xjtu.edu.cn

Abstract. The concept of Bound State in the Continuum (BIC) describes a phenomenon where a state remains confined within the radiation continuum of a system. In the context of the terahertz frequency range, BIC offers a promising approach for creating devices with exceptionally high quality factors (Q-factors) and enabling strong interactions between terahertz radiation and matter, contributing significantly to the advancement of terahertz science and technology. In this work, we propose a chiral metasurface design for terahertz frequencies that achieves significant circular dichroism (CD) of 0.944. This is accomplished by disrupting the symmetry within the plane, which leads to the excitation of quasi-bound states in the continuum (quasi-BICs), resulting in a high Q-factor.

Keywords: Chiral metasurface · Circular dichroism · Bound-states In the Continuum

1 Introduction

The chiral metasurface, is a state that could absorb one spin state of light and at the same time could reflect the other spin state of light. Furthermore, it also has a high Q-factor. Recently attracted much attention and research [6–8, 11]. The bounded continuum (BIC) is a localized electromagnetic eigenstate, and it exists inside the continuum spectrum of radiation. It is interested by public because of its strong interactions between light and matter and high Q factors [1–3, 9, 10]. Theoretically, these states could have an infinite Q-factor, The ideal BIC has perfect energy confinement but is not detectable in the spectrum due to its zero resonance linewidth. Therefore, a common way to observe these states is to convert the ideal BIC into a Quasi-BIC, which is usually done by breaking the in-plane symmetry [4, 5]. Another concept is chiral BICs, which are characterised by the ability to uniquely decouple from circularly polarised light in one spin state, while in the opposite spin direction they exhibit extremely high Q factors and a large number of interactions with circularly polarised light, leading to significant circular dichroism (CD).

In this work, we present a design scheme for a novel chiral metasurface with high Q factor and high CD.

C. Chang et al. (Eds.): YTHZ 2024, SPPHY 400, pp. 57–60, 2024.
https://doi.org/10.1007/978-981-97-3786-4_11

2 Structure Design

The structure of the unit cell is shown in Fig. 1 and Fig. 2. First, a structure of perfect symmetry is constructed, followed by the introduction of a horizontal displacement, denoted as Δx, to disrupt the symmetry, thereby transforming the structure into a quasi-bound state in the continuum (quasi-BIC) configuration. We decide to use only silicon to build this structure.

Fig. 1. Schematic diagram of the asymmetric all-dielectric metasurface.

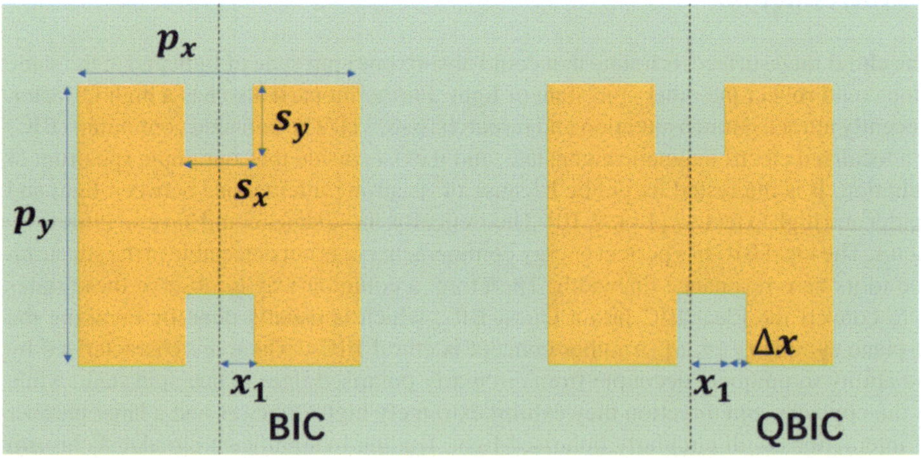

Fig. 2. Top view of the asymmetric metasurface, where $p_x = p_y = 220$ μm, $s_x = 25$ μm, $s_y = 30$ μm.

In order to investigate the chiroptical responses that induced by QBIC in the chiral metasurface, we use CST Microwave Studio, a finite element software to calculate its reflection spectra at normal incidence.

3 Results

To further investigate the performance of this design, we did numerical simulations to see its transmission spectra. In the simulation setup, we set the x and y direction both periodic boundary conditions to simulate a periodic array. The z direction is set open boundary conditions to prevent multiple reflections. The figure shows simulated incident polarized light through the metasurface with perturbation factor $\Delta x = 16\ \mu m$ are shown in the figure. In the figure t_{RL} represents the incident light is right-handed polarized and the transmitted light is left-handed polarized, in other words RCP incident and LCP transmitted. We performed several simulations with different perturbation factors Δx. We found that the peak frequency continuously increases with increasing perturbation factor Δx, and as for the peak value of CD, it keeps decreasing with increasing perturbation factor when it is greater than 16 μm and decreases with decreasing perturbation factor when it is less than 16μm. Thus, it can be concluded that the maximum value of CD appears when the perturbation factor Δx is equal to 16 μm (Fig. 3).

$$CD = \frac{|(t_{RR} + t_{LR}) - (t_{RL} + t_{LL})|}{(t_{RR} + t_{LR}) + (t_{RL} + t_{LL})} \tag{1}$$

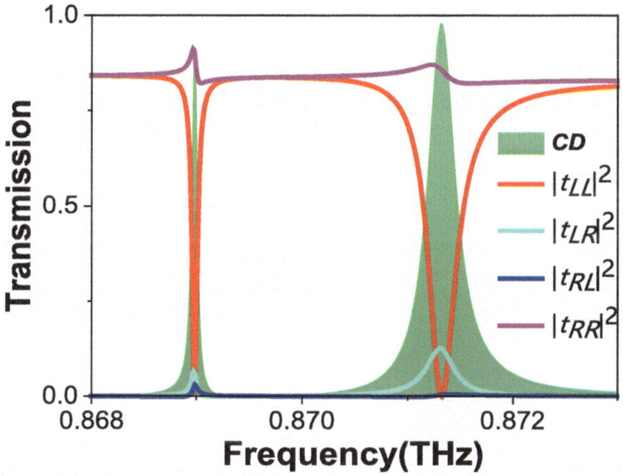

Fig. 3. Transmission spectra of the chiral metasurfaces at = Δx 16 μm.

As for the Q-factor, it is calculated by:

$$Q = \frac{\omega_0}{2\gamma} \tag{2}$$

When the $\Delta x = 16\mu m$, the theoretical Q-factor is calculated as 2625.

4 Summary

In this paper, we introduced a novel chiral metasurface design that disrupts the symmetry of the structure, creating a pathway for radiation in the metasurface device. This design excites the quasi-Bound States in the Continuum (quasi-BIC) resonance mode, characterized by a high Q-factor. The introduction of asymmetry within the plane of the structure enables the achievement of intrinsic planar chirality when exposed to normal incidence, reaching a peak CD value of 0.944 at terahertz frequencies, along with a remarkable Q-factor of 2625.

References

1. Alizadeh, M., Reinhard, B.R.M.: Plasmonically enhanced chiral optical fields and forces in achiral split ring resonators. ACS Photonics **2**, 361–368 (2015)
2. Tang, Y., Cohen, A.E.: Optical chirality and its interaction with matter. Phys. Rev. Lett. **104**, 163901 (2010)
3. Yin, X., Schaferling, M., Metzger, B., Giessen, H.: Interpreting chiral nanophotonic spectra: the plasmonic Born-Kuhn model. Nano Lett. **13**, 6238–6243 (2013)
4. Fan, K., Shadrivov, I.V., Padilla, W.J.: Dynamic bound states in the continuum. Optica **6**, 169–173 (2019)
5. Shen, Z., Fang, X., Li, S., Yin, W., Zhang, L., Chen, X.: Terahertz spin-selective perfect absorption enabled by quasi-bound states in the continuum. Opt. Lett. **47**, 505–508 (2022)
6. Shen, Z., et al.: Chiral metasurfaces with maximum circular dichroism enabled by out-of-plane plasmonic system. Laser Photonics Rev. (2022)
7. Han, D., et al.: 2D-to-3D buckling transformability enabled reconfigurable metamaterials for tunable chirality and focusing effect. Photonics Res. (2023)
8. Han, D., et al.: Kirigami-inspired planar deformable metamaterials for multiple dynamic electromagnetic manipulations. Laser Photonics Rev. (2023)
9. Wang, R., et al.: Ultrasensitive terahertz biodetection enabled by quasi-BIC-based metasensors. Small **19**, 2301165 (2023)
10. Wang, R., et al.: Electric Fano resonance-based terahertz metasensors. Nanoscale **13**(44), 18467–18472 (2021)
11. Sun, L., et al.: A pixelated frequency-agile metasurface for broadband terahertz molecular fingerprint sensing. Nanoscale **14**(27), 9681–9685 (2022)

Injection Locking Study of Terahertz Quantum Cascade Laser Frequency Combs

Yu Ma[1,2], Rusong Li[1], Shichen Zhang[1], Binru Zhou[1], Quanyong Lu[1](✉), Jinchuan Zhang[2], Shenqiang Zhai[2], Junqi Liu[2], Shuman Liu[2], Ning Zhuo[2], and Fengqi Liu[2]

[1] Division of Quantum Materials and Devices, Beijing Academy of Quantum Information Sciences, Beijing 100193, China
luqy@baqis.ac.cn

[2] Key Laboratory of Semiconductor Materials Science, Beijing Key Laboratory of Low Dimensional Semiconductor Materials and Devices, Institute of Semiconductors, Chinese Academy of Sciences, P.O. Box 912, Beijing 100083, China

Abstract. We prepare the side absorbers with Ni on both sides of the double-metal waveguide device. Meanwhile, ground backed coplanar waveguides are used to improve the high-speed characteristic of the device for the RF injection locking. Locking bandwidth up to 50 MHz at RF injection power of 0 dBm. RF modulation not only locks the phases but also modifies the laser spectral bandwidth by up to 660 GHz.

Keywords: Terahertz · Quantum Cascade Laser · Optical Frequency Comb

1 Introduction

Optical frequency combs (OFCs) have revolutionary applications in trace gas-sensing, remote-sensing, laser spectroscopy and other related fields, due to their high stability and low phase noise [1, 2]. Especially in the mid-infrared (MIR) and terahertz (THz) regions, due to the fundamental roto-vibrational bands of a variety of molecules lie with absorption strength higher than those in the visible or near-infrared, intense development of OFCs is mainly driven by their applications in high-resolution spectroscopy [3–7]. Quantum cascade laser (QCL) [8–10] is a semiconductor laser based on interband transitions. Because of its strong nonlinear effect——four wave mixing, QCLs can be longitudinal-mode-locked. Due to the advantages of wavelength adjustment, high power, electrically pumped and on-chip integration, QCLs have become ideal light sources of optical frequency comb in the middle and terahertz spectral regions. But QCLs frequency comb due to internal dispersion, can not fully show the advantages of broadband, high precision, high stability. RF injection can broaden the laser spectrum by exciting more longitudinal modes beyond the threshold, and the repetition frequency can be locked by an external RF signal to maintain full intermode coherence [11]. This is of great significance for terahertz laser spectroscopy.

C. Chang et al. (Eds.): YTHZ 2024, SPPHY 400, pp. 61–64, 2024.
https://doi.org/10.1007/978-981-97-3786-4_12

2 Result and Discussion

As shown in Fig. 1(a), a device with a 2-mm-long cavity and 90-μm-wide ridge was fabricated by double-sided metal waveguide with the side absorbers. The device was bonded to a wideband coplanar waveguide transition on AlN substrate, which was used as an interposer connection between mini SMP cables and our laser. In contrast to conventional structures, we added hollow plated vias along the ground traces passing through the substrate, which connected the ground traces on the top to the ground plane on the bottom. The threshold current of the device is 0.5 A and the maximum output power is 12.5 mW, as shown in Fig. 1(b). I: 0.53–0.61 A, II: 0.75–0.91 A, III: 1.07–1.16 A represent three different frequency comb areas with a single beatnote signal in Fig. 1(c).

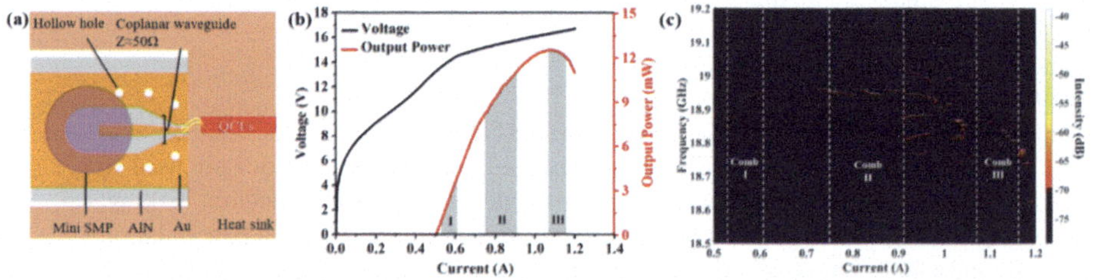

Fig. 1. (a) THz QCLs high-speed package schematic. (b) The P-I-V characteristics of the 90-μm-wide and 2-mm-long device at 10K in the CW mode. I, II, III represent the area of the optical frequency comb. (c) Free running beatnote mapping as a function of drive current measured at 10K in CW mode.

When the frequency of the RF signal gradually approached the frequency of the beat signal, the beat signal was pulled towards the RF signal. A further increase in frequency caused the beat signal to resonate with the RF signal, which locked both signals. Finally, when the injection power was 0 dBm, the lock bandwidth was 50 MHz. After that,

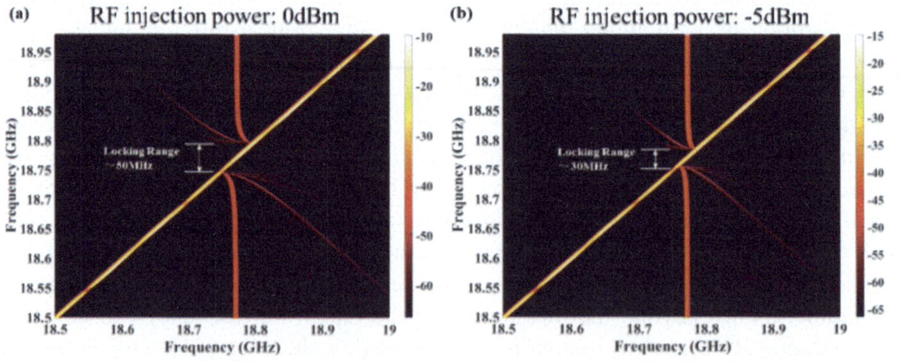

Fig. 2. Evolution of the beatnote of the THz QCLs as a function of the injected RF frequency. (a) The injection power is 0 dBm. (b) The injection power is −5 dBm.

we reduced the power of the RF signal (−5 dBm) and repeated the experiment again, finding that the lock bandwidth was reduced to 30 MHz, as shown in Fig. 2. Meanwhile, the phenomenon of injection locking also appeared when RF injection intensity and RF injection frequency were changed in Fig. 3.

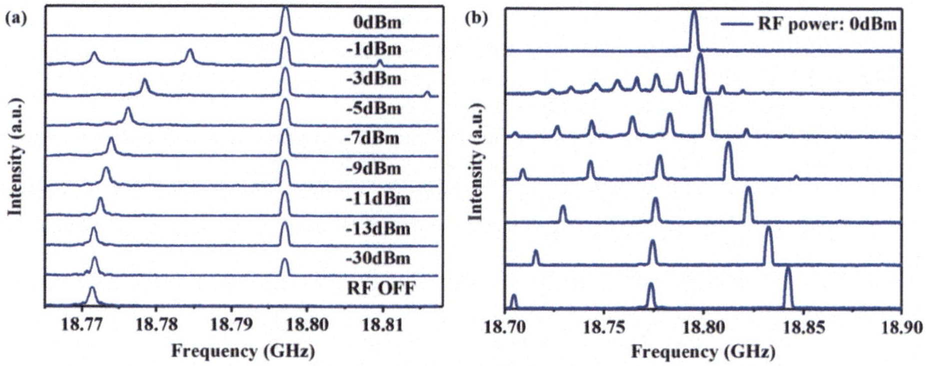

Fig. 3. Pulling and locking of the beat signal.

Figure 4 showed that together with an injection power of 0 dBm, the operating state fundamentally changed. The missing modes were excited under RF injection, which increased the range of comb operation by more than 660 GHz.

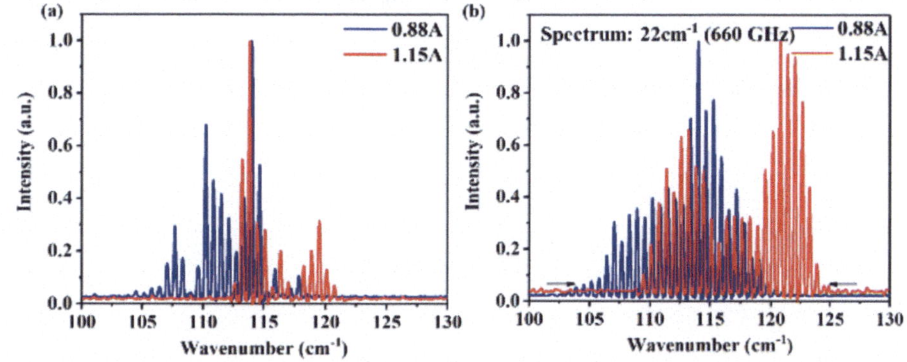

Fig. 4. (a) The CW emission spectra of the device without RF injection. (b) The CW emission spectra of the device with RF injection intensity of 0 dBm.

3 Conclusion

We prepare the side absorbers with Ni on both sides of the double-metal waveguide device. In order to improve the high-speed characteristics of the device, a coplanar waveguide transition way is used to achieve impedance matching for RF signal injection.

The effects of the RF injection not only lock the phases within a locking range of 50 MHz but also modified the laser spectral bandwidth up to 660 GHz.

References

1. Hillbrand, J., Andrews, A.M., Detz, H., et al.: Coherent injection locking of quantum cascade laser frequency combs. Nat. Photonics **13**(2), 101–104 (2019)
2. Ma, Y., Li, W., Li, Y., et al.: Near-full current dynamic range THz quantum cascade laser frequency comb. Micromachines **14**(2), 473 (2023)
3. Ma, Y., et al.: Flat top optical frequency combs based on a single-core quantum cascade laser at wavelength of ~8.7 μm. Chin. Phys. Lett. **40**(1), 014201 (2023)
4. Garrasi, K., et al.: High dynamic range, heterogeneous, terahertz quantum cascade lasers featuring thermally tunable frequency comb operation over a broad current range. ACS Photonics **6**(1), 73–78 (2019)
5. Li, Y., et al.: Terahertz wave enhances permeability of the voltage-gated calcium channel. J. Am. Chem. Soc. **143**(11), 4311–4318 (2021)
6. Xiang, Z., et al.: A primary model of THz and far-infrared signal generation and conduction in neuron systems based on the hypothesis of the ordered phase of water molecules on the neuron surface. Sci. Bull. **65**, 308 (2020)
7. Zhang, J., et al.: Non-invasive, opsin-free mid-infrared modulation activates cortical neurons and accelerates associative learning. Nat. Commun. **12**, 2730 (2021)
8. Liu, J., et al.: Terahertz quantum cascade laser operating at 2.94 THz. Chin. Phys. Lett. **27**(10) (2010)
9. Kumar, S., Hu, Q., Reno, J.L.: 186 K operation of terahertz quantum-cascade lasers based on a diagonal design. Appl. Phys. Lett. **94**(13), 131105 (2009)
10. Wang, T., et al.: Continuous-wave operation of terahertz quantum cascade lasers at 3.2 THz. Chin. Phys. Lett. **30**(6) (2013)
11. Li, Z., Wan, W., Zhou, K., et al.: On-chip dual-comb source based on terahertz quantum cascade lasers under microwave double injection. Phys. Rev. Appl. **12**(4), 044068 (2019)

Metasurface-Based Terahertz Biosensors

Ziqin Meng[1,2](✉)

[1] School of Physics, Peking University, Beijing 100871, China
ziqin.meng@stu.pku.edu.cn
[2] Innovation Laboratory of Terahertz Biophysics, National Innovation Institute of Defense Technology, Beijing 100071, China

Abstract. Metasurface-based terahertz (THz) sensors are a developing technology for label-free, non-destructive and rapid sensing. Recent research has confirmed that the sensors own prominent advantages compared with traditional methods in cell monitoring, biomolecular identification and biochemical trace detection realms. The impressive progress of metasurface-based THz biosensors, such as modulation methods, sensing principles, cell monitoring and biomolecular recognition, as well as the prospects, is reviewed and discussed in this article.

Keywords: Terahertz · Metasurface · Biosensor

1 Introduction

Recently, a class of small metasurface-based terahertz sensors has demonstrated the possibility of highly sensitive detection in the biomedical domain. Metasurface is the planarization of metamaterials, which is composed of artificial periodic structure array with a unit size smaller than the wavelength of the excitation electromagnetic field. Its electromagnetic properties are determined by the design of spatial structure and can also be controlled by external stimulation. Thus, the modulation of the amplitude, phase, and polarization state of the THz waves can be conveniently realized by using the metasurfaces.

However, due to the lack of high-intensity THz radiation sources and the mismatch of the wavelength, the signal of the light-matter interactions is relatively weak, which greatly limits the development of THz sensing [1]. In addition, the existing THz sensors are short of the capability of flexible and dynamic modulation, leading to the difficulty of undertaking complex analysis. These are the key issues that must be overcome for the larger-scale applications of the THz sensors.

Here, the focus is on describing the cutting-edge progress of metasurface-based THz biosensors, especially the breakthrough in the above problems, and revealing the applications in biomedical detection realm. Current challenges and future development prospects are also presented in this review.

C. Chang et al. (Eds.): YTHZ 2024, SPPHY 400, pp. 65–68, 2024.
https://doi.org/10.1007/978-981-97-3786-4_13

2 Progress Review

2.1 Modulation Methods and Sensing Principles

Most common THz metadevices are driving by a single external field but may easily lose their operational stability under some extreme conditions. To improve it, Lou *et al.* [2] raised an optically controlled ultrafast THz metadevice with an electrically modulated pump threshold. The strategy of adding current-bias enhances the interaction between the electro-induced carriers and the photogenerated carriers, which decreases the pump threshold to an ultralow value. Under dual-stimulus, the THz transmission change is increased by 56.3% compared with the single stimulus, providing a significant breakthrough in designing ultrafast THz metadevices with flexible performance.

Using a similar strategy, Zhao *et al.* [3] reported an optically and thermally controlled hybrid metasurface based on silicon and metallic oxide, reaching a dynamical control at the bandwidth of 1.4 THz. It achieves a maximum modulation depth (MD) of 97.2% under thermal tuning, while achieving an MD of 91.5% at 0.87 THz under optical tuning. The testing result revealed its impressive tunable and stable function. This device successfully utilizes the respective characteristics of two different materials, providing an effective strategy for further design of metasurface-based THz devices controlled by dynamic multi-field.

Generally, we use the transmission spectra to characterize the sensors' functions while being less concerned about the phase change. The phase shift also contains abundant information and could be a shortcut to improve the sensitivity, in fact. Recent work by Zhang *et al.* [4] showed a creative method using the significant phase shift to achieve high-sensitive THz sensing. They designed a metadevice based on the Pancharatnam Berry (PB) beta-atoms to collect the surface waves (SWs) created by the incident THz waves, obtaining a refractive sensing performance up to 215.5°/RIU (RIU = refractive index unit) and the phase shift is up to 4 times higher compared to traditional transmission metasurface. Besides, the capability of simultaneous detection of the multiple fingerprint spectra within a continuous spectrum is exhibited. This result demonstrates the potential value of using phase mobility as a means of characterization of metasurface performance, expanding the testing methods of high-sensitive THz metasurfaces.

2.2 Cells Monitoring and Biomolecular Recognition

Compared with the traditional detection methods, metasurface-based THz sensors show promising potential with their label-free, non-destructive, and rapid-detection capabilities. Through well designed experiments, Lou *et al.* [5] proposed a calibration-free and high-precision biosensor. They eliminated the environmental measurement errors, successfully realizing a high-sensitive tool to monitor the cancerous gastric cells. Another electromagnetically induced transparency (EIT) resonance-based THz metasurface biosensor for cancer diagnosis was introduced by Wang *et al.* [6], which dramatically reduced the cells and time consumption to 1/30 and 1/20 of the pathology testing, respectively. Besides, Ma *et al.* [7] present a THz metasensor to improve the sensitivity of quantitative detection. It achieved the high sensitivity of 378 GHz/RIU as well as the Q factor of 21.28 when detecting two different pathogenic bacteria.

For specific identification of biomolecules, Wang *et al.* [8] keenly designed a splitring resonator metasurface that supports magnetic dipole quasi-BIC combining functionalized gold nanoparticles (AuNPs) conjugated with the specific antibody. Enhanced by the unique near-field effect of the quasi-BIC structure, the device's sensitivity is up to 674 GHz/RIU. Furthermore, they tested the performance of the biosensor through detecting the ultralow concentration of two specific proteins, respectively, down to 1 pM, which is unattainable for traditional sensing methods. This study combines the structure of high-sensitive metasurface with biomolecular sensitive materials, and successfully realizes the sensitivity and detection limit beyond traditional metasurface sensing.

The THz biosensors introduced above all operate in a single resonance mode, while Jiao *et al.* [9] demonstrated an optically controlled metasurface-based THz biosensor, which can work in both the Lorentz-lattice mode and the EIT mode at different band in sensing bio-matter. For Lorentz-lattice mode, the strong interaction between the subtle analyte and the biosensor makes a greatly sensitive signal that can be captured directly from the time-domain spectra without using the Fourier transform. And for the EIT mode, like other hybrid metasurface biosensors, it operates in frequency domain and achieves a sensitivity of 118.4 GHz/RIU. As an application, the biosensor is successfully offered for the dynamic identification of colorectal cells between different states in experiments, and its reliability was verified by comparing with the biological detection results. It provides a new strategy to design versatile active metasurface-based THz biosensors.

In addition, the highly efficient THz sources are always key points [11, 12]. THz science and technology are quickly developed for biophysics research [13–15] and applied in bioeffects [16–23], putting forward the development of biomedicine.

3 Conclusion

In summary, we introduce the latest progress of the metasurface-based THz sensors from the aspects of modulation methods and sensing principles, biomolecular recognition and cell monitoring, aiming at demonstrating the great potential and application value of the biosensors as a label-free, non-destructive and rapid detection means. Meanwhile, we should notice that the sensors are still facing many challenges. The recent research focuses on achieving multi-function and multi-mode sensing and expanding the working band as far as possible. Altogether, we expect large applications of metasurface-based THz biosensors in biomedical and other fields, which will undoubtedly promote the development of advanced testing technology.

References

1. Wang, R., et al.: Electric Fano resonance-based terahertz metasensors. Nanoscale **13**, 18467–18472 (2021)
2. Lou, J., et al.: Optically controlled ultrafast terahertz metadevices with ultralow pump threshold. Small **17**, 2104275 (2021)
3. Zhao, X., et al.: Multifield controlled terahertz modulator based on silicon-vanadium dioxide hybrid metasurface. Adv. Optical Mater **10**, 2102589 (2022)
4. Zhang, Z., et al.: Advanced terahertz refractive sensing and fingerprint recognition through metasurface-excited surface waves. Adv. Mater., 2308453 (2024)

5. Lou, J., et al.: Calibration-free, high-precision, and robust terahertz ultrafast metasurfaces for monitoring gastric cancers. Proc. Natl. Acad. Sci. **119**, 43 (2022)
6. Wang, P., et al.: An ultra-sensitive metasurface biosensor for instant cancer detection based on terahertz spectra. Nano Res. **16**, 7304–7311 (2023)
7. Ma, Z., et al.: Identification and quantitative detection of two pathogenic bacteria based on a terahertz metasensor. Nanoscale **15**, 515–521 (2023)
8. Wang, R., et al.: Ultrasensitive terahertz biodetection enabled by Quasi-BIC-based metasensors. Small **19**, 2301165 (2023)
9. Jiao, Y., et al.: Photoactive terahertz metasurfaces for ultrafast switchable sensing of colorectal cells. Mater. Horiz. **9**, 2984–2992 (2022)
10. Sun, L., et al.: A pixelated frequency-agile metasurface for broadband terahertz molecular fingerprint sensing. Nanoscale **14**, 9681–9685 (2022)
11. Chang, C., et al.: High-gain Thompson-scattering X-ray free-electron laser by time-synchronic laterally tilted optical wave. Phys. Rev. Lett. **110**, 064802 (2013)
12. Gao, M., et al.: Enhanced and frequency-tunable terahertz emission from two crossed air-plasmas. Phys. Rev. Res. **5**, 023091 (2023)
13. Wang, P., et al.: Progress on cutting-edge infrared-terahertz biophysics. IEEE Trans. Microw. Theory Tech. **70**(11), 5117–5140 (2022)
14. Xiang, Z., et al.: A primary model of THz and far-infrared signal generation and conduction in neuron systems based on the hypothesis of the ordered phase of water molecules on the neuron surface. Sci. Bull. **65**, 308 (2020)
15. Liu, G., et al.: Myelin sheath as a dielectric waveguide for signal propagation in the mid-infrared to terahertz spectral range. Adv. Funct. Mater. **29**(7), 1807862 (2019)
16. Li, Y., et al.: Terahertz wave enhances permeability of the voltage-gated calcium channel. J. Am. Chem. Soc. **143**(11), 4311–4318 (2021)
17. Liu, X., et al.: Nonthermal and reversible control of neuronal signaling and behavior by midinfrared stimulation. Proc. Natl. Acad. Sci. **118**(10), e2015685118 (2021)
18. Zhang, J., et al.: Non-invasive, opsin-free mid-infrared modulation activates cortical neurons and accelerates associative learning. Nat. Commun. **12**, 2730 (2021)
19. Yu, Y., et al.: Terahertz photons improve cognitive functions in posttraumatic stress disorder. Research **6**, 0278 (2023)
20. Peng, W., et al.: High-frequency terahertz waves disrupt Alzheimer's amyloid fibril formation. eLight **3**, 18 (2023)
21. Tan, X., et al.: Neuromodulation of chemical synaptic transmission driven by THz photons. Research 0010 (2022)
22. Li, Y., et al.: Physicochemical insights on terahertz wave diminished side effects of drugs from slow dissociation. ACS Nano 16, 5, 8419 (2022)
23. Wu, K., et al.: Terahertz wave accelerates DNA unwinding: a molecular dynamics simulation study. J. Phys. Chem. Lett. **11**(17), 7002–7008 (2020)

Rapid Detection of Paclitaxel-Induced Changes in Cervical Cancer Cells Using an Ultrasensitive Biosensor

Liwen Zhang[1], Gan Chen[2], Yating Hao[2], and Yan Peng[2](✉)

[1] Department of Gynecology, Yueyang Hospital of Integrated Traditional Chinese and Western Medicine, Shanghai University of Traditional Chinese Medicine, Shanghai, China
[2] Terahertz Technology Innovation Research Institute, Terahertz Spectrum and Imaging Technology Cooperative Innovation Center, Shanghai Key Lab of Modern Optical System, University of Shanghai for Science and Technology, Shanghai 200093, China
py@usst.edu.cn

Abstract. In this paper, we used a surface plasmon resonance (SPR) biosensor combined with terahertz (THz) spectroscopy to investigate the effect of paclitaxel on cervical cancer cells (Hela cell) and successfully obtain the terahertz absorbance spectra. The results showed that as the concentration of paclitaxel increased, The spectra show a blue shift, increased, and the inhibition of the cells increased, The cell proliferation levels were assessed using the gold standard of drug assessment, and the results obtained were the same. This study shows that the combination of terahertz spectroscopy and SPR biosensor is a promising approach to achieve a simple, rapid, and low-cost method for approaches to Chemistry-Cell Relationships.

Keywords: terahertz spectroscopy · cervical cancer cells · paclitaxel

1 Introduction

Globally, cervical cancer remains one of the most common cancers among women and is the fourth most common cancer after breast, colorectal and lung cancers. Globally, GLOBOCAN 2020 estimates that there are approximately 604,000 new cases of cervical cancer each year [1]. Most new cases and deaths (about 85% and 90%, respectively) occur in low- and middle-income countries, while in developed countries the prevalence and mortality rates of cervical cancer have declined as a result of widespread cervical cancer screening and vaccination programmer, but it still ranks among the three major gynecological malignancies [2, 3]. Currently, the treatment routines for cervical cancer are: surgery, chemotherapy and radiotherapy. Among them, chemotherapy occupies a large proportion, and paclitaxel is recognized as one of the most effective anti-tumor drugs in the world, which can inhibit the proliferation of cancer cells by inducing cell cycle arrest and mitotic failure [4]. Paclitaxel is known to induce multiple cell deaths in cancer. In addition to inducing the mitotic catastrophe that occurs during mitosis, paclitaxel has been shown to induce the expression of several pro-apoptotic mediators,

C. Chang et al. (Eds.): YTHZ 2024, SPPHY 400, pp. 69–74, 2024.
https://doi.org/10.1007/978-981-97-3786-4_14

and it also modulates the activity of anti-apoptotic mediators [5]. Since the discovery of its anti-tumor activity, paclitaxel has been used to treat more than one million patients, making it one of the most widely used anti-tumor agents [6].

In order to study the biochemical characteristics, medical diagnosis and treatment of cancer, cell lines are widely used as common in vitro models in biomedical research and drug development to represent the characteristics of biological tissues in cancer research [7, 8]. In particular, Hela cervical cancer cells have been used as a classical human immortalized cell line in various areas of cancer research [9, 10]. To date, the main methods of cell identification include morphology, immunohistochemistry, isoenzyme assays, karyotyping, etc. [11].However, all these methods require multiple pre-treatments and additional markers or reagents, which makes the identification of cancer cells cumbersome and inefficient. In contrast, techniques based on terahertz (THz, 1 THz = 1012 Hz) waves, which are located between the microwave and infrared regions and have a frequency range of 0.1–10 THz, have shown promise in biomedical research. Compared with other optical and infrared methods, terahertz waves are ideal for non-destructive detection of human tissues and cells [12–19]. Our approach integrates terahertz spectroscopy with a metamaterial-based terahertz biosensor to detect cancer cell quantitatively [20, 21]. The biosensor's surface dielectric constant changes when a sample is applied, altering the resonance frequency correspondingly. [22, 23] The frequency shift of the resonance varies with the content of the sample due to changes in the effective dielectric constant [24–27]. This enables the quantitative detection and analysis of target substances by mapping the frequency shift patterns across different sample contents.In this study, we investigated the interaction between paclitaxel on Hela cells using SPR biosensors combined with THz-TDS, and using terahertz technology we obtained the survival rate of Hela cells after the effect of different concentrations of paclitaxel. In order to explore the underlying mechanism, we measured the survival rate of Hela using CCK-8 assay, while using MTT and tunnel apoptosis assay for parallel study, this study provides a new method to study drug-cell interactions using terahertz spectroscopy.

2 Result and Discussion

2.1 Viability of Hela Cells

Absorption coefficients in terahertz are related to cellular status. According to literature studies with a concentration range of 5–100 nM. we found that cell viability decreased significantly with increasing paclitaxel concentration in a concentration-dependent manner, while Hela cells exposed to 20–100 nM concentrations of paclitaxel significantly reduced cell viability compared to the normal group. To ensure the reliability of our results we used two methods to assess the effect of paclitaxel on Hela cell viability. We found that the results of both methods were similar and were concentration-dependent, while suggesting an effect on viability over a range of concentrations.CCK-8 (Cell Counting Kit-8) and MTT (3-(4,5-Dimethylthiazol-2-yl)-2,5-diphenyltetrazolium bromide) Both are experimental methods used to determine cell proliferation and cell activity, with CCK-8 being more sensitive than MTT. Currently, these two methods are commonly used to assess the effects of drugs on cells [28–30], but due to the limitations of the experiment, other media need to be added to complete the experiment, and the results of the reaction process are affected by cellular or operational problems.

2.2 Terahertz Spectroscopy Analysis of Paclitaxel Effect on HeLa Cells

In this experiment, in order to explore the effect of paclitaxel on HeLa cells, we configured different concentrations of paclitaxel solution for administration (0 nM, 5 nM, 10 nM, 20 nM, 40 nM, 80 nM, 100 nM). A 5 μL drop of uniformly blown cell solution was taken on the SPR biosensor, and the absorbance of the HeLa cell solution was measured by THz-TDS over the frequency range of 0.5–4 THz. The comparison of the absorbance of the cytosol after administration of different concentrations of paclitaxel is shown in Fig. a. In order to capture the changes of the measurement results more intuitively, we inserted small graphs in Fig. a for magnification to better demonstrate the changes of the peaks, and as shown in the figure, the average frequency points of different concentrations were 0 nM (1.7209), 5 nM (1.7252), 10 nM (1.7314), 20 nM (1.7343), 40 nM (1.7424), 80 nM (1.7452), 100 nM (1.7471). Figure 2b shows the relationship curve between different cytosolic samples and the frequency shift of the hypersurface resonance peak. From the figure, it can be seen that there is a non-linear relationship between the amount of frequency shift corresponding to the hypersurface resonance peak and the number of cells after paclitaxel administration at different concentrations; as the concentration of paclitaxel increases, the number of cells decreases, and the resonance peak of the SPR biosensor is shifted to a higher frequency. We extracted the frequency shifts of different cytosolic samples to establish the correlation as a function of concentration. The fitted equations are as follows.

$$\mathbf{y = -Vmax{*}x/(Km + x) + a}$$

where x is the concentration of paclitaxel administered, y is the frequency shift induced by the sample on the terahertz SPR biosensor, Vmax is the maximum frequency shift, and Km is the value of x corresponding to half of the maximum frequency shift (Figs. 1 and 3).

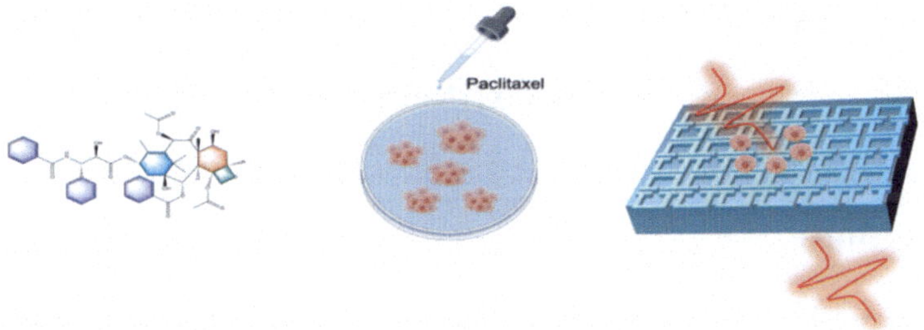

Fig. 1. Diagram of terahertz workflow and structural formula of paclitaxel

In order to make the results more intuitive and clearer, the average frequency points of different concentrations we extracted separately to do the fitting, and the results are

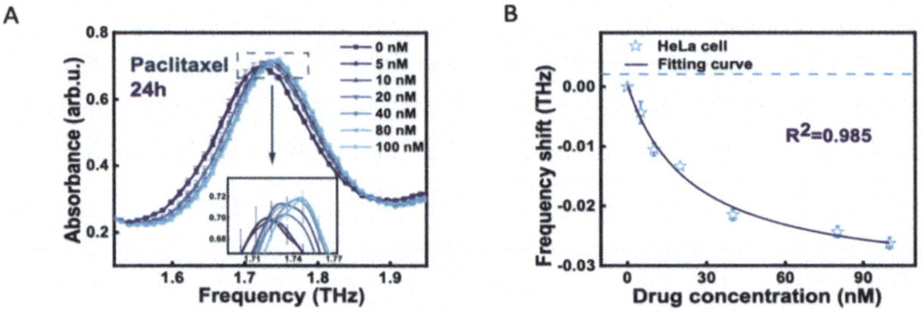

Fig. 2. Absorbance spectra (A), partial amplification of absorbance spectra (B)

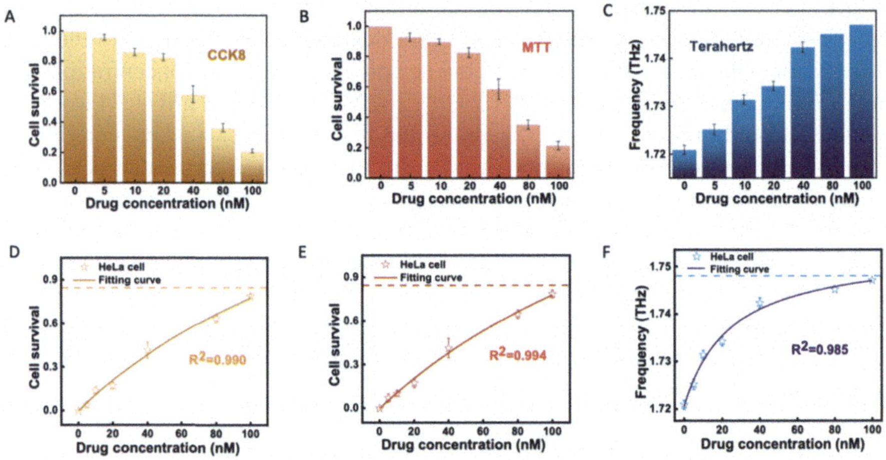

Fig. 3. Cell viability assay (A,B), the effect of drug concentration on the viability of HeLa cells (D,E) corresponding frequency shift as a function (C) column chart of the corresponding frequency shifts (F) of different paclitaxel concentrations in Hela

shown in Fig. 2c. Figure 2d We extracted the frequency point data to do the fitting to establish the correlation with the paclitaxel drug concentration as a function. The fitting formula is as follows.

$$\mathbf{y = Vmax*x/(Km + x) + a}$$

3 Conclusion

In summary, we have investigated the interaction between paclitaxel and Hela and demonstrated that the effects of paclitaxel on Hela cells can be detected in a labelling-free, non-invasive manner by terahertz spectroscopy. With the help of cell viability, tunnel researchers explored the intrinsic mechanism of terahertz detection. The results showed that paclitaxel affected the proliferation of Hela cells in different dose-dependent ways. As the concentration of paclitaxel increased, paclitaxel inhibited the proliferation of Hela cells. The effect of paclitaxel on Hela cells was detected by terahertz absorbance extracted from terahertz TD-ATR spectroscopic measurements. This coefficient contains comprehensive information about the cells measured. Based on the terahertz absorption coefficients, the survival of tumor cells after the action of paclitaxel can be appropriately assessed. The results of the CCK-8 assay, MTT, tunnel apoptosis analysis support the feasibility of using terahertz spectroscopy in conjunction with SPR biosensors to study drug-Hela interactions. It also provides a simple, rapid and cost-effective new method for subsequent drug-cell line studies.

Acknowledgments. This work was supported in part by National key research and development program of China (2022YFA1404004), National Natural Science Foundation of China (61988102, 62335012), High-level Chinese Medicine Key Discipline Construction Project (ZYYZOKK-2023065) State Administration of Traditional Chinese Medicine.

References

1. Sung, H., Ferlay, J., et al.: Global Cancer Statistics 2020: GLOBOCAN estimates of incidence and mortality worldwide for 36 cancers in 185 countries. CA Cancer J. Clin. **71**(3), 209–249 (2021). https://doi.org/10.3322/caac.21660
2. Revising the WHO classification: female genital tract tumours - Cree - 2020 - Histopathology - Wiley Online Library **76**(1), 151–156 (2020). https://onlinelibrary.wiley.com/doi/abs/10.1111/his.13977. Accessed 17 Feb 2024
3. Lowy, D.R., et al.: Human papillomavirus infection and the primary and secondary prevention of cervical cancer. Cancer **1980**, 113 (2008)
4. Javeed, A., et al.: Paclitaxel and immune system. Eur. J. Pharm. Sci. Off. J. Eur. Feder. Pharm. Sci. **38**(4), 283 (2009)
5. Hong, X., et al.: Disruption of protein neddylation with MLN4924 attenuates paclitaxel-induced apoptosis and microtubule polymerization in ovarian cancer cells. Biochem. Biophys. Res. Commun. **508**(3), 986 (2019)
6. Punzi, S., Meliksetian, M., Riva, L., et al.: Development of personalized therapeutic strategies by targeting actionable vulnerabilities in metastatic and chemotherapy-resistant breast cancer PDXs. Cells **8**(6), 605 (2019)
7. Mouradov, D., Sloggett, C., Jorissen, R.N., et al.: Colorectal cancer cell lines are representative models of the main molecular subtypes of primary cancer. Can. Res. **74**, 3238 (2014)
8. Wilding, J.L., Bodmer, W.F.: Cancer cell lines for drug discovery and development. Can. Res. **74**, 2377 (2014)
9. Kim, H., Lee, J., Oh, C., et al.: Cooperative tumour cell membrane targeted phototherapy. Nat. Commun. **8**, 15880 (2017). https://doi.org/10.1038/ncomms15880
10. Adey, A., Burton, J.N., Kitzman, J.O., et al.: The haplotype-resolved genome and epigenome of the aneuploid HeLa cancer cell line. Nature **500**(7461), 207–211 (2013)
11. Zaynagetdinov, R., et al.: Identification of myeloid cell subsets in murine lungs using flow cytometry. Am. J. Resp. Cell Mol. Biol. **49**(2), 180 (2013)
12. Ang, R., Wu, Q., Zhang, Q., et al.: Conversion from terahertz-guided waves to surface waves with metasurface. Opt. Express **26**(24), 31233–31243 (2018)
13. Liu, G., et al.: Myelin sheath as a dielectric waveguide for signal propagation in the mid-infrared to terahertz spectral range. Adv. Funct. Mater. **29**(7), 1807862 (2019)
14. Li, Y., et al.: Terahertz wave enhances permeability of the voltage-gated calcium channel. J. Am. Chem. Soc. **143**, 9 (2021)

15. Zhang, J.X., et al.: Non-invasive, opsin-free mid-infrared modulation activates cortical neurons and accelerates associative learning. Nature Comm. **12**, 2730 (2021)
16. Wu, K., et al.: Terahertz wave accelerates DNA unwinding: a molecular dynamics simulation study. J. Phys. Chem. Lett. **11**(17), 7002–7008 (2020)
17. Xiang, Z., et al.: A primary model of THz and far-infrared signal generation and conduction in neuron systems based on the hypothesis of the ordered phase of water molecules on the neuron surface. Sci. Bull. **65**, 308 (2020)
18. Lou, J., et al.: Calibration-free, high-precision, and robust terahertz ultrafast metasurfaces for monitoring gastric cancers. PNAS **119**(43) (2022)
19. Wang, Y., Li, S., Wang, H., et al.: Simple terahertz metasurface with broadband and efficient functionality. Opt. Express **30**(25), 45488–45498 (2022)
20. Sun, Y., et al.: Quantitative analysis of bisphenol analogue mixtures by terahertz spectroscopy using machine learning method. Food Chem. **352**, 129313 (2021)
21. Shi, W., Wang, Y., Hou, L., et al.: Detection of living cervical cancer cells by transient terahertz spectroscopy. J. Biophoton. **14**(1), e202000237 (2021)
22. Peng, Y.: Three-step one-way model in terahertz biomedical detection. PhotoniX **2**, 1–18 (2021)
23. Peng, Y.: Terahertz spectroscopy in biomedical field: a review on signal-to-noise ratio improvement. PhotoniX **1**, 1–18 (2020)
24. Shou, Y., Wang, Y., Miao, L., et al.: Realization of all-optical higher-order spatial differentiators based on cascaded operations. Opt. Lett. **47**, 5981 (2022)
25. Liu, S., Xu, F., Zhan, J., et al.: Terahertz liquid crystal programmable metasurface based on resonance switching. Opt. Lett. **47**(7), 1891–1894 (2022)
26. Wang, R., et al.: Ultrasensitive terahertz biodetection enabled by Quasi-BIC-based metasensors. Small **19**, 2301165 (2023)
27. Zhang, Z., et al.: Advanced terahertz refractive sensing and fingerprint recognition through metasurface-excited surface waves. Adv. Mater. **36**, 2308453 (2024)
28. Braun, K., et al.: Comparison of different cytotoxicity assays for in vitro evaluation of mesoporous silica nanoparticles. Toxicol. Vitro BIBRA **52**, 214–221 (2018)
29. Zhou, S., et al.: An electrochemical method for evaluation the cytotoxicity of fluorene on reduced graphene oxide quantum dots modified electrode. Sens. Actuators B Chem. **255**, 2595–2600 (2018). https://doi.org/10.1016/j.snb.2017.09.066
30. Qin, H., Liu, J., Zhang, Z., et al.: In situ electrochemical assessment of cytotoxicity of chlorophenols in MCF-7 and HeLa cells. Anal. Biochem. **462**, 60–66 (2014)

Permittivity-Asymmetric Quasi-Bound States in the Continuum of Localized Spoof Surface Plasmons

Si-Qi Li, Feng Zhang, Yue-Yi Zhang, Yu-Lu Lei, Ji-Tao Yang, and Chao-Hai Du(✉)

State Key Laboratory of Advanced Optical Communication Systems and Networks, School of Electronics, Peking University, Beijing 100871, China
duchaohai@pku.edu.com

Abstract. Quasi-bound states in the continuum (quasi-BICs) of localized spoof surface plasmons (LSSPs) have high quality factors (Q-factors) because of low radiation loss. In this paper, permittivity-asymmetric quasi-BICs of LSSPs are proposed. The unit cell of the metasurface contains a circular grating and two dielectric sheets. The quasi-BICs of the LSSPs are realized by breaking the symmetry of the permittivity of the dielectric sheets. This paper provides a new way to construct the quasi-BICs of the LSSPs.

Keywords: Quasi-Bound States in the Continuum · Localized Spoof Surface Plasmons · Permittivity

1 Introduction

Localized spoof surface plasmons (LSSPs) have been hot topics in Terahertz technologies [1–9]. They have favorable characteristics of subwavelength confinement and nearfield enhancement, but their quality factors (Q-factors) are low, partly because of radiation loss. Therefore, quasi-bound states in the continuum (quasi-BICs) of the LSSPs have been proposed, which have the advantage of high Q-factors because of low radiation loss [10]. Based on the quasi-BICs of the LSSPs, a displacement sensor was designed and reported at the ICMMT 2023 [11]. Inspired by reference [12], permittivity-asymmetric quasi-BICs of the LSSPs are proposed in this paper.

This work was supported in part by the National Natural Science Foundation of China (no. 62271011), the Beijing Natural Science Foundation for Distinguished Young Scholars (no. JQ21011), and the National Key Research and Development Program of China (no. 2021YFA1600302). This work was supported by State Key Laboratory of Advanced Optical Communication Systems and Networks, China. The authors are grateful to Zi-Wen Zhang and Jin Zhao from School of Electronics, Peking University for useful discussions.

C. Chang et al. (Eds.): YTHZ 2024, SPPHY 400, pp. 75–79, 2024.
https://doi.org/10.1007/978-981-97-3786-4_15

2 Results and Discussions

2.1 Structure Design

The schematic of the unit cell of the metasurface is shown in Fig. 1(a), which contains a circular grating and two dielectric sheets. The permittivity of the dielectric sheet in green is 1 and the permittivity of the dielectric sheet in blue is ε. Here, $R_1 = 5$ mm, $R_2 = 15$ mm, $R_3 = 16$ mm, $b = 1$ mm, $h = 10$ mm, $a/d = 0.4$, and $N = 60$. The period of the unit cell is 42 mm, the thickness of the dielectric sheets is 1 mm, and the asymmetry parameter for the metasurface is defined as $\Delta\varepsilon = \varepsilon - 1$.

2.2 Simulation Configuration

The simulation was carried out with CST Microwave Studio, and the frequency domain solver was selected. A linearly polarized plane wave was incident normally on the metasurface, as shown in Fig. 1(b). The material loss was neglected.

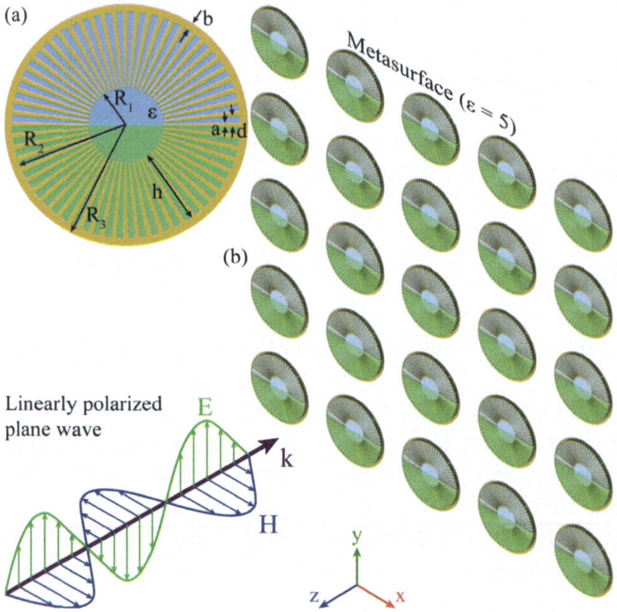

Fig. 1. (a) Schematic of the unit cell of the metasurface. The size information is labelled next to the structures with different parameters. (b) Simulation configuration of the metasurface.

2.3 Simulation Results

The S-parameters spectra are shown in Fig. 2. When ε is 1, only $TE_{1,1}$ can be excited, $TE_{2,1}$ and above cannot be excited, because they are true-BICs protected by rotational symmetry. Therefore, we break the symmetry of the permittivity of the dielectric sheets and obtain the permittivity-asymmetric circular grating metasurface. Here, we select $\varepsilon = 5$ as an example. $TE_{2,1}$ and above transform into quasi-BICs from the true-BICs and can be excited. Their Q-factors are high because of low radiation loss.

The electromagnetic field distributions are shown in Fig. 3. For the metasurface with $\varepsilon = 5$, they are deformed due to the asymmetry of the permittivity of the dielectric sheets. According to the minimum energy principle, they are more likely to stay in the dielectric sheet with $\varepsilon = 5$.

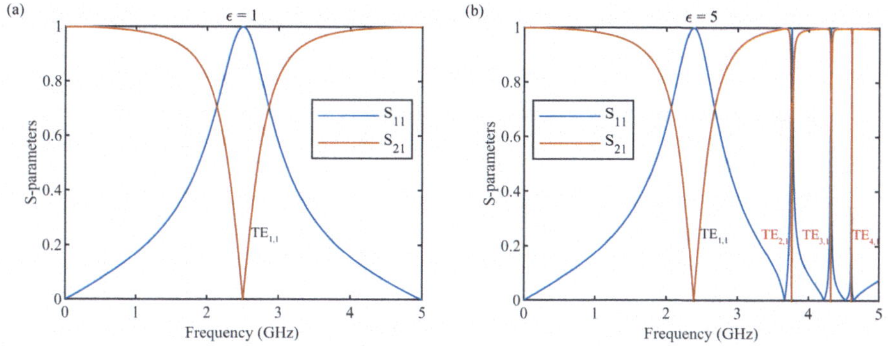

Fig. 2. Simulated S-parameter spectra of the metasurface with (a) $\varepsilon = 1$ and (b) $\varepsilon = 5$.

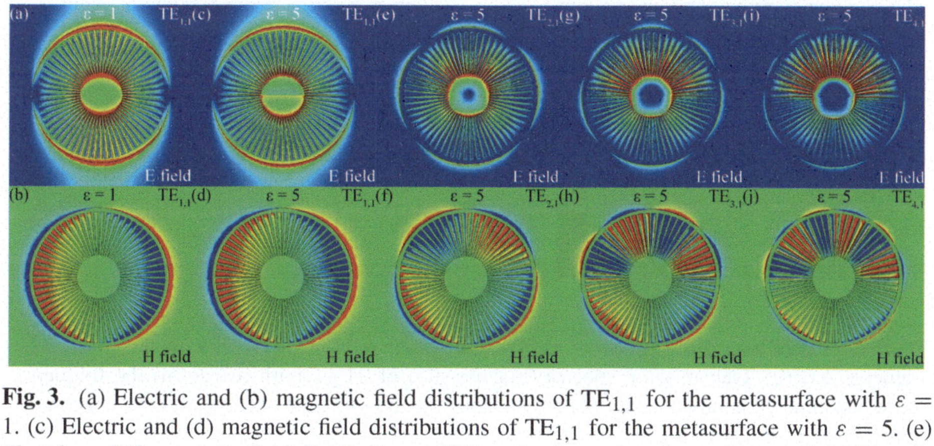

Fig. 3. (a) Electric and (b) magnetic field distributions of $TE_{1,1}$ for the metasurface with $\varepsilon = 1$. (c) Electric and (d) magnetic field distributions of $TE_{1,1}$ for the metasurface with $\varepsilon = 5$. (e) Electric and (f) magnetic field distributions of $TE_{2,1}$ for the metasurface with $\varepsilon = 5$. (g) Electric and (h) magnetic field distributions of $TE_{3,1}$ for the metasurface with $\varepsilon = 5$. (i) Electric and (j) magnetic field distributions of $TE_{4,1}$ for the metasurface with $\varepsilon = 5$.

2.4 Discussions

The behavior for the Q-factor of a quasi-BIC as a function of the asymmetry parameter is given in

$$Q = Q_0 \times \alpha^{-2} \tag{1}$$

where α is the asymmetry parameter and is defined as $\Delta\varepsilon = \varepsilon$ - 1 for the permittivity-asymmetric circular grating metasurface. The formula has been verified on the eccentric circular grating metasurface in reference [3]. The formula remains valid for the permittivity-asymmetric circular grating metasurface in this paper, as shown in Fig. 4.

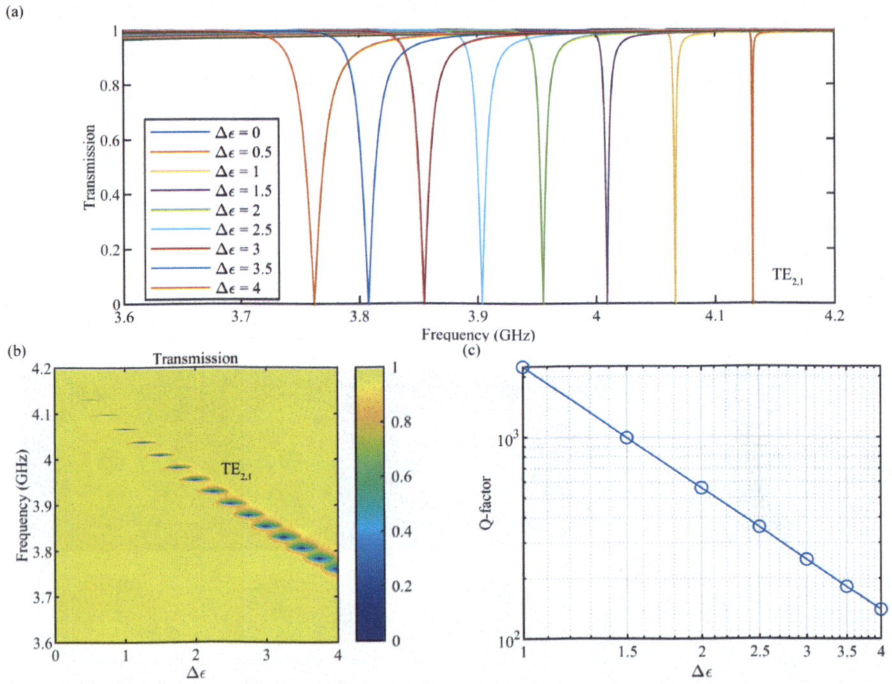

Fig. 4. (a) Evolution of the transmission spectra (linear scale) of $TE_{2,1}$ with the asymmetry parameter $\Delta\varepsilon$. (b) Transmission spectra (linear scale) of $TE_{2,1}$ with respect to the frequency and the asymmetry parameter $\Delta\varepsilon$. (c) Dependence of the Q-factor of $TE_{2,1}$ on the asymmetry parameter $\Delta\varepsilon$ (log-log scale).

3 Conclusion

In this paper, permittivity-asymmetric quasi-BICs of the LSSPs are proposed, which have promising applications in sensors, filters, invisible cloaks, and other functional devices.

References

1. Pors, A., Moreno, E., Martin-Moreno, L., Pendry, J.B., Garcia-Vidal, F.J.: Localized spoof plasmons arise while texturing closed surfaces. Phys. Rev. Lett. **108**(22), 223905 (2012)
2. Li, S.-Q., et al.: Localized spoof surface plasmon skyrmions excited by guided waves. Adv. Func. Mater. **33**(47), 2305789 (2023)
3. Zhang, Z., et al.: Advanced terahertz refractive sensing and fingerprint recognition through metasurface-excited surface waves. Adv. Mater. **36**, 2308453 (2024)
4. Lou, J., et al.: Calibration-free, high-precision, and robust terahertz ultrafast metasurfaces for monitoring gastric cancers. PNAS **119**(43) (2022)
5. Liu, G., et al.: Myelin sheath as a dielectric waveguide for signal propagation in the mid-infrared to terahertz spectral range. Adv. Funct. Mater. **29**(7), 1807862 (2019)
6. Lou, J., et al.: Optically controlled ultrafast terahertz metadevices with ultralow pump threshold. Small **17**, 2104275 (2021)
7. Wang, R., et al.: Ultrasensitive terahertz biodetection enabled by Quasi-BIC-based metasensors. Small **19**, 2301165 (2023)
8. Zhang, J.X., et al.: Non-invasive, opsin-free mid-infrared modulation activates cortical neurons and accelerates associative learning. Nature Comm. **12**, 2730 (2021)
9. Li, Y., et al.: Terahertz wave enhances permeaility of the voltage-gated calcium channel. J. Am. Chem. Soc. **143**, 9 (2021)
10. Li, S.-Q., Du, C.-H., Yin, L.-Z., Gao, Z.-C., Li, F.-H.: Quasi-bound states in the continuum of localized spoof surface plasmons. ACS Photon. **9**(10), 3333–3340 (2022)
11. Li, S.Q., Li, F.H., Zhu, Y., Du, C.H.: Displacement sensor based on quasi bound states in the continuum of localized spoof surface plasmons. In: 2023 International Conference on Microwave and Millimeter Wave Technology, ICMMT, Qingdao, China, pp. 1–3 (2023)
12. Berté, R., et al.: Permittivity-asymmetric Quasi-bound states in the continuum. Nano Lett. **23**(7), 2651–2658 (2023)

Ultra-Wide Dual-Band Fingerprint Spectral Sensing

Yifang Yuan(✉) and Yiping Han

School of Physics, Xidian University, Xi'an 700071, China
yyf_lalala@126.com

Abstract. Metasurface devices for biological sensing have been used to design a variety of terahertz (THz) biosensors. However, due to the fixed nature of the metasurface structure, these sensor devices are limited to tuning at a single frequency point or in a narrow range, making it difficult to achieve a wide range of frequency modulation. In this paper, a novel design concept for optically controlled frequency modulation is proposed to enable fingerprint spectral sensing over a very large resonant frequency shift range of 0.86 to 1.81 THz. At the same time, transitions on the order of picoseconds are achieved based on the excitation and relaxation times of the carriers. It provides an example for large-scale applications of tunable devices and ultra-wideband biosensing.

Keywords: Dual-band · fingerprint spectral · sensing

1 Abstract

THz waves are placed in the transition region between electronics and photonics research, possessing many unique properties such as low photon energy and strong penetrating power. At present, THz waves are widely used in biochemical detection, broadband communication, medical imaging, electromagnetic metamaterial [1, 2]. The modulation effect of THz waves on biological macromolecules has been demonstrated by numerous simulation studies [3–6]. Meanwhile, THz waves have shown great application potential in enhancing brain cognition and modulation of biological macromolecules [7–10], with the development of THz sources [11, 12]. These studies and applications have further promoted the research on the generation, modulation and detection of THz waves.

Metamaterials are periodic or aperiodic artificial structures with sub-wavelength size whose electromagnetic properties are mainly determined by the element structure. They are free to design their electromagnetic response to overcome the shortcomings of natural materials. Once the element structure of traditional metamaterials is determined, its characteristics are immutable, which limits the application scenarios. Thus, in order to promote the practical application of tunable THz metasurfaces, various external excitation methods have been tried to dynamically control the resonance characteristics of the metasurfaces. Due to the high sensitivity of Fano resonance to the change of the refractive index of the surrounding medium, it has also been widely studied in fundamental research such as biodetection and biosensors [13]. In recent years, electromagnetically

C. Chang et al. (Eds.): YTHZ 2024, SPPHY 400, pp. 80–83, 2024.
https://doi.org/10.1007/978-981-97-3786-4_16

induced transparency (EIT), a special case of Fano resonance, has also become a branch of related research, which shows great application prospects in ultra-sensitive biosensing [14–19], enhanced nonlinear effects and optical storage. In these emerging devices, the wide range of adjustable resonant frequency of metasurface devices not only plays a crucial role in extending the device frequency operating range, but also may help to study the biophysics [20, 21]. This makes wider and faster frequency shift particularly important.

In this work, we achieve resonance mode transformation by changing the conductivity properties of the semiconductor, which enables fingerprint spectral sensing in a very large resonance frequency shift range of 0.86 to 1.81 THz. It greatly extends the frequency range of fingerprint spectrum detection. It provides a paradigm for large-scale applications of tunable devices and ultra-wideband biosensing.

2 Results and Discussion

Firstly, a terahertz photomodulated resonator is constructed, which completes the switching of the coupling mode by changing the conduction of the silicon bridge, as shown in Fig. 1(a). Next, the frequency shift of the resonator is achieved by changing the structural parameters of the open ring, and multiple modules are constructed to realise the sensing of the fingerprint spectrum, as shown in Fig. 1(b). The specific structure is shown in Fig. 1(c,d).

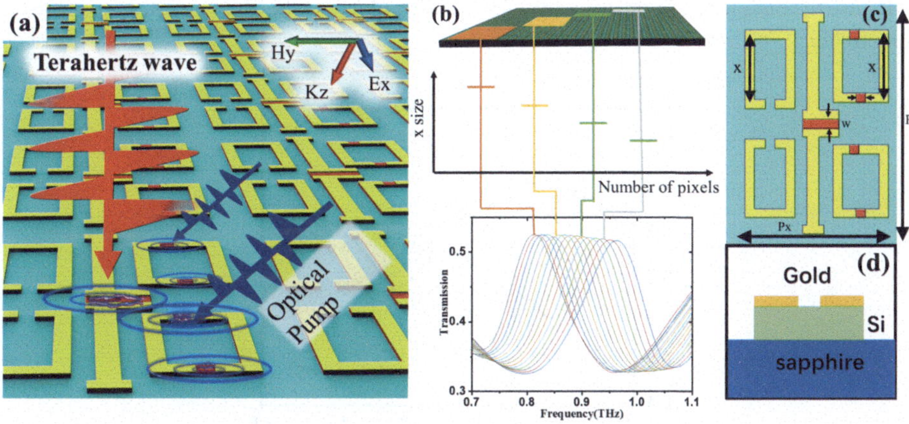

Fig. 1. Schematic diagram of Fano ultra-wide frequency adjustment. (a) Schematic of the working principle of the device. (b) Fingerprint spectrum sensing schematic. (c) 3D schematic of the structural elements. The detailed parameters are given as follows: Px = 56 μm, Py = 80 μm, w = 3 μm, x is a variable parameter. (d) Structural section.

The fingerprint spectrum obtained by changing the structural parameters when the silicon bridge is conducting in EIT mode is shown in Fig. 2(a). Figure 2(b) shows the sensing curve of the Fano mode fingerprint spectrum when the silicon bridge is not conducting. It can be seen that an ultra-wide bandwidth shift of 0.95 THz is achieved as the conductivity of silicon changes from 0 to 15000 S/m. According to the definition of the frequency tuning range $\Delta f_{I-II}/f_I$, the modulator exhibits a resonance frequency tuning range of 52.5%. By varying the structural parameters of x (22–29 um), we obtained the fingerprint spectrum detection model in two resonance modes respectively. The fingerprint spectrum of EIT is in the range of 0.81–0.96 THz and that of Fano is in the range of 1.61–1.91 THz.

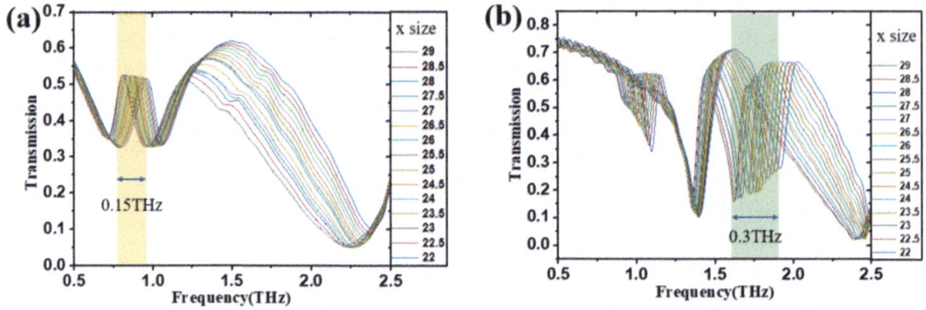

Fig. 2. (a) EIT mode fingerprint spectrum sensing curve. (b) Fano mode fingerprint spectrum sensing curve.

As show in Fig. 3, to quantify the attenuation changes detected by the substance fingerprint spectrum, the substance spectrum signal is defined as $T = R_s/R_b$, where R_s and R_b are the envelopes of the transmittance of the substance before and after absorption, respectively.

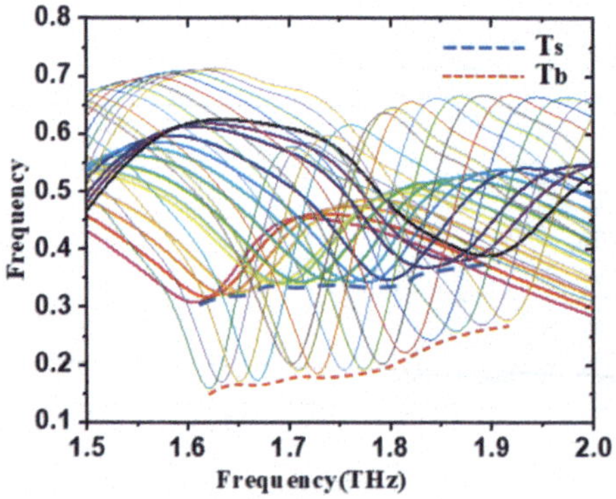

Fig. 3. Fingerprint spectrum detection process diagram

3 Conclusions

In summary, we propose an ultra-wideband detection of fingerprint spectra based on resonant mode conversion for dual-band fingerprint spectrum detection. Our scheme provides a new method for large-scale application of bio-detection. Due to biological interests involving the full spectral range from THz waves to mid-infrared (MIR), our design concept is applicable to a wider range of application scenarios.

References

1. Lou, J., et al.: Calibration-free, high-precision, and robust terahertz ultrafast metasurfaces for monitoring gastric cancers. PNAS **119**(43), e220921811 (2022)
2. Lou, J., et al.: Optically controlled ultrafast terahertz metadevices with ultralow pump threshold. Small **17**(44), e2104275 (2021)
3. Li, Y., et al.: Terahertz wave enhances permeability of the voltage-gated calcium channel. J. Am. Chem. Soc. **143**, 4311–4318 (2021). https://doi.org/10.1021/jacs.0c09401
4. Wu, K., et al.: Terahertz wave accelerates DNA unwinding: a molecular dynamics simulation study. J. Phys. Chem. Lett. **11**(17), 7002–7008 (2020)
5. Yin, J., et al.: Terahertz photons inhibit cancer cells long term by suppressing nano telomerase activity. ACS Nano (2024). https://doi.org/10.1021/acsnano.3c09216
6. Sun, L., et al.: Inhibition of cancer cell migration and glycolysis by terahertz wave modulation via altered chromatin accessibility. Research, 9860679 (2022)
7. Liu, X., et al.: Nonthermal and reversible control of neuronal signaling and behavior by midinfrared stimulation. Proc. Natl. Acad. Sci. U. S. A. **118**(10), e2015685118 (2021)
8. Zhang, J., et al.: Non-invasive, opsin-free mid-infrared modulation activates cortical neurons and accelerates associative learning. Nat. Commun. **12**, 2730 (2021)
9. Yu, Y., et al.: Terahertz photons improve cognitive functions in posttraumatic stress disorder. Research **6**, 0278 (2023). https://doi.org/10.34133/research.0278
10. Zhang, C., et al.: Driving DNA origami assembly with a terahertz wave. Nano Lett. **22**, 468–475 (2022)
11. Chang, C., et al.: High-gain thompson-scattering X-ray free-electron laser by time-synchronic laterally tilted optical wave. Phys. Rev. Lett. **110**, 064802 (2013)
12. Gao, M., et al.: Enhanced and frequency-tunable terahertz emission from two crossed air-plasmas. Phys. Rev. Res. **5**, 023091 (2023)
13. Sun, L., et al.: A pixelated frequency-agile metasurface for broadband terahertz molecular fingerprint sensing. Nanoscale **14**, 9681–9685 (2022)
14. Wang, R., et al.: Ultrasensitive terahertz biodetection enabled by Quasi-BIC-based metasensors. Small **19**, e2301165 (2023)
15. Zhang, Z., et al.: Advanced Terahertz refractive sensing and fingerprint recognition through metasurface-excited surface waves. Adv. Mater., 2308453 (2024)
16. Wang, P., et al.: An ultra-sensitive metasurface biosensor for instant cancer detection based on terahertz spectra. Nano Res. **16**, 7304–7311 (2023)
17. Ma, Z., et al.: Identification and quantitative detection of two pathogenic bacteria based on a terahertz metasensor. Nanoscale **15**, 515–521 (2023)
18. Wang, R., et al.: Electric Fano resonance-based terahertz metasensors. Nanoscale **13**, 18467–18472 (2021)
19. Jiao, Y., et al.: Photoactive terahertz metasurfaces for ultrafast switchable sensing of colorectal cells. Mater. Horiz. **9**, 2984–2992 (2022)
20. Liu, G., et al.: Myelin sheath as a dielectric waveguide for signal propagation in the mid-infrared to terahertz spectral range. Adv. Funct. Mater. **29**, 1807862 (2019)
21. Xiang, Z., et al.: A primary model of THz and far-infrared signal generation and conduction in neuron systems based on the hypothesis of the ordered phase of water molecules on the neuron surface I: signal characteristics. Sci. Bull. **65**, 308–317 (2020)

Photoresponse Characteristics of Quantum Ratchet Detectors Grown by MOCVD

Yi Wang[1,2], Peng Bai[1,2](✉), Ning Yang[1,2], Weidong Chu[1,2], Yan Xie[3], Yingxin Wang[3], Meng Chen[3], and Ziran Zhao[3]

[1] Institute of Applied Physics and Computational Mathematics, Beijing 100088, China
bai_peng@iapcm.ac.cn

[2] National Key Laboratory of Computational Physics, Beijing 100088, China

[3] Department of Engineering Physics, Tsinghua University, Beijing 100084, China

Abstract. Terahertz detection has important application in space exploration, biomedicine, and national defense security. At present, although terahertz thermal detectors can operate at room temperature, their response speed is relatively slow, and the detection band coverage is narrow. Photon-type terahertz detectors have the potential for high-speed response, and the response band is flexible and adjustable, but their operating temperature is generally low (all working in the liquid helium temperature range). Developing a terahertz detection device with high speed, high sensitivity, high temperature operation, wide band response, simple fabrication process, and relatively low cost has become a focus and frontier issue in terahertz technology research. The quantum ratchet detector proposed by us is expected to achieve high-temperature, high-speed, and broadband terahertz detection. In this paper, we use MOCVD technology for material epitaxial growth and fabricate quantum ratchet detectors. We studied their photoresponse characteristics and ratchet characteristics, and discussed the possibility of high-temperature operation.

Keywords: Quantum ratchet · MOCVD · Broadband detector

1 Introduction

Quantum ratchet is a nonequilibrium system with broken spatial inversion symmetry, whose typical characteristic is to form directional movement of carriers in the presence of external random perturbations (such as electromagnetic waves, heat, alternating electric fields, etc.) [1]. This concept originated from people's understanding of symmetry and the process of knowing the second law of thermodynamics. In recent years, the quantum ratchet effect has been widely studied in the fields of solar photovoltaic [2] effects and molecular motors [3–9]. Using the quantum ratchet effect to realize infrared and terahertz detection is also an important research direction with promising prospects. In 2022, we realized an ultra-broadband quantum ratchet terahertz detector based on the GaAs/AlGaAs material system [10]. On this basis, an ultra-broadband quantum ratchet upconversion device was further realized, which is the widest response band semiconductor upconversion device so far [11].

C. Chang et al. (Eds.): YTHZ 2024, SPPHY 400, pp. 84–88, 2024.
https://doi.org/10.1007/978-981-97-3786-4_17

The previous quantum ratchet detector research was based on the GaAs/AlGaAs material system grown by molecular beam epitaxy (MBE), and the related experimental results were in line with theoretical expectations. In this article, we studied the infrared/terahertz spectral response characteristics of the quantum ratchet detector fabricated from GaAs/AlGaAs materials grown by Metal Organic Chemical Vapor Deposition (MOCVD)technology, analyzed its photoelectric response microscopic mechanism, and presented the photovoltaic response characteristics and high-temperature operating characteristics of the device.

2 Result and Discussion

As shown in Table 1, the detailed parameters of the quantum ratchet structure grown by MOCVD are presented. The doping concentrations of the upper and lower contact layers, the carrier injection layer, and the absorption layer are all 1 × 1019 cm^{-3}. The number of quantum ratchet light absorption periods is 15. The P-type doping impurities are all C, which is the only difference from the previous MBE growth (doping impurities are Be).

Table 1. The wafer details of the ratchet detector. The period, thickness, and doping concentration of the epitaxy structure of wafers are given.

Layer ID	Period	Material	Composition	Thickness (Å)	Dopant	Conductive type	Concentration (cm^{-3})
10	--	GaAs	--	1000	C	p	1E+19
9	15	AlxGa1-xAs	x = 0.1	800	--	--	--
8	15	GaAs	--	200	C	p	1E+19
7	15	AlxGa1-xAs	x = 0 >> 0.2	800	--	--	--
6	15	GaAs	--	200	C	p	1E+19
5	--	AlxGa1-xAs	x = 0.1	800	--	--	--
4	--	GaAs	--	200	C	p	1E+19
3	--	AlxGa1-xAs	x = 0 >> 0.2	800	--	--	--
2	--	GaAs	--	7000	C	p	1E+19
1	--	GaAs	--	200	--	--	--
3-in., GaAs (100) 625 μm, semi-insulating substrate							

The photocurrent spectrum measurement is realized by a Fourier Transform Infrared Spectrometer (FTIR Bruker VERTEX 80 IFS 66 v/s) as shown in Fig. 1. The light emitted by the light source is first collimated into parallel light by an off-axis parabolic mirror, and then passes through the Michelson interferometer. The parallel light after the interference is converged on the sample by a parabolic mirror. The sample is placed in a cryostat. A current amplifier provides bias voltage and amplifies the current signal. After the amplification, the current signal is converted into a voltage signal and input into the FTIR analog-to-digital converter. Finally, the photocurrent response spectrum is obtained through the inverse Fourier transform in the computer.

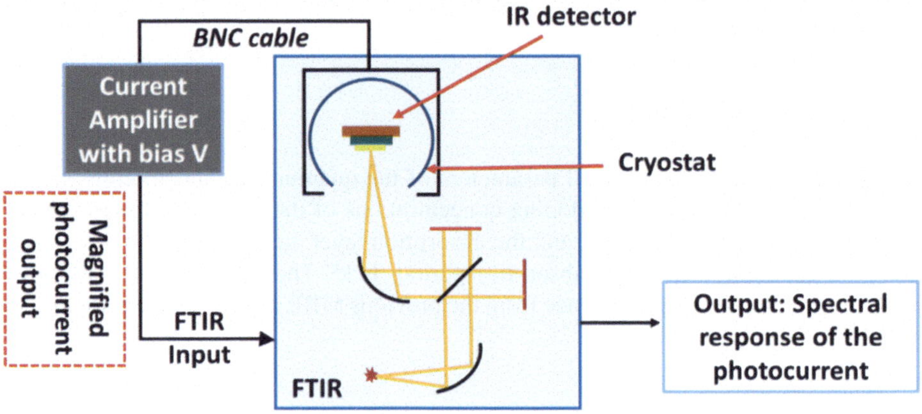

Fig. 1. Schematic diagram of the infrared detector photocurrent spectrum measurement experimental setup.

As shown in Fig. 2(a), the infrared photocurrent response spectrum of the quantum ratchet detector (beam splitter: KBr, cryostat window: ZnSe) is presented. We can find that the detector response band covers from 500 cm^{-1} to 5000 cm^{-1}. The ultra-broadband response is due to the simultaneous presence of free carrier absorption and SO-band absorption inside the detector. The multiple peaks and valleys within the range of 1000 cm^{-1}–2000 cm^{-1} are mainly caused by the multiple phonon absorption of GaAs and AlGaAs materials, and the two peaks within the range of 3000 cm^{-1}–4000 cm^{-1} are caused by the SO-Band absorption in the detector.

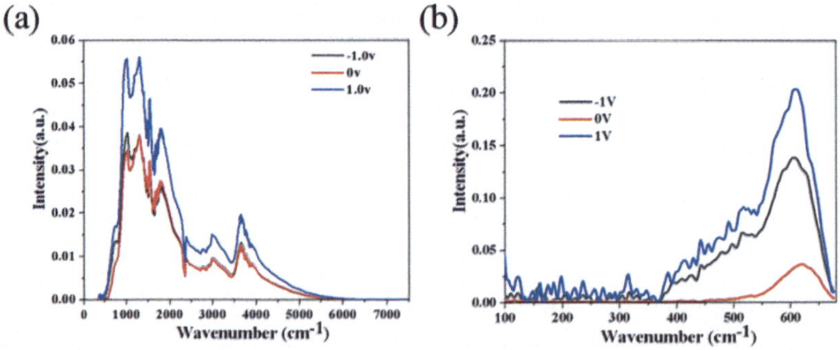

Fig. 2. (a) Infrared photocurrent response spectrum of the quantum ratchet detector, (b) THz band photocurrent response spectrum.

The terahertz band photocurrent response spectrum (beam splitter: Mayler film, cryostat window: HDPE) is shown in Fig. 2(b). Within the tested band and bias range, the cutoff wavelength of the detector response is 370 cm^{-1}, and the spectral coverage of the bias response is almost the same as that of the negative bias response (similar characteristics are also exhibited in the infrared band). This is completely inconsistent with the performance of the sample grown by MBE. There are two possible reasons for this situation. One possible reason is that within the tested bias range, the detector does not reach the electric field conditions for hot carrier injection, resulting in no obvious expansion of long-wave detection. Another possible reason is that the interface characteristics of the MOCVD material growth are not particularly good, and the truncation of the GaAs/AlGaAs heterojunction interface does not conform to the design parameters, resulting in less asymmetry inside the device than expected, thereby resulting in a situation where the positive and negative bias responses are consistent at low bias.

Although the photoelectric response characteristics of the detector under positive and negative biases are almost the same, we still measured a significant photocurrent at 0 bias, which proves that the asymmetry inside the device does exist. The response photocurrent at zero bias is the ratchet photocurrent, which is mainly caused by the asymmetry of the potential barrier inside the device. We measured the total ratchet photocurrent of the detector at different temperatures, and the results are shown in Fig. 3. The detectable temperature of the detector can be increased to 50K, which is nearly twice the operating temperature of the same type of ratchet detector [12]. There is still a lot of room for optimization of the ratchet detector, and it is expected to achieve a higher operating temperature.

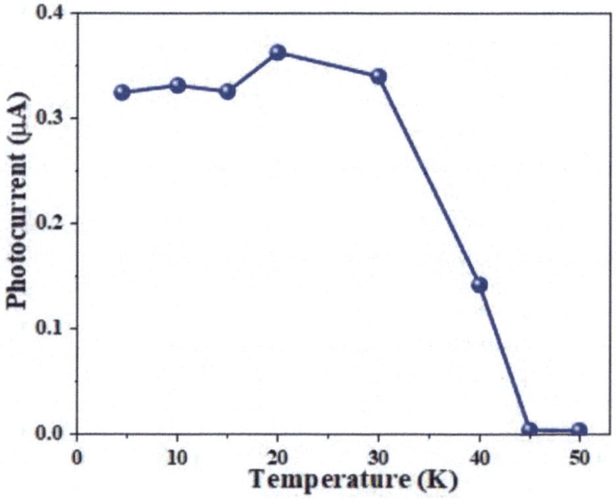

Fig. 3. The ratchet photocurrent measured at different temperatures.

3 Conclusion

In this paper, we experimentally studied the infrared/terahertz spectral response characteristics of MOCVD-grown quantum ratchet detectors. The results show that the detector has an ultra-wide spectral response capability, but the performance in the long-wave expansion is not particularly ideal. This may be because the MOCVD interface truncation characteristics are poor, resulting in a large deviation between the asymmetry of the ratchet barrier and the theoretical expectation. In addition, we also gave the ratchet photocurrent of the detector. At zero bias, the detector can generate a significant photocurrent under the action of infrared terahertz radiation. And the operating temperature of this ratchet photovoltaic effect can be as high as 50 K, which proves that the detector has great potential in high-temperature operation.

References

1. Kedem, O., Lau, B., Ratner, M.A., et al.: Light-responsive organic flashing electron ratchet. Proc. Natl. Acad. Sci. **114**(33), 8698–8703 (2017)
2. Lau, B., Kedem, O., Kodaimati, M., et al.: A silicon ratchet to produce power from below-bandgap photons. Adv. Energy Mater. **7**(22), 1701000 (2017)
3. Lau, B., Kedem, O.: Electron ratchets: state of the field and future challenges. J. Chem. Phys. **152**(20) (2020)
4. Xiang, Z., et al.: A primary model of THz and far-infrared signal generation and conduction in neuron systems based on the hypothesis of the ordered phase of water molecules on the neuron surface. Sci. Bull. **65**, 308 (2020)
5. Wu, K., et al.: Terahertz wave accelerates DNA unwinding: a molecular dynamics simulation study. J. Phys. Chem. Lett. **11**(17), 7002–7008 (2020)
6. Li, Y., et al.: Terahertz wave enhances permeability of the voltage-gated calcium channel. J. Am. Chem. Soc. **143**(11), 4311–4318 (2021)
7. Liu, G., et al.: Myelin sheath as a dielectric waveguide for signal propagation in the mid-infrared to terahertz spectral range. Adv. Funct. Mater. **29**(7), 1807862 (2019)
8. Zhang, J., et al.: Non-invasive, opsin-free mid-infrared modulation activates cortical neurons and accelerates associative learning. Nat. Commun. **12**, 2730 (2021)
9. Liu, X., et al.: Nonthermal and reversible control of neuronal signaling and behavior by midinfrared stimulation. Proc. Natl. Acad. Sci. **118**(10), e2015685118 (2021)
10. Bai, P., Li, X., Yang, N., et al.: Broadband and photovoltaic THz/IR response in the GaAs-based ratchet photodetector. Sci. Adv. **8**(21), eabn2031 (2022)
11. Bai, P., et al.: Ultra-broadband THz/IR upconversion and photovoltaic response in semiconductor ratchet-based upconverter. Appl. Phys. Lett. **119**(24) (2021)
12. Li, X.H., Huang, S.H., Yu, Q., et al.: High-temperature photon-type ultra-broadband detectors based on ratchet structure. Appl. Phys. Lett. **123**(6) (2023)

Recent Progress on Terahertz Biosensors

Xudong Zhang[1,2(✉)] and Longji Duan[1]

[1] School of Physics, University of Electronic Science and Technology of China, Chengdu 611731, China
zhangxudong@std.uestc.edu.cn
[2] Sun Yat-Sen University, Guangzhou 510275, China

Abstract. Terahertz (THz) sensors are novel approaches to investigate biological processes and biomolecular recognition non-destructively in a label-free way. Recent research endeavors have focused on enhancing the sensitivity and practicality of THz sensors. This paper introduces the cutting-edge design methods of THz sensors, such as surface plasmon resonance, bound states in the continuum, electromagnetically induced transparency and so on. Some typical terahertz sensor designs are summarized to illustrate how to realize intelligent and multi-mode sensing.

Keywords: Terahertz Biosensor · Metasurface · Multi-mode

1 Introduction

Positioned within the frequency spectrum between the microwave and infrared regions, terahertz (THz) waves exhibit promise potentials in matter [1–3] and biological studies [4–10], owing to satisfactory penetration, low energy, high resolution, transient performance and advancements in THz generation technology [11, 12]. THz biosensors represent innovative means for exploring biological processes and biomolecular recognition in a label-free and non-destructive method. They provide exceptional opportunities for biochemical analysis aimed at disease detection and diagnosis [13].

The effectiveness of terahertz biosensing is often hampered by insufficient interaction between analyte and THz field, resulting in low sensitivity [14]. Additionally, terahertz sensing encounters challenges related to the complexity of structure and the need for rapid multi-biomolecular detection. Addressing these issues will enhance the competitiveness of terahertz sensors across various biosensing applications.

This review article endeavors to offer comprehensive insight into recent advances in the design strategy of terahertz biosensors. Some typical terahertz sensor designs and principles will be summarized to realize multi-function and multi-mode sensing. An exploration of the future development prospects for THz biosensors are also presented here.

C. Chang et al. (Eds.): YTHZ 2024, SPPHY 400, pp. 89–93, 2024.
https://doi.org/10.1007/978-981-97-3786-4_18

2 Progress Review

2.1 Metasurface-Based Terahertz Biosensors

In the development of highly sensitive terahertz sensors, resonant metasurfaces with high-quality (Q) factors have attracted considerable attention. This is attributed to their ability to greatly enhance local optical near-fields and facilitate interactions between light and matter at specific frequencies in the subwavelength scale. Consequently, various sensor mechanisms have been proposed, such as surface plasmon resonance (SPR) [1], bound states in the continuum (BICs) [2, 3], electromagnetically induced transparency (EIT) [15, 16] and so on. Overall, these approaches offer low device-level complexity and eliminate the need for bulky and expensive spectrometers to access optical changes induced by biomolecules.

SPR techniques operate by resonantly exciting electromagnetic surface waves at the interface between a metal and dielectric material. This resonance is highly sensitive to the refractive index of the surrounding analyte. Etezadi *et al.* presented a label-free biosensing platform constituted of gold nanoantenna arrays providing localized surface plasmon resonances, which enables the dynamic monitoring of protein conformational changes [17].

By introducing asymmetry to the system or applying perturbations, an ideal BIC has the capability to be converted into a quasi-BIC. This offers an innovative approach for controlling light–matter interactions, as their customizable Q-factor and near-field enhancement can be adjusted depending on the level of introduced perturbations. Wang *et al.* proposed an on-chip ultra-sensitive THz plasmonic metasensor utilizing magnetic dipole quasi-BICs. Remarkably, the sensitivity of this device reaches up to 674 GHz/RIU [18].

In contrast to conventional electric and magnetic metasurfaces, toroidal metasurfaces are distinguished by poloidal currents circulating on the surface of a torus. EIT stands out as one of the most effective toroidal metasurfaces for achieving a high Q-factor response in toroidal metasurfaces by utilizing subwavelength meta-atoms periodically arranged on the substrate surfaces. Wang *et al.* demonstrated an EIT resonance-based metasurface biosensor for identifying cancer cells, significantly reduced the cost of detection to 1/30 and time consumption to 1/20 of traditional pathology testing methods. The sensor achieves a refractive index sensitivity of up to 504 GHz/RIU at 2.05 THz [19].

2.2 Multi-mode Terahertz Sensor

Conventional metasurface sensors usually rely on periodic structures with fixed resonant modes or frequencies to gather analyte information. However, this approach limits the practicality and flexibility of sensing applications. Exploring methods to enhance the acquisition and accumulation of information within a single device would significantly advance the development of THz sensors.

Leveraging special material properties, such as photosensitive materials, thermal materials, and anisotropic media, alongside novel methodologies, has jointly propelled the advancement of THz sensors toward reconfigurable and multi-mode capabilities [20–22].

Jiao *et al.* demonstrated a phototunable THz metasurface sensor [23]. By enhancing conductivity of the silicon bridge through the addition of an optical pump, two pairs of split-ring resonators establish an EIT resonance. The transmission spectra of the THz metasurfaces were measured under different pump fluence conditions. The optically controlled active metasurfaces achieve switchable sensing functions from a Lorentz-lattice to EIT. The frequency sensitivities are observed to be 118.4 GHz/RIU and 96.1 GHz/RIU, with a response time as short as 32 picoseconds. This work introduces a new approach for constructing dependable and adaptable metasurface sensors. Additionally, advancements are made in the development of terahertz ultrafast metasurfaces, which are both calibration-free and characterized by high precision and robustness [26, 27].

Angle multiplexing enables encoding the integration of various optical parameters, including polarization or phase characteristics, onto a single metasurface, expanding the design possibilities for nanophotonic devices. Aleksandrs Leitis *et al.* implemented a dielectric metasurface comprising an array of germanium resonators arranged in an anisotropic zigzag pattern, achieved a plethora of ultra-sharp and highly surface-sensitive resonances spanning from 1100 cm^{-1} to 1800 cm^{-1}, with spectral tunability characterized by increments smaller than 1.5 cm^{-1}. The schematic diagram of the working mechanism can be seen in Fig. 1. Within the operating range of the spectrum, each frequency can be uniquely associated with a specific angular position. This angle-multiplexed approach offers numerous customizable resonances in the mid-infrared region from a single metasurface chip, constrained solely by the incidence angle spectrum [24].

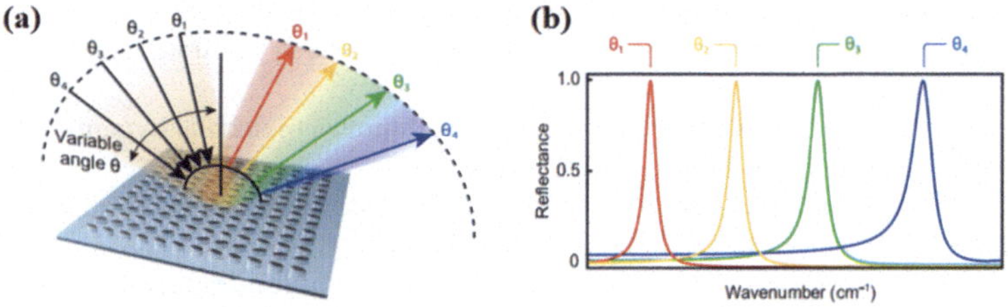

Fig. 1. Angle-multiplexed broadband fingerprinting. (a) A germanium-based high-Q all-dielectric metasurface. (b) Ultrasharp, angle-Specific resonances across broad spectral Range (reprinted from [24], Open Access Journal).

Numerous metasurface sensors have been proposed to leverage optical changes induced by bio-molecules, such as transmission or reflection spectra. However, the accompanying phase change information is often ignored. Z. Zhang *et al.* presented a novel paradigm utilizing significant phase shifts of surface waves, the schematic of the fabricated metasurfaces is illustrated in Fig. 1a. A metadevice utilizing Pancharatnam-Berry meta-atoms is presented, achieving sensitive detection for both transmission phase and amplitude across a continuous spectrum. Their results exhibit sensitivity up to 215.5°/RIU, marking a fourfold enhancement in phase-shift compared to conventional transmission-type metasurfaces.

3 Conclusion

In conclusion, we summarize the basic principle and several strategies of designing terahertz sensor, including surface plasmon resonance, bound states in the continuum, electromagnetically induced transparency and so on. Besides, some typical experiments are introduced to demonstrate how to conduct multi-function and multi-mode sensing. It will be essential to provide terahertz sensors of flexible, cost-effective, and suitable for deployment in complex detection environments in future research. As terahertz sensors become increasingly utilized in biological detection, we anticipate unprecedented breakthroughs in disease treatment, which will undoubtedly drive the advancement of biomedicine.

References

1. Kavungal, D., et al.: Artificial intelligence-coupled plasmonic infrared sensor for detection of structural protein biomarkers in neurodegenerative diseases. Sci. Adv. **9**(28), 9644 (2023)
2. Jahani, Y., et al.: Imaging-based spectrometer-less optofluidic biosensors based on dielectric metasurfaces for detecting extracellular vesicles. Nat. Commun. **12**(1), 3246 (2021)
3. Tittl, A., et al.: Imaging-based molecular barcoding with pixelated dielectric metasurfaces. Science **360**(6393), 1105–1109 (2018)
4. Li, Y., et al.: Terahertz wave enhances permeability of the voltage-gated calcium channel. J. Am. Chem. Soc. **143**(11), 4311–4318 (2021)
5. Liu, G., et al.: Myelin sheath as a dielectric waveguide for signal propagation in the mid-infrared to terahertz spectral range. Adv. Funct. Mater. **29**, 1807862 (2019)
6. Yu, Y., et al.: Terahertz photons improve cognitive functions in posttraumatic stress disorder. Research **6**, 0278 (2023)
7. Wu, K., et al.: Terahertz wave accelerates DNA unwinding: a molecular dynamics simula-tion study. J. Phys. Chem. Lett. **11**(17), 7002–7008 (2020)
8. Liu, X., et al.: Nonthermal and reversible control of neuronal signaling and behavior by mid-infrared stimulation. Proc. Natl. Acad. Sci. **118**(10), e2015685118 (2021)
9. Zhang, J., He, Y., Liang, S., et al.: Non-invasive, opsin-free mid-infrared modulation activates cortical neurons and accelerates associative learning. Nature Comm. **12**, 2730 (2021)
10. Yin, J., et al.: Terahertz photons inhibit cancer cells long-term by suppressing nano telomerase activity. ACS Nano (2024). https://doi.org/10.1021/acsnano.3c09216
11. Gao, M., et al.: Enhanced and frequency-tunable terahertz emission from two crossed air-plasmas. Phys. Rev. Res. **5**, 023091 (2023)
12. Chang, C., et al.: High-gain thompson-scattering X-ray free-electron laser by time-synchronic laterally tilted optical wave. Phys. Rev. Lett. **110**(6), 064802 (2013)
13. Wang, P., et al.: Progress on cutting-edge infrared-terahertz biophysics. IEEE Trans. Microwave Theory Tech., 5117–5140 (2022)
14. Wang, R., et al.: Electric Fano resonance-based terahertz metasensors. Nanoscale **13**, 18467–18472 (2021)
15. Zhang, S., et al.: Plasmon-induced transparency in metamaterials. Phys. Rev. Lett. **101**, 047401 (2008)
16. Limonov, M.F., et al.: Fano resonances in photonics. Nat. Photonics **11**, 543 (2017)
17. Etezadi, D., et al.: Real-time in situ secondary structure analysis of protein monolayer with mid-infrared plasmonic nanoantennas. ACS Sens. **3**(6), 1109–1117 (2018)
18. Wang, R., et al.: Ultrasensitive terahertz biodetection enabled by Quasi-BIC-based metasensors. Small **19**, 2301165 (2023)

19. Wang, P., et al.: An ultra-sensitive metasurface biosensor for instant cancer detection based on terahertz spectra. Nano Res. **16**, 7304–7311 (2023)
20. Fan, K., et al.: Phototunable dielectric huygens' metasurfaces. Adv Mater. **30**(22) (2018)
21. Hu, Y., et al.: Spatiotemporal terahertz metasurfaces for ultrafast all-optical switching with electric-triggered bistability. Laser Photonics Rev. **15** (2021)
22. Hu, Y., et al.: Bifunctional spatiotemporal metasurfaces for incident angle-tunable and ultrafast optically switchable electromagnetically induced transparency. Small **17**(21) (2021)
23. Jiao, Y., et al.: Photoactive terahertz metasurfaces for ultrafast switchable sensing of colorectal cells. Mater. Horiz. **9**, 2984–2992 (2022)
24. Leitis, A., et al.: Angle-multiplexed all-dielectric metasurfaces for broadband molecular fingerprint retrieval. Sci. Adv. **5**(5) (2019)
25. Zhang, Z., et al.: Advanced terahertz refractive sensing and fingerprint recognition through metasurface-excited surface waves. Adv. Mater. **36**, 2308453 (2024)
26. Lou, J., et al.: Calibration-free, high-precision, and robust terahertz ultrafast metasurfaces for monitoring gastric cancers. Proc. Natl. Acad. Sci. **119**, e2209218119 (2022)
27. Lou, J., et al.: Optically controlled ultrafast terahertz metadevices with ultralow pump threshold. Small **17**, 2104275 (2021)

Power Detection of Solid-State Terahertz Transmitters: Terahertz Induced Thermoacoustic Signal and Its Characteristics

Weipeng Wang[1,2,3], Lin Huang[1,2,3](✉), Hongji Zhou[1,2,3], Sen Gong[1,2,3], Hongxin Zeng[1,2,3], Jun Zhou[1,2,3], Huajie Liang[1,2,3], Dan Liang[1,2,3], Tao Jiang[1,2,3], Cong Dai[1,2,3], Ziqiang Yang[1,2,3], and Yaxin Zhang[1,2,3](✉)

[1] School of Electronic Science and Engineering, University of Electronic Science and Technology of China, Chengdu 611731, China
{lhuang,zhangyaxin}@uestc.edu.cn

[2] Yangtze Delta Region Institute, University of Electronic Science and Technology of China, Huzhou 313001, China

[3] Engineering Center of Integrated Optoelectronic and Radio Meta-Chips, University of Electronic Science and Technology, Chengdu, China

Abstract. We propose a real-time terahertz (THz) power detection method based on the terahertz-thermoacoustic (THz-TA) effect. Our self-made THz transmitter operates at a center frequency of 0.17 THz with a maximum output power of 92.8 mW. The test results reveal a linear relationship between the amplitude of THz-TA signals and the output power of the THz transmitter within the range of 2.8–92.8 mW (Pearson correlation coefficient, R = 0.99571). Additionally, we observe that the time interval of the TA signal evoked by THz is consistent with the pulse width of the THz signal (0.5, 1, 2, and 5 ms). Thus, the THz-TA technology enables real-time THz power detection and provides insights into the pulse width of amplitude-modulated terahertz signals.

Keywords: Power detection · Solid-state terahertz (THz) transmitter · Thermoacoustic (TA)

1 Introduction

In 2019, the Federal Communications Commission initiated research on terahertz (THz) spectrum services within the frequency range of 95 GHz to 3 THz and issued a 10-year trial spectrum license for network services. Concurrently, the definition of the sixth-generation (6G) communication network began [1]. To enable the deployment of THz base stations in the 6G network, the development of cost-effective THz power detection technology is crucial for rapid testing of base station transmission power. While existing methods such as Golay cells, bolometers, pyroelectric detectors, and plasmonic detectors offer wide detection bandwidth, high sensitivity, and accuracy, they suffer from limitations in portability and cost [2–15].

C. Chang et al. (Eds.): YTHZ 2024, SPPHY 400, pp. 94–98, 2024.
https://doi.org/10.1007/978-981-97-3786-4_19

To reduce costs, we use THz waves to excite TA signals. Determine the power of THz waves by measuring the amplitude of the TA signals. Compared to other researchers' exploration of THz photoacoustic signals [16, 17], we explore THz-thermoacoustic (THz-TA) signals excited by a solid-state THz transmitter (center frequency = 0.17 THz, maximum output power = 92.8 mW) for the first time, providing valuable insights into this novel detection approach.

2 Results

Figure 1 (a) illustrates a compact THz-TA test platform utilizing a self-made solid-state THz transmitter. The setup includes the following components:

1. A signal generator (E8257D, Keysight, USA) produces a pulse-modulated signal with the following parameters: carrier frequency = 14.1666667 GHz, pulse repetition frequency = 20 kHz. This signal is input to the solid-state THz transmitter via a flexible coaxial cable. The synchronous signal from the generator serves as a trigger for the oscilloscope (RTO1014, Rohde & Schwarz, Germany).
2. The solid-state THz transmitter multiplies the frequency of the pulse-modulated signal by 12 (resulting in 170 GHz), amplified the power, and radiated THz signal through a WR-5 waveguide.
3. The ultrasonic signal is transmitted to the transducer's surface and converted into an electrical signal. The electrical signal is amplified by a pre-amplifier (PREAMP2-D9V, Ultratek, USA) with a gain of 57 dB. Then it is processed by an ultrasonic pulser/receiver (DPR300, JSR Ultrasonics Inc, USA) with a gain of 19 dB and a filter range of 1 to 5 MHz. After input into the oscilloscope (sampling rate = 2.5 GSa/s, impedance = 50 Ω) for screening and data acquisition, it is averaged over 5000 measurements (taking approximately 0.25 s).

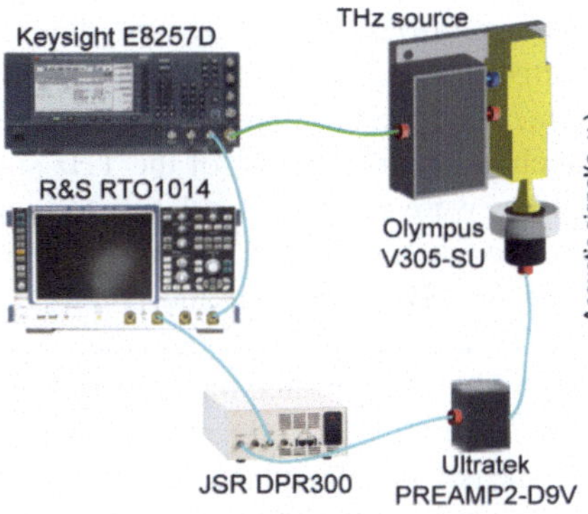

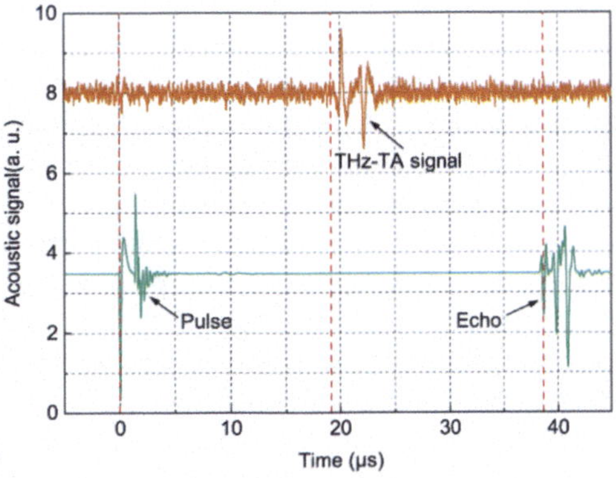

Fig. 1. (a) The schematic depicts the THz-TA test platform. At the front end of the focused transducer, a 25.4 mm long plastic sleeve filled with soy sauce is mounted. This configuration ensures effective reception of THz-TA signals. (b) The time-domain waveforms are shown for both the THz-TA signal (magnified 3000 times) and the pulse-echo signal.

In Fig. 1 (b), the time-domain waveform of the THz-TA signal is prominently displayed at the top, magnified 3000 times for ease of observation. To validate that this signal indeed corresponds to the thermoacoustic (TA) effect, we conducted a pulse-echo test. Notably, the THz-TA signal occurred at approximately 19.2 μs, while the echo signal occurred at approximately 38.7 μs. The observed time difference confirms the association of the top signal with the THz-TA effect.

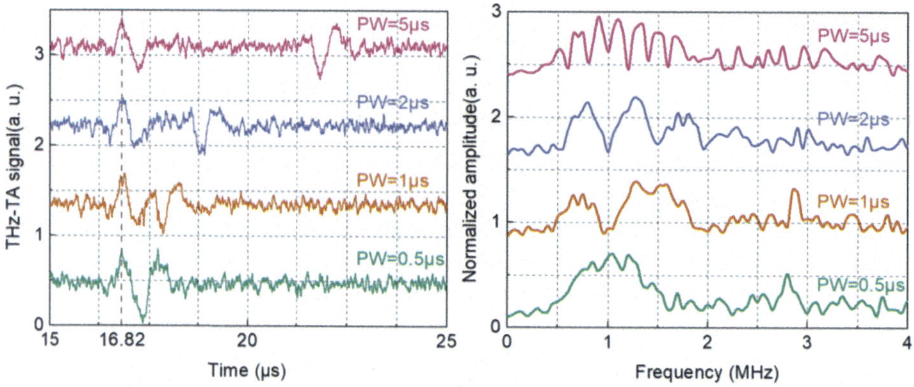

Fig. 2. (a) The time-domain waveforms of THz-TA signals are presented for four different pulse widths (PW). Notably, the first spike of THz-TA signals occurs at 16.82 μs. These waveforms provide insights into the behavior of the THz-TA effect as a function of pulse width. (b) The corresponding frequency-domain representation of the THz-TA signals from (a) is shown. This transformation allows us to analyze the spectral characteristics and energy distribution of the THz-TA effect.

We adjusted the pulse width of the pulse-modulated THz signal and meticulously analyzed the time interval between positive and negative spikes of the THz-TA signal. When the two edges of the signal waveform exhibit symmetry, the corresponding TA signals possess equal amplitude and inverse phase. Figure 2 (a) illustrates the time-domain waveforms of THz-TA signals for four pulse widths (0.5, 1, 2, and 5 μs). Notably, the first spike of THz-TA signals appeared at 16.82 μs. We observed that the time interval between the first and second spikes precisely matched the pulse width of the THz signal. Additionally, the waveforms of these two spikes were symmetric relative to their horizontal baselines. As the pulse width narrowed, the first and second spikes gradually overlapped, impacting the peak-to-peak voltage of the THz-TA signal and potentially influencing the sensitivity of our THz-TA test platform. Figure 2 (b) depicts the corresponding spectrum, revealing that narrower pulse widths concentrated the energy of the THz-TA signal at lower frequencies.

To achieve optimal sensitivity, we measured the peak-to-peak voltage corresponding to different pulse widths of THz-TA signals. Figure 3 (a) presents the arithmetic mean results after three tests. Initially, as the pulse width narrowed, the peak-to-peak voltage of the THz-TA signal increased. However, beyond a certain point, the voltage began to decrease due to the overlapping of the two spikes in opposite polarity. Notably, fluctuations occurred when adjusting the pulse width with the minimum step of the signal generator (0.01 μs). These fluctuations were attributed to the limited passband of the transducer and uneven amplifier gain following changes in the THz-TA signal spectrum. Intriguingly, the THz-TA signal abruptly disappeared when the pulse width transitioned from 0.42 to 0.41 μs. Unfortunately, further investigation was hindered by the signal generator's limited minimum step.

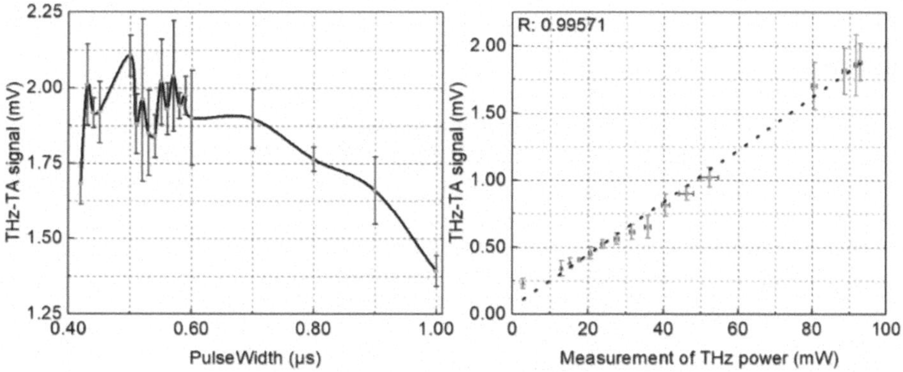

Fig. 3. (a) The plot illustrates the relationship between the peak-to-peak voltage of the THz-TA signal and the pulse width of the THz signal. The data points represent the mean peak-to-peak voltage, with error bars indicating the standard deviation. The solid line depicts the changing trend of the THz-TA signal's peak-to-peak voltage. (b) This graph shows the functional relationship between the peak-to-peak voltage of the THz-TA signal and the transmitting power of the solid-state THz transmitter (measured using a THz power meter). The data points represent the mean peak-to-peak voltage, with x and y error bars indicating the standard deviations. The dotted black line represents the best-fit line, with a Pearson's R value of 0.99571.

Then, we selected a pulse width of 0.5 μs for THz power testing based on the results from Fig. 3 (a). By adjusting the output power of the signal generator, we recorded the peak-to-peak voltage of the THz-TA signal until it became unreadable on the oscilloscope. Subsequently, we connected a commercial THz power meter (PM5, Erickson, USA) to the WR-5 waveguide. The minimum detectable THz power by our THz-TA test platform is 2.8 mW, while the maximum measurable THz power is 92.8 mW. The linear dependence between THz-TA signal and THz power, as indicated by a Pearson's R value of 0.99571, aligns with theoretical expectations [18].

3 Summary

We constructed a compact THz-TA test platform using a self-made solid-state THz transmitter and explored the interplay between THz power, pulse width, and THz-TA signals. Our findings reveal that the THz-TA test platform can detect a minimum THz power of 2.8 mW, with a rapid response time of approximately 0.25 s (averaging over 5000 measurements at a 20 kHz pulse repetition rate). Within the power range of 2.8–92.8 mW, the output power of the THz transmitter exhibits a linear relationship with the peak-to-peak voltage of the THz-TA signal (Pearson's R = 0.99571).

Acknowledgment. This work was supported by the National Key Research and Development Program of China (2021YFA1401000); National Natural Science Foundation of China (61931006, 62131007, U20A20212, 61921002, U1930127); Sichuan Science and Technology Program (2020JDRc0028); the Fundamental Research Funds for the Central Universities (ZYGX2020ZB011, ZYGX2019J013); The China Postdoctoral Science Foundation (2021M700706, 2020M683285).

References

1. Zhang, Z., et al.: IEEE Veh. Technol. Mag. **14**, 28 (2019)
2. Corcos, D., et al.: IEEE Trans. Terahertz Sci. Technol. **5**, 902 (2015)
3. Kuznetsov, S.A., Paulish, A.G., Navarro-Cía, M., Arzhannikov, A.V.: Sci. Rep. **6**, 21079 (2016)
4. Otsuji, T.: IEEE Trans. Terahertz Sci. Technol. **5**, 1110 (2015)
5. Liu, G., et al.: Adv. Funct. Mater. **29**(7), 1807862 (2019)
6. Li, Y., et al.: J. Am. Chem. Soc. **143**(11), 4311–4318 (2021)
7. Xiang, Z., et al.: Sci. Bull. **65**, 308 (2020)
8. Wu, K., et al.: J. Phys. Chem. Lett. **11**(17), 7002–7008 (2020)
9. Zhang, J., et al.: Nat. Commun. **12**, 2730 (2021)
10. Lou, J., et al.: Proc. Natl. Acad. Sci. **119**(43), e2209218119 (2022)
11. Zhang, Z., et al.: Adv. Mater., 2308453 (2024)
12. Zhu, Z., et al.: Nano Lett. **24**(10), 3243–3248 (2024)
13. Wang, R., et al.: Small **19**, 2301165 (2023)
14. Zhu, Z., et al.: Analyst **144**, 2504–2510 (2019)
15. Wang, P., et al.: IEEE Trans. Microwave Theory Tech. **70**(11), 5117–5140 (2022)
16. Tsubouchi, M., Hoshina, H., Nagai, M., Isoyama, G.: Sci. Rep. **10**, 18537 (2020)
17. Li, J., et al.: Adv. Photon. **3** (2021)
18. Xu, M., Wang, L.V.: IEEE Trans. Med. Imaging **21**, 814 (2002)

Enhanced Fano Metasurfaces for Tailored Spectroscopic Detection and Exploration of Planar Structure Generality

Hao Ruan and Yifei Feng(✉)

Innovation Laboratory of Terahertz Biophysics, National Innovation Institute of Defense Technology, Beijing 100071, China
rh06293055@163.com

Abstract. Metasurfaces are composite materials composed of subwavelength structures arranged periodically. By designing the structural shape and excitation methods, it can exhibit fascinating optical and physical properties not present in natural materials. This property has facilitated the application of terahertz technology in sensing, gradually enriching research in the terahertz spectral range. In this work, we proposed a terahertz plasmonic biosensor based on Fano resonance, and achieved fingerprint spectroscopy detection of L-glutamine by adjusting the structural parameters of the metasurface. Moreover, we investigated the generality of high-quality factor Fano resonances in planar structures and explored the control of the number of resonant eigenpeaks by varying the number of antennas. Our approach advances the application of terahertz metasurfaces in biosensing, stimulating continuous exploration of metamaterial structure design.

Keywords: Metasurfaces · Terahertz · Fano resonance · Biosensing

1 Introduction

Terahertz (THz) spectroscopy plays a crucial role in label-free and non-destructive sensing due to its ability to capture the unique biomolecular absorption characteristics associated with molecular rotational and vibrational modes [1–7]. Terahertz biosensors based on meta-surface have made some progresses [8–14]. However, the mismatch between THz wavelengths and the absorption cross-section of analytes results in weak interactions between light and matter at the micrometric scale [15]. This limitation poses a significant obstacle to the practical application of THz technology in biochemical molecular recognition. To overcome this challenge, the proposed micro/nanophotonics approach leverages subwavelength resonators to greatly enhance the near-field interactions [16]. This advancement addresses specific practical applications of THz technology and empowers unprecedented control over light. If the resonance and absorption fingerprints align in the spectrum, the intensified near-field will enhance the coupling between molecules and resonators. This, in turn, impacts the linewidth and intensity of the resonance, allowing for the extraction of molecular fingerprints from the resulting changes.

C. Chang et al. (Eds.): YTHZ 2024, SPPHY 400, pp. 99–102, 2024.
https://doi.org/10.1007/978-981-97-3786-4_20

Here, we designed a subwavelength antenna structure with simple structural parameters, which greatly facilitates the control of resonance peak center frequency and achieves broadband enhancement for terahertz fingerprint detection of two-dimensional materials and biomolecules. The designed sensor achieves Fano resonance by introducing structural asymmetry of microstructure [17]. The configuration is ingeniously designed to enhance biosensing capabilities, facilitating the detection of molecular absorption signatures at multiple spectral points through robust near-field coupling. This chemical specific method allows for the precise resolution of absorption fingerprints, eliminating the need for frequency scanning or mechanical motion. By enhancing the performance of Terahertz sources [18, 19], as a result, it opens doors to highly sensitive and versatile miniaturized THz spectroscopy devices, useful for detecting bio-molecules [20–22].

2 Method

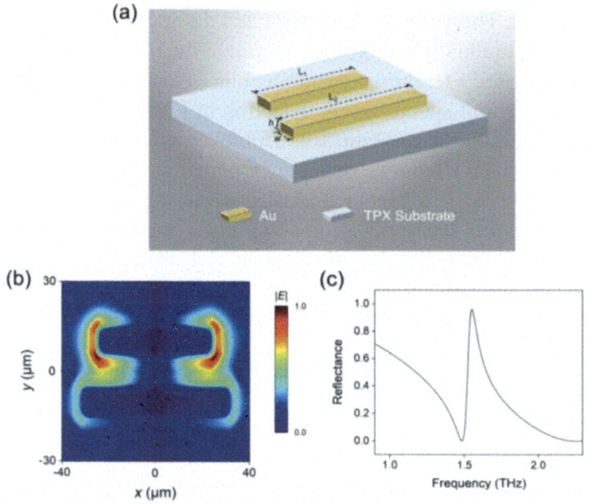

Fig. 1. (a) Sketch of the unit cell with geometric parameters. (b) The electric field energy distribution corresponds to Fano resonance in the reflection spectrum. (c) The reflectance spectrum of the metasurface.

The unit cell with a meta-atom on the substrate is outlined in the Fig. 1(a). In order to authenticate the metasurface's response, we employed numerical computations using 3D finite-difference time-domain (FDTD) solution software. In the simulations, we represented the conductivity of the gold material using a perfect conductor, while the substrate employed TPX material known for its high THz transmittance. The THz beam is incident vertically upwards at the bottom and acts on the antenna structure at the top, then detects the reflection spectrum of the metasurface. The unit cell consisted of two parallel linear structures with lengths denoted as $L_1 = 50$ μm and $L_2 = 65$ μm, each antenna having a width of $w = 10$ μm, height of $h = 5$ μm, and a separation of 20 μm. Furthermore, to elucidate the resonant mechanism of this structure, the corresponding electric field distribution at the resonant frequencies is provided in the figure. Figure 1(b) shows that typical electric dipolar mode are excited. Notably, the typical electric field concentrated mainly in the gaps and at the end of antennas. Figure 1(c) indicate that the asymmetry and sharp increase in slope of the spectral lines are achieved in the Fano resonance due to the unequal lengths of closely spaced subwavelength antennas, disrupting the symmetry of the structure.

3 Result

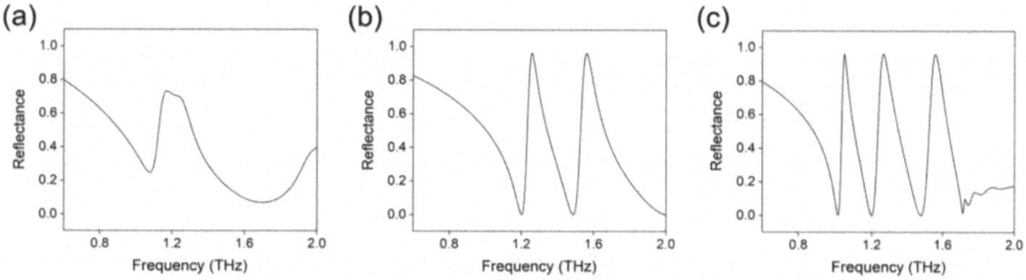

Fig. 2. (a) Simulated reflection spectra of the metasensor for measuring L-glutamine. (b) Simulated reflection spectra obtained from three metal antennas with lengths of $L_1 = 50$ μm, $L_2 = 65$ μm, and $L_3 = 80$ μm. (c) Simulated reflection spectra obtained from four metal antennas with lengths of $L_1 = 50$ μm, $L_2 = 65$ μm, $L_3 = 80$ μm, and $L_4 = 95$ μm.

To validate the sensor's ability to recognize molecular fingerprints, we chose the common biological sample L-glutamine as the analyte. Based on prior research, the absorption fingerprint of L-glutamine in the terahertz frequency band is identified at 1.23 THz. In the recorded reflectance spectrogram, depicted in Fig. 2(a), distinct fingerprint spectra emerged around 1.23 THz. This outcome signifies the commendable performance of our sensor in molecular fingerprint recognition.

The geometric parameters of the unit cell structure play a crucial role in resonance effects. To investigate this phenomenon, Fig. 2(b) and (c) display simulated reflection spectra obtained with different numbers of metal strips. When the lengths of the strips are set to $L_1 = 50$ μm, $L_2 = 65$ μm, and $L_3 = 80$ μm, two resonance peaks appear at 1.26 THz and 1.56 THz. Introducing a fourth strip with a length of $L_4 = 95$ μm results in three resonance peaks at 1.05 THz, 1.56 THz, and 1.28 THz. Similarly, by adding corresponding strips, more multispectral metamaterials can be achieved. Therefore, compared to metamaterials composed of complex unit cells, the proposed structure can be more easily manufactured, especially in achieving multispectral devices.

In summary, we investigated plasma biosensors utilizing metal metamaterials. The efficacy of the designed metasurface in fingerprint recognition of biomolecules was confirmed by analyzing spectral positions. Subsequent investigations revealed that the augmentation of resonance peaks could be achieved straightforwardly by incorporating additional gold strips within each unit cell. This finding holds promise for the exploration of planar or integrated microstructures characterized.

References

1. Li, Y., et al.: Terahertz wave enhances permeability of the voltage-gated calcium channel. J. Am. Chem. Soc. **143**(11), 4311–4318 (2021)
2. Liu, X., et al.: Nonthermal and reversible control of neuronal signaling and behavior by midinfrared stimulation. Proc. Natl. Acad. Sci. **118**, 10, e2015685118 (2021)
3. Zhang, J., et al.: Non-invasive, opsin-free mid-infrared modulation activates cortical neurons and accelerates associative learning. Nat. Comm. **12**, 2730 (2021)
4. Liu, G., et al.: Myelin sheath as a dielectric waveguide for signal propagation in the mid-infrared to terahertz spectral range. Adv. Funct. Mater. **29**(7), 1807862 (2019)
5. Yu, Y., et al.: Terahertz photons improve cognitive functions in posttraumatic stress disorder. Research **6**, 0278 (2023)
6. Xiang, Z., et al.: A primary model of THz and far-infrared signal generation and conduction in neuron systems based on the hypothesis of the ordered phase of water molecules on the neuron surface. Sci. Bull. **65**(308) (2020)
7. Wu, K., et al.: Terahertz wave accelerates DNA unwinding: a molecular dynamics simulation study. J. Phys. Chem. Lett. **11**(17), 7002–7008 (2020)
8. Lou, J., et al.: Calibration-free, high-precision, and robust terahertz ultrafast metasurfaces for monitoring gastric cancers. PNAS **119**(43), e2209218119 (2022)
9. Sun, L., et al.: A pixelated frequency-agile metasurface for broadband terahertz molecular fingerprint sensing. Nanoscale **14**(27), 9681–9685 (2022)
10. Wang, R., Xu, L., Huang, L., et al.: Ultrasensitive terahertz biodetection enabled by quasi-BIC-based metasensors. Small **19**(35), 2301165 (2023)
11. Zhang, Z., et al.: Advanced terahertz refractive sensing and fingerprint recognition through metasurface-excited surface waves. Adv. Mater., 2308453 (2024)
12. Wang, P., et al.: Progress on cutting-edge infrared-terahertz biophysics. IEEE Trans. Microw. Theory Tech. **70**(11), 5117–5140 (2022)
13. Ma, Z., et al.: Identification and quantitative detection of two pathogenic bacteria based on a terahertz metasensor. Nanoscale **15**(2), 515–521 (2023)
14. Jiao, Y., et al.: Photoactive terahertz metasurfaces for ultrafast switchable sensing of colorectal cells. Mater. Horiz. **9**(12), 2984–2992 (2022)
15. Wang, R., Xu, L., Wang, J., et al.: Electric Fano resonance-based terahertz metasensors. Nanoscale **13**(44), 18467–18472 (2021)
16. Lou, J., et al.: Optically controlled ultrafast terahertz metadevices with ultralow pump threshold. Small **17**, 2104275 (2021)
17. Wang, P., et al.: An ultra-sensitive metasurface biosensor for instant cancer detection based on terahertz spectra. Nano Res. **16**, 7304–7311 (2023)
18. Chang, C., et al.: High-gain thompson-scattering X-ray free-electron laser by time-synchronic laterally tilted optical wave. Phys. Rev. Lett. **110**, 064802 (2013)
19. Gao, M., et al.: Enhanced and frequency-tunable terahertz emission from two crossed air-plasmas. Phys. Rev. Res. **5**, 023091 (2023)
20. Zhu, Z., et al.: Broadband Terahertz signatures and vibrations of dopamine. Analyst **145**(18), 6006–6013 (2020)
21. Zhu, Z., et al.: Characteristic fingerprint spectrum of neurotransmitter norepinephrine with broadband terahertz time-domain spectroscopy. Analyst **144**, 2504 (2019)
22. Tang, C., et al.: Integrating terahertz metamaterial and water nanodroplets for ultrasensitive detection of amyloid β aggregates in liquids. Sens. Actuators B Chem. **329**, 129113 (2021)

Analysis of Argon Gas Plasma Density and Transmittance Characteristics by Terahertz Time-Domain Spectroscopy

Haitao Tao[1,2], Ming Fang[1], Haiqing Liu[2,3(✉)], Susu Hu[3], Yinxian Jie[3], and Chun Zhou[3]

[1] School of Electronic and Information Engineering, Anhui University, Hefei 230039, China
mingfang@ahu.edu.cn
[2] Insitute of Energy, Hefei Comprehensive National Science Center, Hefei 230031, Anhui, China
[3] Institute of Plasma Physics, Hefei Institutes of Physics Science, Chinese Academy of Sciences, Hefei 230031, China
hqliu@ipp.ac.cn

Abstract. The diagnostics of plasma density is of great significance. In this paper, a plasma diagnostic system by terahertz time-domain spectroscopy (THz-TDS) based on fiber laser is introduced. The plasma discharge is carried out by an inductively coupled plasma source (ICP). The corresponding electron density and the transmittance of terahertz wave through plasma are measured under different radio frequency (RF) power and different Argon gas pressure in the cavity. The electron density increases when the RF power is increased or the Argon gas pressure is increased. Furthermore, the transmittance of terahertz wave decreases when the RF power or the Argon gas pressure is increased. This experiment further expands the application ability of THz-TDS technology in the diagnostics of gas discharge plasma parameters, and provides an important reference for future applications in magnetic confinement nuclear fusion plasma.

Keywords: THz-TDS · ICP · Plasma Density · Transmittance

1 Introduction

Plasma is a kind of gas with obvious electromagnetic properties and electrical neutrality. Plasma is widely used in biomedical sterilization, industrial machining, integrated circuit etching and slag metallurgy [1, 2]. As one of the important core parameters of plasma, it is of great significance to study electron density. The common diagnostic methods of plasma density include Langmuir probe method, Thomson scattering method, microwave interference method and spectroscopy method. Langmuir probe method requires direct contact with plasma and is greatly affected by plasma temperature. The signal-to-noise ratio of Thomson scattering method is poor. Microwave interference method requires different frequency light sources and the optical path is complex, and the data processing of spectroscopy method is difficult. Therefore, a more compact and convenient non-contact diagnostic method is urgently needed.

C. Chang et al. (Eds.): YTHZ 2024, SPPHY 400, pp. 103–107, 2024.
https://doi.org/10.1007/978-981-97-3786-4_21

Thanks to the maturity of photoconductive antenna technology and the development of femtosecond laser in terahertz frequency domain technology, terahertz time-domain spectroscopy (THz-TDS) diagnostics of plasma density has received extensive attention in the past 20 years. Since the refractive index of the plasma is not 1, the phase of the terahertz wave will change after passing through the plasma. The terahertz time-domain spectroscopy technology obtains the absorption coefficient and refractive index of the plasma by obtaining the amplitude and phase information of the terahertz wave [3–5]. In 2003, D. R. Jones *et al.* first used THz-TDS technology to diagnose the characteristics of gas discharge plasma [6]. Later, Ayumi Ando *et al.* [7] and Y. Zhao *et al.* [8] compared THz-TDS with Langmuir probe method and Thomson scattering method respectively, which confirmed the reliability of the method. Subsequently, some scholars used THz-TDS with asynchronous optical sampling technology and single measurement technology to shorten the sampling time to millisecond level. In 2023, Ding *et al.* combined traditional terahertz time-domain spectroscopy with asynchronous optical sampling technology to provide an effective guarantee for the reliability of the system [9]. However, at present, terahertz time-domain spectroscopy is mainly used to analyze the characteristics of gas discharge plasma density.

This paper introduces the terahertz time-domain spectroscopy system based on fiber laser. The plasma is generated by an inductively coupled plasma source (ICP), and the pressure of the gas entering the plasma source cavity and radio frequency (RF) power are changed respectively. The phase and amplitude information of the terahertz wave are obtained by the terahertz time-domain spectroscopy system, and then the information of the plasma density and the transmittance of the terahertz wave through the plasma is analyzed.

2 Experimental Setup

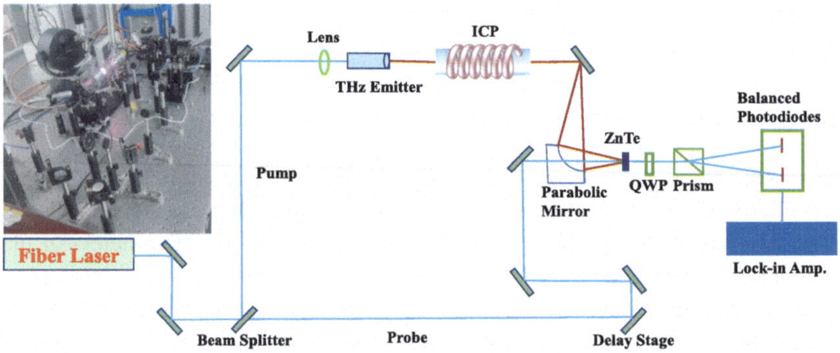

Fig. 1. The diagram of plasma diagnostic system by terahertz time-domain spectroscopy

The schematic diagram of the experimental system is shown in Fig. 1. A fiber femtosecond laser (center wavelength 780 nm, repetition rate 70 MHz, pulse width 150 fs) is used as the pump source for the terahertz pulse. Two reflectors are used to collimate the output beam from the fiber laser. After passing through a beam splitter (transmission 20%, reflection 80%), it is divided into a pump beam and a probe beam. Then the probe beam is passed by two optical reflectors on the translation stage to reach the detector. The power of the pump light after the beam splitter is 24 mW. The pump beam is focused on the gap of the photoconductive antenna through an aspheric lens (f = 8 mm). The photoconductive antenna acts as a transmitter of the terahertz wave, which is excited under an alternating voltage with a repetition frequency of 10 kHz and peak value of 20 V. The terahertz wave is reflected by a gold-coated mirror onto a gold-coated off-axis parabolic mirror (f = 50.8 mm) through an inductively coupled plasma source. The probe light and the terahertz wave arrive at the ZnTe crystal simultaneously after passing through the off-axis parabolic mirror. After passing through the quarter-wave plate and the Wollaston prism, two orthogonal beams are extracted and received by the two receivers of the balanced detector, and the instantaneous electric field of the terahertz wave is recorded by the lock-in amplifier.

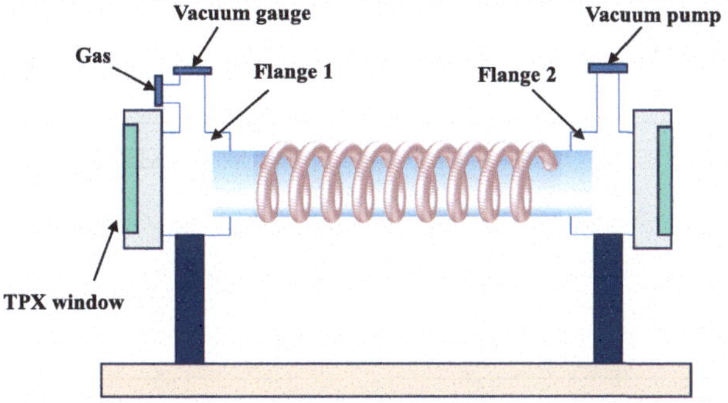

Fig. 2. Diagram of Inductively coupled plasma source (ICP)

The inductively coupled plasma source used in this experiment is shown in Fig. 2. It is composed of two flanges (Flange 1, Flange 2) and a quartz tube (length 160 mm, outer diameter 30 mm, inner diameter 28 mm). Both flanges are equipped with a self-made TPX window (diameter 50.8 mm, thickness 3 mm). Flange 1 is connected to the vacuum gauge (ZDO-53Bv01) and the air inlet to ensure real-time monitoring of the pressure in the cavity. Flange 2 is connected to the vacuum pump (DUO 20M) system, and the limit vacuum is 5×10^{-2} Pa. The RF power supply (AG-1305, RF power 0–500 W, frequency 13.56 MHz) matches the RF source to the copper ring antenna through a matcher, and is connected to a water-cooling system at both ends of the antenna to ensure the long-time stable operation of the plasma.

3 Experimental Result

Detailed calculation of electron density can be seen in reference [10]. When the terahertz wave passes through the plasma, the terahertz wave will be dispersed, resulting in a phase difference, and the refractive index of the plasma can be obtained by the phase shift to obtain the electron density. The THz-TDS signals at time domain with and without plasma are shown in Fig. 3(a). Fourier transform is performed to obtain the THz signal at frequency domain, as shown in Fig. 3(b).

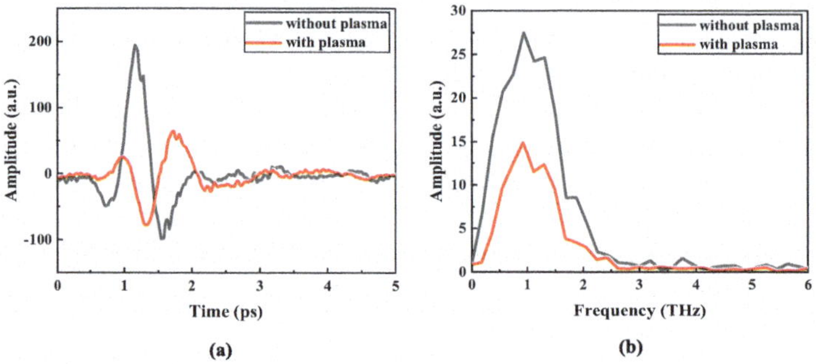

Fig. 3. (a) THz-TDS signal at time domain with and without plasma; (b) The corresponding THz signal at frequency domain.

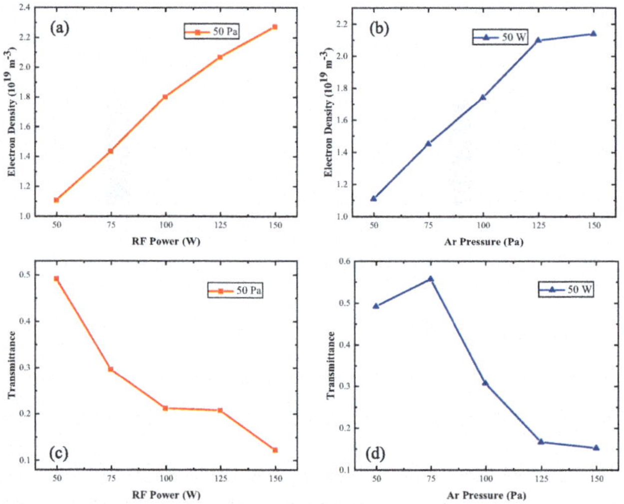

Fig. 4. (a) Plasma density versus RF power; (b) Plasma density versus intracavity Ar pressure; (c) Terahertz wave transmittance versus RF power; (d) Terahertz wave transmittance versus intracavity Ar pressure.

It is easy to obtain the plasma density from the terahertz time domain signals with and without plasma by the method in reference [6]. By calculating the change in the amplitude of the terahertz wave in frequency domain, the transmittance of the terahertz wave through the plasma can also be obtained.

By changing the RF power of the plasma source and the Ar gas pressure in the cavity, the plasma density changes are shown in Fig. 4 (a) and Fig. 4(b), and the transmittances of the terahertz wave through the plasma are shown in Fig. 4 (c) and Fig. 4(d). It can be seen that the electron density increases when the RF power is increased or the Argon gas pressure is increased. Furthermore, the transmittance of terahertz wave decreases when the RF power is increased or the Argon gas pressure is increased. This indicating that the higher the plasma density, the lower the transmittance, which is in good agreement with the theoretical calculation.

4 Conclusion

In this paper, a plasma diagnostic system utilizing terahertz time-domain spectroscopy based on fiber laser has been developed. The plasma density and the transmittance of terahertz wave through plasma are studied with a self-made inductively coupled plasma source. The experimental results show that when the RF power increases or the Ar gas pressure increases, the plasma density will increase and the transmittance of terahertz wave will decrease. It further improves the application prospect of THz-TDS technology in the diagnostics of gas discharge plasma parameters, and provides an important reference for future applications in magnetic confinement nuclear fusion plasma.

Acknowledgements. This work was supported by the Institute of Energy, Hefei Comprehensive National Science Center (Grant No. 21KZZ517, 22KZZ514, 24JYJB01); the University Synergy Innovation Program of Anhui Province (No. GXXT-2022-014, GXXT-2021-028); the HFIPS Director's Fund (Grant No. YZJJKX202301); and the CAS Project for Young Scientists in Basic Research (Grant No. YSBR-059).

References

1. Park, G.Y., et al.: Atmospheric-pressure plasma sources for biomedical applications. Plasma Sources Sci. Technol. **21**, 043001 (2012)
2. Samal, S.: Thermal plasma technology: the prospective future in material processing. J. Clean. Prod. **142**, 3131–3150 (2017)
3. Gao, M., et al.: Enhanced and frequency-tunable terahertz emission from two crossed air-plasmas. Phys. Rev. Res. **5**, 023091 (2023)
4. Huang, Y., et al.: Localized-plasma-assisted rotational transitions in the terahertz region. Phys. Rev. A **103**(3), 033109 (2021)
5. Xu, X., et al.: Laser-chirp controlled Terahertz wave generation from air plasma. Chin. Phys. Lett. **40**, 045201 (2023)
6. Jamison, S.P., et al.: Plasma characterization with terahertz time–domain measurements. J. Appl. Phys. **93**, 4334 (2003)
7. Ando, A., et al.: Electron density measurement of inductively coupled plasmas by terahertz time-domain spectroscopy (TH z -TDS). J. Appl. Phys. **110**, 073303 (2011)
8. Zhao, Y., et al.: Comparison of Terahertz time domain spectroscopy and laser Thomson scattering for electron density measurements in inductively coupled plasma discharges. J. Inst. **14**, C12019–C12019 (2019). https://doi.org/10.1088/1748-0221/14/12/C12019
9. Ding, J., et al.: Integrated system of traditional THz time-domain spectroscopy and asynchronous optical sampling. Chin. Phys. B **32**, 048702 (2023)
10. Chen, K., et al.: Application of terahertz time-domain spectroscopy in atmospheric pressure plasma jet diagnosis. Results Phys. **16**, 102928 (2020)

Terahertz Biosensing Detections Based on Metasurfaces

Zihao Zhang[1,2] and Yubin Gong[1(✉)]

[1] University of Electronic Science and Technology of China, Chengdu 611721, China
zhzhang@std.uestc.edu.cn, ybgong@uestc.edu.cn
[2] Innovation Laboratory of Terahertz Biophysics, National Innovation Institute of Defense Technology, Beijing 100071, China

Abstract. Terahertz (THz) biosensing have become the advanced bioanalysis technology with advantages of low-cost, non-invasive and label-free. Here, some progresses of terahertz biosensing detections based on metasurfaces are discussed. Aiming at different biological analytes and diverse sensing requirements, metasurface biosensors were proposed to significantly enhance the light-matter interactions and confine the localization of electromagnetic energy through designed metal-atoms, and integrated various materials. The characteristics of ultra-sensitive, frequency-agile and calibration-free make these biosensors beneficial for clinical diagnosis and inspire more advanced applications in THz domain.

Keywords: Biosensor · Cancer detection · Metasurface · Terahertz

1 Introduction

Terahertz (THz) radiation, residing between microwave and infrared in frequency domain, is a promising technology for the Terahertz modulation on ion channels [1], neurons [2–4], learning and brain function [5–9], considering the intra- and inter-molecular vibrations and rotations such as van der Waals force and hydrogen bonds occurring in the THz band [10, 11]. Based on the identification of biomolecules and measurements of changing refractive indexes, THz biosensors are potential alternatives of conventional biological detection methods. Compared to traditional biosensing methods, including optical dye-binding assay and pathological biopsy, which are authoritative but high-cost, time-consumption, and poor-early-diagnosis, terahertz biosensors pave a new way for the label-free, rapid-response, and low-cost biosensing detection. However, the relatively long wavelengths of THz waves make them insensitive to small volume or low concentration analytes, which block the widespread applications in biosensing. To address this issue, meta-surfaces, two-dimension structures consisting of subwavelength meta-atoms and nano-antennas, are adopted to THz biosensors [12–18]. By designing and engineering metal-atoms, metasurfaces can enhance the light-matter interactions and confine electromagnetic energy to obtain ultra-sensitivity and specific THz biosensors. As a result, metasurface biosensors hold the potential to become the future of high-performance THz biosensors.

C. Chang et al. (Eds.): YTHZ 2024, SPPHY 400, pp. 108–111, 2024.
https://doi.org/10.1007/978-981-97-3786-4_22

2 Metasurface Terahertz Biosensor

2.1 Cancer Detection

Cancer continues to pose a significant threat to human lives. To enhance early cancer detection and diagnosis, various meta-surface THz biosensors are proposed, aiming to reduce mortality rates.

Wang et al. demonstrated a high-sensitivity (504 GHz/RIU) and polarization-insensitivity meta-surface biosensor to distinguish cancer cells from other normal lung and brain cells [12]. The sensor is based on the quartz substrate (1500 μm), with titanium (10 nm) and gold (200 nm) deposited by magnetron sputtering. The unit meta-atom of the sensor is made of a pair of Au, consisting four C-shape structures surrounded in a square ring structure, and all of them are placed in a symmetrical manner, which ensures that the response to THz radiation is independent of polarization. Experiments were carried out with different densities of normal and cancerous lung cells which were radiated by 2.07 THz waves. Results reveal that the frequency shifts (49.6 GHz max.) as the density of cancer cells increase, while it differs in normal cells. Compared with clinical detection technologies, the required number of cells is 30-fold reduced and the time consumption is only 5 min.

To avoid invasive colonoscopy-guided biopsy and achieve early diagnosis of colorectal adenoma, which is a precancerous lesion, an ultrafast switchable biosensor was demonstrated, utilizing a phototunable silicon-based THz metasurface [13]. Its THz response can shift from Lorentz-lattice mode (1.215 THz) to electromagnetism-induced transparency (EIT) mode (0.777 THz) within 32 ps under the radiation of 800 nm wave-length optical pumping. The designed unit meta-atom consists of two pairs of splitting-ring resonators and two wires separated by a silicon bridge, which is constructed by embedding a silicon epilayer between metallic structures and a sapphire substrate. Experimental results illustrate the sensing outcomes of normal colorectal epithelial cells (NCECs), colorectal adenoma cells (CACs) and colorectal cancer cells (CCCs) by different transmission loss for THz waves. This depiction highlights the capability to discern distinct stages of cancerous colorectal cancers.

In addition, a calibration-free and robust metasurface was designed to discriminate the categories of gastric cells, including normal epithelial, adenoma and cancerous cells, based on distinct net sizes of nucleus and cell fluid [14]. Environment noise is eliminated by normalizing the transmission results with two different THz spectra, thus dispensing with extra calibration. It's worth noting that the response of the sensor to 0.825 and 1.385 THz is facilitated by an optical pump and silicon. Meanwhile, three types of gastric cells exhibit phase variation at 1.11 THz, providing a novel avenue for obtaining more sensing information of analytes.

2.2 Biomolecule Detection

Similar with the reflection of cell morphological changes to cancer cases, specific biomolecules are clinically relevant to some diseases, such as inflammation and viral infection. Targeting inflammatory factors C-reactive protein (CRP) and Serum Amyloid

A (SAA), a chip-based portable metasurface biosensor was reported by exploring quasi-bound states in the continuum (quasi-BICs) [15]. The unit meta-atom is in the shape of crossed bi-directional arrows, which was designed asymmetrically to achieve ultrahigh-Q resonances to enhanced the near field with-in the metasurface. With increasing of the asymmetric factor *d*, a typical sharp resonance feature appeared, suggesting the reinforced light-matter interaction. Experiments were carried out by dropping the hybrid solution of different concentration of antigens (CRP and SAA). The results demonstrate that this metasurface biosensor has the ability of identifying the solution with a solubility difference of 1 pM, with frequency shifting 17 GHz. Besides resolving ultra-low concentration solutions, the biosensor is ultra-sensitive to the refractive index of analytes (674 GHz/RIU).

For sensing biomolecular fingerprints, a pixelated frequency-agile broadband metasurface THz biosensor was demonstrated [16]. Symmetric C-shape rings of different radius were adopted, enhancing flexibility and compactness within the same bandwidth, in combination with graphene micro-ribbon arrays on the quartz substrate. This structure establishes a one-to-many mapping between spatial and spectral information. The experimental results on glucose demonstrate that the absorptance is improved by nearly 5 times with the metasurface.

3 Conclusion

Through the utilization of designed unit metal-atoms and phototunable materials, researches achieved ultra-sensitive, ultrafast switchable and label-free metasurface biosensors. The detection results of different cancer cells and biomolecules align with traditional biological experiments, suggesting the potential of metasurface THz biosensors to impact clinical diagnosis and treatments.

References

1. Li, Y., et al.: Terahertz wave enhances permeability of the voltage-gated calcium channel. J. Am. Chem. Soc. **143**(11), 4311–4318 (2021)
2. Liu, X., et al.: Nonthermal and reversible control of neuronal signaling and behavior by midinfrared stimulation. Proc. Natl. Acad. Sci. **118**(10), e2015685118 (2021)
3. Wu, K., et al.: Terahertz wave accelerates DNA unwinding: a molecular dynamics simulation study. J. Phys. Chem. Lett. **11**(17), 7002–7008 (2020)
4. Li, Y., et al.: Physicochemical insights on terahertz wave diminished side effects of drugs from slow dissociation. ACS Nano **16**(5), 8419–8426 (2022)
5. Zhang, J.X., et al.: Non-invasive, opsin-free mid-infrared modulation activates cortical neurons and accelerates associative learning. Nat. Comm. **12**, 2730 (2021)
6. Yu, Y., et al.: Terahertz photons improve cognitive functions in posttraumatic stress disorder. Research **15**(6), 0278 (2023)
7. Xiang, Z., et al.: A primary model of THz and far-infrared signal generation and conduction in neuron systems based on the hypothesis of the ordered phase of water molecules on the neuron surface. Sci. Bull. **65**(4), 308–317 (2020)
8. Liu, G., et al.: Myelin sheath as a dielectric waveguide for signal propagation in the mid-infrared to terahertz spectral range. Adv. Funct. Mater. **29**(7), 1807862 (2019)

9. Peng, W., et al.: High frequency terahertz waves disrupt Alzheimer's amyloid fibril formation. eLight **3**, 18 (2023)
10. Wang, P., et al.: Progress on cutting-edge infrared-terahertz biophysics. IEEE Trans. Microw. Theory Tech. **70**(11), 5117–5140 (2022)
11. Chang, C., et al.: High-gain Thompson-scattering X-ray free-electron laser by time-synchronic laterally tilted optical wave. Phys. Rev. Lett. **110**(6), 064802 (2013)
12. Wang, P., et al.: An ultra-sensitive metasurface biosensor for instant cancer detection based on terahertz spectra. Nano Res. **16**(5), 7304–7311 (2023)
13. Jiao, Y., et al.: Photoactive terahertz metasurfaces for ultrafast switchable sensing of colorectal cells. Mater. Horiz. **9**(12), 2984–2992 (2022)
14. Lou, J., et al.: Calibration-free, high-precision, and robust terahertz ultrafast metasurfaces for monitoring gastric cancers. Proc. Natl. Acad. Sci. **119**(43), e2209218119 (2022)
15. Wang, R., et al.: Ultrasensitive terahertz biodetection enabled by quasi-BIC-based metasensors. Small **19**(35), 2301165 (2023)
16. Sun, L., et al.: A pixelated frequency-agile metasurface for broadband terahertz molecular fingerprint sensing. Nanoscale **14**(27), 9681–9685 (2022)
17. Zhang, Z., et al.: Advanced terahertz refractive sensing and fingerprint recognition through metasurface-excited surface waves. Adv. Mater. Early View 2308453 (2024)
18. Lou, J., et al.: Optically controlled ultrafast terahertz metadevices with ultralow pump threshold. Small **17**(44), 2104275 (2021)

Refractive Index Sensing Based on Metasurface Coupled Surface Wave

Zeyan Zhang[1,2](✉)

[1] Peking University, Beijing 100871, China
2201110344@stu.pku.edu.cn
[2] National Innovation Institute of Defense Technology, Beijing 100071, China

Abstract. Sensitivity is the most important character of metasurface-based sensors in effective substance detection and comprehensive bio-interaction studies. Demonstrating our experimental verification, successful excitation and propagation of surface waves (SWs) occur on a metasurface, benefitting from both long-range transmission and strong confinement. Integrated sensing capabilities are enabled by the proposed metasurface, which supports SWs. Notable is the refractive sensing capability, which extends up to 215.5°/RIU. Experimental results reveal maximum phase changes of 91.1°, 101.8°, and 126.4°, respectively, with the coverage of 5 μm thickness of polyimide, quartz, and silicon nitride layers within the THz band.

Keywords: metasurface · surface waves · refractive sensing

1 Introduction

Metasurfaces, with their subwavelength structures and enhanced field properties, possess sensitivity in detecting environmental changes in biological and chemical analysis [1–12]. Nevertheless, traditional metasurfaces rely on measuring transmission or reflection spectra, which constrains methods for enhancing the interaction between analytes and light, thus limiting sensitivity improvement. Terahertz waves, situated within the molecular fingerprint spectra [9–12] offer potential to augment interaction for sensing, particularly in bioeffects [13–21].

Surface waves (SWs), electromagnetic wave modes confined at the metal-medium interface, have drawn considerable attention. However, prevailing research has largely focused on crafting elaborate metasurfaces to intricately manipulate SWs wavefronts rather than emphasizing their potential for sensitive sensing applications. In this study, we pivot the focus towards the often-neglected long-range propagation capability of SWs, intending to delve into its unexplored potential in biosensing.

In this investigation, we effectively exploit the long-range propagation characteristic of SWs, presenting sensitive sensing in the transmission phase for refractive sensing. Experimental results vividly depict propagation phase disparities of 91.1°, 101.8°, and 126.4° within 5 μm thicknesses of polyimide, quartz, and silicon nitride layers, respectively.

C. Chang et al. (Eds.): YTHZ 2024, SPPHY 400, pp. 112–116, 2024.
https://doi.org/10.1007/978-981-97-3786-4_23

2 Results and Discussions

Let's start by considering the design of Pancharatnam–Berry (PB) meta-atoms. Upon configuring the metasurface with PB meta-atoms, we can then deduce the condition for maximum efficiency [13]:

$$|r_{xx}| = |r_{yy}| = 1,\ \Delta\varphi = \varphi_{xx} - \varphi_{yy} = \pm\pi \tag{1}$$

This condition implies that the PB meta-atoms must serve as reflective half plates for achieving maximum efficiency. Illustrated in Fig. 1a, the designed PB meta-atoms comprise a gold resonator antenna, a quartz substrate, and a gold reflective layer. The metasurface's capability to excite SWs is depicted in Fig. 1b, validated through both simulation and experimental methods. Critical parameters defining the PB meta-atom include $p = 277.6$ μm, $r = 126.6$ μm, $w = 20$ μm, $h = 100$ μm and $\alpha = 33°$. The PB meta-atoms, when integrated with a waveguide, collectively form the entire metasurface.

Fig. 1. (a) Depiction of the structured layout for the Pancharatnam–Berry (PB) meta-atom [14]; (b) Comparative evaluation contrasting simulated outcomes with experimental fitting of the SWs metasurface [14].

In investigating the sensing prowess and effectiveness of the metasurface in detecting fingerprint spectra, we use some analyte layers with specific refractive index n to coat the metasurface. By varying the thickness of the analyte layers within the decay length range of SWs in air, we acquire comprehensive data. Defining the propagation distance as the length of the metasurface, configured as N_x times to p, we observe alterations in the SWs dispersion curve when the analyte is applied. Consequently, both the equivalent wavevector and phase of propagation undergo modifications, as depicted in Fig. 2a.

Exploring further, we scrutinize the impact of the propagation distance N_x. When a thickness of $t = 10$ μm is considered, metasurfaces are individually coated with analyte layers possessing refractive indices $n = 1$ and $n = 2$. Subsequently, we analyze variations in the propagation phase between $n = 1$ and $n = 2$ concerning changes in N_x. As the propagation distance escalates, so does the propagation phase difference of SWs, particularly evident under the coverage of analyte layers with $n = 1$ and $n = 2$, as illustrated in Fig. 2b. Employing a fitting analysis based on five uniformly selected frequency points within the operational frequency range, we delineate trends in transmission phase variation between $n = 1$ and $n = 2$, as portrayed in Fig. 2c. The influence of the real part of the refractive index on the transmission phase intricately correlates with the equivalent wavenumber. Consequently, the variation in transmission phase aligns with the operating frequency, facilitating a process of linear fitting. Leveraging the propagation properties of SWs, we achieve sensitive refractive sensing, exemplified by an impressive 215.5°/RIU at $N_x = 35$, with potential for further enhancement through increased propagation length.

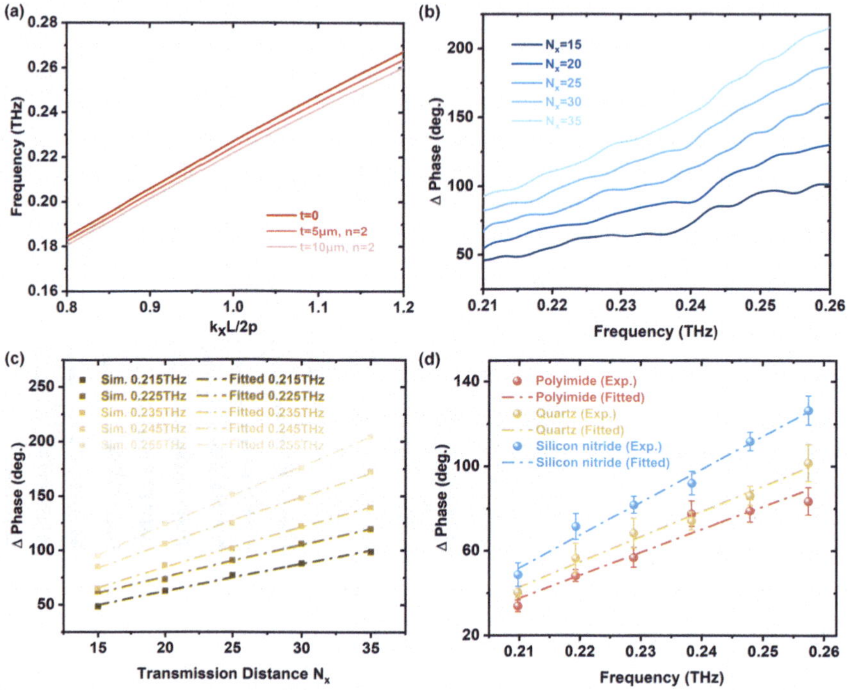

Fig. 2. (a) Movement of the SWs dispersion curve under varying conditions of covered analyte layers. (b) Examination of transmission phase variations at $t = 10$ μm and different N_x for two distinct analyte layers with refractive indices $n = 1$ and $n = 2$, respectively [14]. (c) Utilization of linear fitting to establish a relationship between transmission phase difference and N_x. (d) At $N_x = 30$, measurement and subsequent linear fitting of differences in SWs propagation phase across different analyte layers and in the absence of an analyte layer are conducted [14].

In strengthening the evidence supporting the sensing application of the metasurface, our investigation concentrates on stable materials—polyimide ($n = 1.8$), quartz ($n = 1.95$), and silicon nitride ($n = 2.7$). Each material exhibits a consistent refractive index and a low absorption coefficient around the operating frequency [22]. At a fixed propagation distance, we experimentally gauge and fit the propagation phase differences induced by the analyte layers, as illustrated in Fig. 2d. It becomes evident that the presence of analyte layers accelerates the phase change of SWs compared to the condition of an uncovered device, with these phase differences accentuated with increasing frequency and refractive index. These findings manifest as a well-established linear relationship, corroborating Fig. 2c. Upon covering the metasurface with 5 μm thicknesses of polyimide, quartz, and silicon nitride layers, we achieve respective maximum phase changes of 91.1°, 101.8°, and 126.4°. Notably, these values surpass those achieved in previous works, underscoring the unique capabilities of our approach.

3 Conclusion

In conclusion, our study presents a fresh approach to refractive sensing employing SWs. Through experimental validation, we showcase the metasurface's capability to efficiently excite and propagate SWs across a considerable range within the THz frequency spectrum. This departure from conventional metasurface sensing methodologies introduces novel techniques to enhance the interaction between electromagnetic waves and target analytes.

References

1. Wang, R., et al.: Electric Fano resonance-based terahertz metasensors. Nanoscale **13**, 18467–18472 (2021)
2. Lou, J., et al.: Optically controlled ultrafast terahertz metadevices with ultralow pump threshold. Small **17**, 2104275 (2021)
3. Lou, J., et al.: Calibration-free, high-precision, and robust terahertz ultrafast metasurfaces for monitoring gastric cancers. PNAS **119**(43) (2022)
4. Wang, P., et al.: An ultra-sensitive metasurface biosensor for instant cancer detection based on terahertz spectra. Nano Res. **16**, 7304–7311 (2023)
5. Ma, Z., et al.: Identification and quantitative detection of two pathogenic bacteria based on a terahertz metasensor. Nanoscale **15**, 515–521 (2023)
6. Wang, R., et al.: Ultrasensitive terahertz biodetection enabled by quasi-BIC-based metasensors. Small **19**, 2301165 (2023)
7. Jiao, Y., et al.: Photoactive terahertz metasurfaces for ultrafast switchable sensing of colorectal cells. Mater. Horiz. **9**, 2984–2992 (2022)
8. Sun, L., et al.: A pixelated frequency-agile metasurface for broadband terahertz molecular fingerprint sensing. Nanoscale **14**, 9681–9685 (2022)
9. Wang, P., et al.: Progress on cutting-edge infrared-terahertz biophysics. IEEE Trans. Microw. Theory Tech. **70**(11), 5117–5140 (2022)
10. Liu, G., et al.: Myelin sheath as a dielectric waveguide for signal propagation in the mid-infrared to terahertz spectral range. Adv. Funct. Mater. **29**(7), 1807862 (2019)
11. Chang, C., et al.: High-gain Thompson-scattering X-ray free-electron laser by time-synchronic laterally tilted optical wave. Phys. Rev. Lett. **110**, 064802 (2013)
12. Xiang, Z., et al.: A primary model of THz and far-infrared signal generation and conduction in neuron systems based on the hypothesis of the ordered phase of water molecules on the neuron surface. Sci. Bull. **65**, 308 (2020)
13. Liu, X., et al.: Nonthermal and reversible control of neuronal signaling and behavior by midinfrared stimulation. Proc. Natl. Acad. Sci. **118**(10), e2015685118 (2021)
14. Zhang, J.X., et al.: Non-invasive, opsin-free mid-infrared modulation activates cortical neurons and accelerates associative learning. Nature Comm. **12**, 2730 (2021)
15. Yu, Y., et al.: Terahertz photons improve cognitive functions in posttraumatic stress disorder. Research **6**, 0278 (2023)
16. Li, Y., et al.: Terahertz wave enhances permeaility of the voltage-gated calcium channel. J. Am. Chem. Soc. **143**, 9 (2021)

17. Peng, W., et al.: High frequency terahertz waves disrupt Alzheimer's amyloid fibril formation. eLight **3**, 18 (2023)
18. Li, Y., et al.: Physicochemical insights on terahertz wave diminished side effects of drugs from slow dissociation. ACS Nano **16**(5), 8419 (2022)
19. Wu, K., et al.: Terahertz wave accelerates DNA unwinding: a molecular dynamics simulation study. J. Phys. Chem. Lett. **11**(17), 7002–7008 (2020)
20. Tan, X., et al.: Neuromodulation of chemical sysnaptic transmission driven by THz photons. Research **2022**, 0010 (2022)
21. Yin, J., et al.: Terahertz wave as a non-drug and non-invasive strategy for long-term inhibition of cancer cells through telomerase activity suppression. ACS Nano (2024). https://doi.org/10.1021/acsnano.3c09216
22. Zhang, Z., et al.: Advanced terahertz refractive sensing and fingerprint recognition through metasurface-excited surface waves. Adv. Mater., e2308453 (2024)
23. Cai, T., et al.: Experimental realization of a superdispersion-enabled ultrabroadband terahertz cloak. Adv. Mater. **34**, e2205053 (2022)

Preliminary Study of Intraocular Tissues Based on Infrared-Terahertz

Xin Yan[1], Bei Tian[1,2](✉), and Ride Wang[1,2](✉)

[1] Capital Medical University, Beijing Tongren Hospital, Beijing 100730, China
tianbei@ccmu.edu.cn, wangride@mail.nankai.edu.cn
[2] Beijing Tongren Eye Center, Beijing Tongren Hospital, Capital Medical University, Beijing 100730, China

Abstract. Terahertz (THz) technology has been widely used in biomedical applications because of its unique characteristics, with significant promise in the field of ophthalmology. In this research, the Infrared-terahertz technique was used to characterize the human anterior lens capsule, the epiretinal membrane, and rat retinal tissues to obtain a characteristic correlation between terahertz absorption spectra and the structure and protein composition of intraocular tissues. In this experiment, we used infrared confocal microimaging method with wavelengths in the range of 5–11 μm for different scanning ranges. The results of this research indicate that the Infrared-terahertz method can realize the imaging of intraocular aqueous tissue. Also, the Infrared-terahertz imaging results show the same sulcus-like structure in both the human anterior lens capsule and the human pathologic epiretinal membrane the imaging features of the membranous structure of the rat retina differ significantly from those of the transparent capsule in infrared-terahertz imaging. It is concluded that Terahertz infrared imaging of different pathophysiological states can obtain images with different structural features, which is worth further study.

Keywords: THz · Infrared-terahertz · intraocular tissue · Near-field imaging · Infrared imaging

1 Introduction

Currently, THz technology is mainly applied to the following aspects of biomedicine [1–9]: Identify and classify biomolecules by acquiring THz characteristics of substances to be measured; Deploying THz imaging technology in the clinic to collect images of tumors and other tissues, and providing information for further treatment; Advanced research on the biological effects of THz and determining the effects of THz on body tissues and cells.

Infrared-terahertz (IRT) resonant absorption complies with unique biomolecule fingerprints, including neurotransmitters, proteins, ribonucleic acid, and deoxyribonucleic acid, rendering it a suitable technology to achieve conformational variations and functional adjustments [10–13]. Because of the specificity of human tissue composition, different materials of THz sensors are needed to assist the measurement and improve

C. Chang et al. (Eds.): YTHZ 2024, SPPHY 400, pp. 117–121, 2024.
https://doi.org/10.1007/978-981-97-3786-4_24

the accuracy of the measurement [14–17]. In this study, preliminary infrared- terahertz spectroscopic imaging was performed on the intraocular clear capsular membrane-like tissues of the anterior lens capsule, the premacular membrane, and rat retinal tissue.

2 THz Imaging of Human Anterior Lens Capsule Tissue

2.1 Sampling of Specimens from Human Anterior Lens Capsule Tissue

Obtained post-surgical discarded anterior capsule membrane, fixed in 4% paraformaldehyde, and stored at 4° cold chain placed in PE tubes for delivery. The samples were laid flat on a gold reflector and tested for imaging after 6 h in a ventilated area. The thickness of the samples was about 2 μm at the thinnest point and about 20 μm at the thickest point, with an average thickness of about 15 μm.

2.2 Infrared-Terahertz Imaging

In the experiment, we use the infrared confocal microimaging method with wavelengths of 5–11 μm [18], the imaging range of near-field imaging is 2 mm * 2 mm. Imaging of the anterior lens capsule reveals a groove-like tissue structure, which is analyzed as characteristic information about the wrinkling of the model structural proteins (see Fig. 1). Enhanced imaging in the imaging range of 0.2 mm * 0.2 mm gives different wavelength absorption peaks (see Fig. 2).

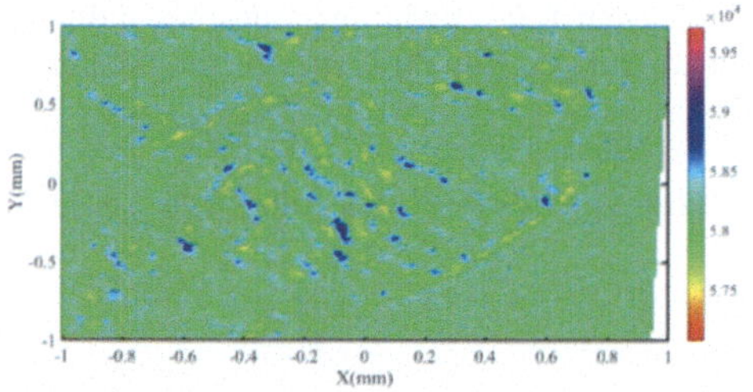

Fig. 1. Lower row for 2 mm * 2 mm, image of the basic institutional organization of the anterior lens capsule, with sulcus-like structural features.

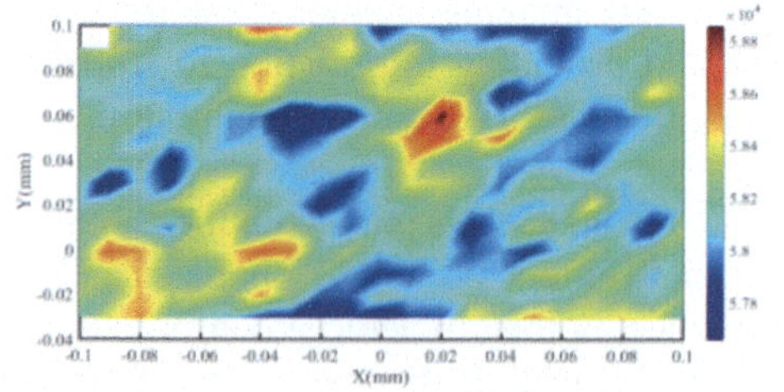

Fig. 2. Enhanced imaging of 0.2 mm * 0.2 mm to see different absorption wavelengths.

3 THz Imaging of Human Retinal Pathological Tissue

3.1 Sampling of Specimens from Human Macular Precursor

Obtained post-surgical discarded macular precursor membranes, stored at 4° cold chain placed in containers for delivery, laid flat on a gold reflector, and tested for imaging after 6 h in a ventilated area, with a sample thickness of approximately 150 μm.

3.2 Infrared-Terahertz Imaging

In the experiment, we use the infrared confocal microimaging method with wave-lengths of 5–11 μm [18]. In the experiment, we used the infrared confocal microimaging method with wavelengths in the range of 5–11 μm to perform spectrum imaging with different wavelength bands to analyze the spectrum absorption information for both protein characterization and characterization without spectral information. Under infrared-terahertz imaging macular precursor membranes show a sulcus-like structure (see Fig. 3). As the samples are wrinkled, the different light signals may respond to two kinds of information: differences in the signals displayed due to light scattering, indicating differences in height, they may also represent differences in the density of the tissue components. The spectral imaging response of the protein component of the pathologic macular precursor tissue is significantly enriched, and the different tissue components require further identification (see Fig. 3a).

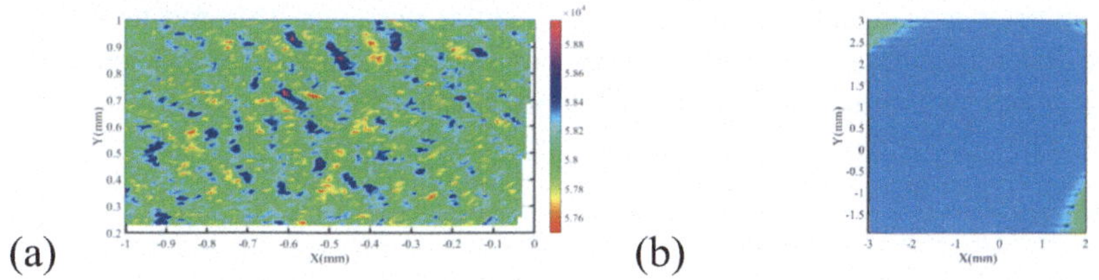

Fig. 3. (a) Imaging range of 1 mm * 1 mm range of infrared-terahertz imaging, a richer optical signal can be seen. (b) Fresh retinal tissue was taken from SD rats

3.3 Sampling of Specimens from Human Macular Precursor

Fresh retinal tissue was taken from SD rats, laid flat on a gold reflector, and tested for imaging after 6 h in a ventilated area, with a sample thickness of about 222 μm.

3.4 Infrared-Terahertz Imaging

In the experiment, we use the infrared confocal microimaging method with wave-lengths of 5–11 μm [18]. The retinal tissue of rats showed a point-like imaging state and clear structure in infrared-terahertz detection (see Fig. 3b).

4 Result

(1) THz technology enables imaging of aqueous tissue in the eyes.
(2) The results of Infrared-terahertz imaging showed that the same sulcus-like structure could be found in the human anterior lens capsule membrane and macular precursor.
(3) There is a significant difference between the imaging features of the membranous structure of the rat retina and the infrared-terahertz imaging features of the clear capsular membrane.

5 Discussion

Terahertz infrared (TIR) detection technology has a broad application in the detection of biological tissues containing water, and our study confirms that TIR can obtain preliminary imaging of water-rich intraocular tissues. The anterior capsule of the lens is a normal human tissue, and the nucleus of the human crystalline lens shows a "groove-like" tissue structure. Infrared-terahertz imaging of the anterior lens capsule reveals a "groove-like" tissue structure that matches the fibrous tissue morphology of the crystalline nucleus. The premacular membrane is an intraocular pathologic tissue, and in the presence of macular degeneration, the retina appears as a "groove-like" fold. Terahertz infrared detection of the premacular membrane showed a "groove-like" morphology that matched the morphology of the retina in the pathologic state. In the rat, the retinal tissue showed a point-like pattern in Terahertz infrared detection, which matched the distribution of RPE and cone and rod cells, and was very different from the morphology of the anterior capsule of the lens and the macular prefovea. Infrared-terahertz imaging for different pathophysiological states can obtain images with different structural characteristics, which is worth further in-depth study.

References

1. Gezimati, M., Singh, G.: Terahertz imaging and sensing for healthcare: current status and future perspectives. IEEE Access (2023)
2. Gezimati, M., Singh, G.: Advances in terahertz technology for cancer detection applications. Opt. Quant. Electron. **55**(2), 151 (2023)
3. Liu, G., et al.: Myelin sheath as a dielectric waveguide for signal propagation in the mid-infrared to terahertz spectral range. Adv. Funct. Mater. **29**(7), 1807862 (2019)
4. Liu, X., et al.: Nonthermal and reversible control of neuronal signaling and behavior by midinfrared stimulation. Proc. Nat. Acad. Sci. **118**(10), e2015685118 (2021)
5. Zhang, J.X., et al.: Non-invasive, opsin-free mid-infrared modulation activates cortical neurons and accelerates associative learning. Nat. Commun. **12**, 2730 (2021)
6. Li, Y., et al.: Terahertz wave enhances permeability of the voltage-gated calcium channel. J. Am. Chem. Soc. **143**(9), 4311 (2021)
7. Xiang, Z., et al.: A primary model of THz and far-infrared signal generation and conduction in neuron systems based on the hypothesis of the ordered phase of water molecules on the neuron surface. Sci. Bull. **65**, 308 (2020)
8. Wu, K., et al.: Terahertz wave accelerates DNA unwinding: a molecular dynamics simulation study. J. Phys. Chem. Lett. **11**(17), 7002–7008 (2020)

9. Yu, Y., et al.: Terahertz photons improve cognitive functions in posttraumatic stress disorder. Research **6**, 0278 (2023)
10. Wang, P., et al.: Progress on cutting-edge infrared-terahertz biophysics. IEEE Trans. Microw. Theory Tech. **70**(11), 5117–5140 (2022)
11. Zhang, Z., et al.: Advanced terahertz refractive sensing and fingerprint recognition through metasurface-excited surface waves. Adv. Mater., 2308453 (2024)
12. Lou, J., et al.: Calibration-free, high-precision, and robust terahertz ultrafast metasurfaces for monitoring gastric cancers. PNAS **119**(43), 2209218119 (2022)
13. Jiao, Y., et al.: Photoactive terahertz metasurfaces for ultrafast switchable sensing of colorectal cells. Mater. Horiz. **9**, 2984–2992 (2022)
14. Wang, R., et al.: Electric Fano resonance-based terahertz metasensors. Nanoscale **13**, 18467 (2021)
15. Sun, L., et al.: A pixelated frequency agile metasurface for broadband terahertz molecular fingerprint sensing. Nanoscale **14**(27), 9681–9685 (2022)
16. Wang, R., et al.: Ultrasensitive terahertz biodetection enabled by quasi-BIC-based metasensors. Small **35**(19), 2301165 (2023)
17. Wang, P., et al.: An ultra-sensitive metasurface biosensor for instant cancer detection based on terahertz spectra. Nano Res. **16**(5), 7304–7311 (2023)
18. Zhu, J., et al.: Mid-infrared multispectral confocal microscope using off-axis parabolic mirrors to study epiretinal membranes. Appl. Opt. **60**(27), 8616–8623 (2021)

Rapid Detection of Pathogenic Microorganisms by Terahertz Spectroscopy

Jun Chen[1], Congyong Li[1,2], Wei Zhang[1], Zhe Luan[1], and Gang Sun[1(✉)]

[1] Department of Gastroenterology and Hepatology, First Medical Center, Chinese PLA Gen-Eral Hospital, Beijing, China
sunok301@126.com

[2] Sixth Health Care Department, Second Medical Center of PLA General Hospital, Beijing, China

Abstract. Prompt identification of infectious pathogenic microorganisms is essential for preventing bacterial growth and spread, and serves as the foundation for implementing appropriate antibiotic therapy in clinical settings. Illogical drug usage can attenuate the therapeutic efficacy of antibiotics and increase the risk of drug resistance in patients with bacterial infections. The traditional methods for the identification of pathogenic microorganisms, such as isolation and cultivation, are inefficient due to the relatively long turn-around time. Physical, immunological, and molecular biological techniques have significantly improved the efficiency of pathogenic microorganisms detection, while there are still some drawbacks. Terahertz technology is a promising new method for pathogenic microorganisms detection, but currently the application of Terahertz is confined partially due to its working mechanism. This study provides a concise summary of the primary methods for detecting pathogenic microorganisms and outlines the advancements in terahertz spectroscopy for pathogenic microorganism identification.

Keyword: Rapid Detection · Infectious bacteria detection · Terahertz waves

1 Introduction

Pathogenic bacteria can invade the human body via the respiratory tract, digestive tract, skin injuries, invasive procedures, blood transfusion and sexual contact, leading to infection, inflammation, allergy, sepsis, and eventually death. Approximately 2.8 million people in America were infected with pathogens annually, causing 35,000 deaths owing to antibiotic resistance [1]. Clinicians usually select broad-spectrum antibiotics empirically until a causal pathogen is identified. Detecting harmful germs quickly and accurately is crucial for effectively treating infectious diseases and lowering the mortality rate. Isolation and culture are the primary methods used for clinical pathogen detection. However, the culture process is relatively long, requires a lot of effort, and is not accurate enough. Performance is influenced by the sample quality and the speed at which microorganisms grow. Alternative physical methods (e.g. spectroscopy and mass spectrometry) [2], immunological detection methods (e.g. ELISA) [3], and molecular biological detection

C. Chang et al. (Eds.): YTHZ 2024, SPPHY 400, pp. 122–126, 2024.
https://doi.org/10.1007/978-981-97-3786-4_25

methods (e.g. PCR, qPCR, multiplex PCR, and LAMP) [4] are limited by cost, sensitivity, and specificity and are unable to substitute isolation and culture procedures. Terahertz (THz) is an electromagnetic wave that ranges in frequency from 0.1 to 100 THz, positioned between microwave and infrared spectra [5–9]. Biomedical applications of Terahertz included identifying neurotransmitters (e.g., dopamine, norepinephrine, γ-aminobutyric acid) [10–12], diagnosing cancer cells, brain damage, and detecting infectious pathogens [13–17]. Nano-telomerase inhibition by Terahertz can even suppress cancer cells long-term [18]. Terahertz spectroscopy utilizes low-energy photons to permeate non-metallic and non-polar material and exhibits sensitive absorption properties towards water molecules. This frequency band exhibits significant absorption and resonance for various biological macromolecules, making it suitable for non-destructive and radiation-free detection. Researchers have utilized terahertz spectroscopy to detect bacteria and viruses by examining bacterial cell structures, internal components, metabolites, and nucleic acids. (Fig. 1).

1.1 Detection of Bacterial Cells via THz Spectroscopy

In 2004, Globus et al. [19] utilized Fourier transform infrared spectroscopy along with high-power noise sources to analyze the projection spectra of freeze-dried cells of Erwinia herbicola and Bacillus Subtilis for the first time. The spectrum disparities between the two strains indicated that terahertz spectroscopy may potentially distinguish different species. Yoon et al. [20] classified molds, yeasts, and bacteria based on the intrinsic dielectric constant in the terahertz frequency range, and confirmed that different cell wall components were responsible for this difference. Jun et al. [21] developed a label-free identification tool for single pathogens such as bacteria, which recorded the spectral characteristics of bacterial growth, thermal inactivation, DNA inactivation, and cell wall destruction, and could distinguish gram negative bacteria from gram positive bacteria. Wang et al. [22] achieved terahertz super-resolution images of microorganisms, distinguishing Escherichia coli and Staphylococcus aureus using image analysis and data processing. Ma et al. [17] used a terahertz hypersensitive sensor based on the coupling of electric dipole and ring dipole to increase the detectable concentration of Escherichia coli and Staphylococcus aureus to 10^4 CFU ml-1, achieving a highly sensitive quantitative detection of pathogenic microorganisms. Terahertz spectroscopy applications is limited because it can identify only a small number of pathogenic microorganisms due to signal superposition from biological macromolecules like proteins, amino acids, DNA, and RNA, as well as distinctions in intracellular water content.

1.2 Detection of Internal Components via THz Spectroscopy

Recently, terahertz spectra of Dipicolinic acid (DPA), DNA, and bacterial metab-olites have been discovered. DPA is a distinctive component found in Bacillus thuringiensis spores. The transmission spectra of DPA exhibits similar traits to those of Bacillus thuringiensis and Bacillus globigii. The transmission spectrum of Aspergillus niger differs significantly from that of Aspergillus Niger, indicating that certain Bacillus spores may be distinguished based on DPA spectral features [23]. Wang et al. [24] employed terahertz time-domain spectroscopy (THz-tds) to analyze the spectra of astaxanthin, riboflavin, and cellular riboflavin metabolism for the purpose of identifying bacteria with elevated riboflavin synthesis. Continuous detection of bacterial metabolites using terahertz spectra would allow for the analysis of distinct terahertz spectra, which can be utilized for identifying specific strains and assessing industrial bacterial metabolites. Yang et al. [25] utilized graphene-mediated THz scattering scanning near-field optical microscopy to directly image individual proteins, capturing the morphology and THz scattering pictures of individual immunoglobulin G (IgG) and ferritin molecules. Detecting pathogenic bacteria may be achievable by identifying certain proteins inside them.

1.3 Detection of Nucleic Acids via THz Spectroscopy

Pathogenic microorganisms utilize nucleic acid as a big molecule to store genetic information. The vibrational and rotational excitation levels associated with the nucleic acid structure are within the terahertz spectrum. Terahertz waves have the potential to be used as tools for identifying and analyzing the breakdown process of nucleic acid production in bacteria. In 2013, Arora et al. [26] utilized terahertz time-domain spectroscopy to analyze the PCR amplification products of DNA and differentiated various nucleic acid solutions based on their spectra. Zhou and colleagues [27] integrated graphene and terahertz metasurfaces into a microfluidic system to decrease sample volume, improve biomolecule-THz wave interaction and identify Escherichia coli DNA sequences. Yang et al. [28] utilized rolling circle amplification (RCA), an isothermal nucleic acid amplification method, to amplify certain bacterial pieces. Following purification of the amplified samples, E. coli, Acinetobacter baumannii, and Pseudomonas aeruginosa were effectively differentiated using terahertz spectroscopy, with a pathogen DNA detection limit of 0.05 ng/μL. Wu et al. [29] demonstrated THz can effectively unwind DNA.

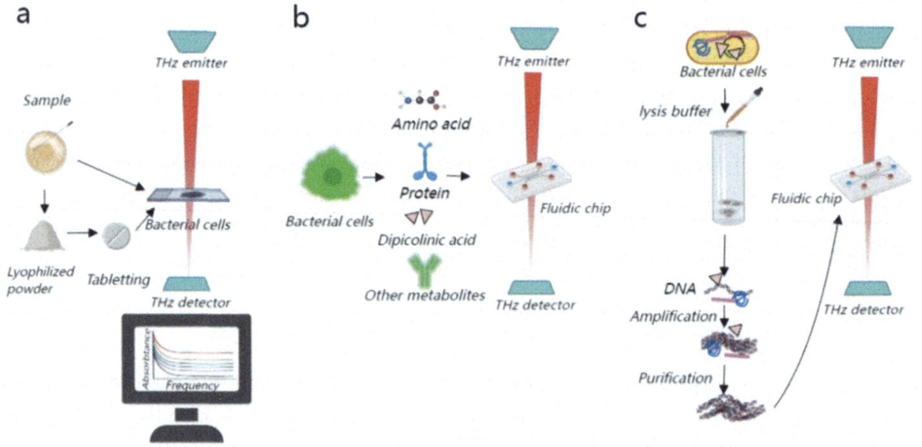

Fig. 1. Schematic representation of bacterial detection via THz spectroscopy. Terahertz spectroscopy was utilized in the following three ways to identify bacteria; a) directly detect the bacterial cell body or powder; b) detect the components of the bacterial cell body or metabolites; and c) detect specific nucleic acid fragments following bacterial amplification.

References

1. Karbelkar, A.A., Furst, A.L.: Electrochemical diagnostics for bacterial infectious diseases. ACS Infect. Dis **6**(7), 1567–1571 (2020)
2. Pahlow, S., Meisel, S., Cialla-May, D., Weber, K., Rosch, P., Popp, J.: Isolation and identification of bacteria by means of Raman spectroscopy. Adv. Drug Deliv. Rev. **89**, 105–120 (2015)
3. Fach, P., Perelle, S., Grout, J., Dilasser, F.: Comparison of different PCR tests for detecting Shiga toxin-producing Escherichia coli O157 and development of an ELISA-PCR assay for specific identification of the bacteria. J. Microbiol. Methods **55**(2), 383–392 (2003)
4. Jin, J., et al.: A real-time LAMP-based dual-sample microfluidic chip for rapid and simultaneous detection of multiple waterborne pathogenic bacteria from coastal waters. Anal. Methods **13**(24), 2710–2721 (2021)
5. Liu, G., et al.: Myelin sheath as a dielectric waveguide for signal propagation in the mid-infrared to terahertz spectral range. Adv. Funct. Mater. **29**(7), 1807862 (2019)
6. Xiang, Z., et al.: A primary model of THz and far-infrared signal generation and conduction in neuron systems based on the hypothesis of the ordered phase of water molecules on the neuron surface. Sci. Bull. **65**, 308 (2020)
7. Liu, X., et al.: Nonthermal and reversible control of neuronal signaling and behavior by midinfrared stimulation. Proc. Natl. Acad. Sci. **118**(10), e2015685118 (2021)
8. Zhang, J., et al.: Non-invasive, opsin-free mid-infrared modulation activates cortical neurons and accelerates associative learning. Nat. Commun. **12**, 2730 (2021)
9. Li, Y., et al.: Terahertz wave enhances permeability of the voltage-gated calcium channel. J. Am. Chem. Soc. **143**(11), 4311–4318 (2021)
10. Zhu, Z., et al.: Broadband terahertz signatures and vibrations of dopamine. Analyst **145**(18), 6006–6013 (2020)
11. Zhu, Z., et al.: Characteristic fingerprint spectrum of neurotransmitter norepinephrine with broadband terahertz time-domain spectroscopy. Analyst **144**(8), 2504–2510 (2019)
12. Cheng, C., et al.: Broadband terahertz recognizing conformational characteristics of a significant neurotransmitter gamma-aminobutyric acid. RSC Adv. **9**(35), 20240–20247 (2019)
13. Lou, J., et al.: Calibration-free, high-precision, and robust terahertz ultrafast metasurfaces for monitoring gastric cancers. Proc. Natl. Acad. Sci. U.S.A. **119**(43), e2209218119 (2022)
14. Jiao, Y., et al.: Photoactive terahertz metasurfaces for ultrafast switchable sensing of colorectal cells. Mater. Horiz. **9**(12), 2984–2992 (2022)
15. Zhang, Z., et al.: Advanced terahertz refractive sensing and fingerprint recognition through metasurface-excited surface waves. Adv. Mater., 2308453 (2024)
16. Wang, Y., et al.: Terahertz spectroscopic diagnosis of early blast-induced traumatic brain injury in rats. Biomed. Opt. Express **11**(8), 4085–4098 (2020)
17. Ma, Z., et al.: Identification and quantitative detection of two pathogenic bacteria based on a terahertz metasensor. Nanoscale **15**(2), 515–521 (2023)
18. Yin, J., et al.: Terahertz photons inhibit cancer cells long term by suppressing nano telomerase activity. ACS Nano **18**(6), 4796–4810 (2024)
19. Globus, T., Woolard, D.L., Khromova, T., Partasarathy, R.: Ediss GJPoS-TISfOE: Terahertz signatures of biological-warfare-agent simulants **5411**, 25–32 (2004)
20. Yoon, S.A., et al.: Identifying different types of microorganisms with terahertz spectroscopy. Biomed. Opt. Express **11**(1), 406–416 (2020)
21. Jun, S.W., Ahn, Y.H.: Terahertz thermal curve analysis for label-free identification of pathogens. Nat. Commun. **13**(1), 3470 (2022)

22. Wang, J., Peng, L., Han, D., Zheng, T., Chang, T., Cui, H.L.: Label-free detection and identification of single bacteria via terahertz near-field imaging. Front. Microbiol. **14**, 1195448 (2023)
23. Fitch, M., Dodson, C., Ziomek, D., Osiander, R.: Time-domain terahertz spectroscopy of bioagent simulants, vol. 5584. SPIE (2004)
24. Wang, C., et al.: Application of terahertz time-domain spectroscopy in intracellular metabolite detection. J. Biophotonics **3**(10–11), 641–645 (2010)
25. Yang, Z.B., et al.: Near-field nanoscopic terahertz imaging of single proteins. Small **17**(3) (2021)
26. Arora, A., et al.: Terahertz-time domain spectroscopy for the detection of PCR amplified DNA in aqueous solution. Analyst **137**(3), 575–579 (2012)
27. Zhou, R., et al.: Label-free terahertz microfluidic biosensor for sensitive DNA detection using graphene-metasurface hybrid structures. Biosens. Bioelectron. **188**, 113336 (2021)
28. Yang, X., et al.: Terahertz spectroscopy for the isothermal detection of bacterial DNA by magnetic bead-based rolling circle amplification. Analyst **142**(24), 4661–4669 (2017)
29. Wu, K., et al.: Terahertz wave accelerates DNA unwinding: a molecular dynamics simulation study. J. Phys. Chem. Lett. **11**(17), 7002–7008 (2020)

Multi-functional High-Efficiency Terahertz Polarization Converter Based on Graphene Metasurface

Yushan Hou and Liuyang Zhang(✉)

School of Mechanical Engineering, Xi'an Jiaotong University, Xi'an 710049, China
liuyangzhang@xjtu.edu.cn

Abstract. In this paper, we propose a multi-functional and switchable broadband polarization converter in the terahertz (THz) band, which can switch between linear-to-linear polarization conversion and linear-to-circular polarization conversion modes via artificially tuning the Fermi energy level of graphene (E_f). When E_f is set to 0 eV, the polarization converter can transform linear polarized (LP) wave into its cross-polarized counterpart within the 0.53–1.41 THz range, exhibiting a remarkable relative bandwidth of 91%. Conversely, when E_f is set to 1 eV, the polarization converter can convert LP wave into right-handed circularly polarized (RCP) wave within the 0.7–1.6 THz band and left-handed circularly polarized (LCP) wave within the 1.8–3.1 THz band, with relative bandwidths of 78% and 53%, respectively. With these exceptional capabilities, this graphene metasurface-based polarization converter holds significant promise for a wide array of THz applications, including optical switches, modulators, communication systems, and sensors.

Keywords: THz metasurface · Polarization converter · Multilayered graphene

1 Introduction

Terahertz (THz) wave, spanning from 0.1 to 10 THz, has drawn significant interest across various domains, including communication, sensing, and imaging [1–3]. Metasurfaces, which generally refers to two-dimensional (2D) artificially designed subwavelength unit cell structures, brings a novel approach to manipulating THz wave due to their extraordinary electromagnetic (EM) performances compared to natural materials [4]. To achieve tunable control over THz wave, researchers integrate active materials such as semiconductors, graphene, liquid crystals, and phase change materials with metasurfaces, allowing regulation through external stimuli like electric fields, thermal effects, photoexcitation, and mechanical stretching [5–7]. Among these materials, graphene stands out due to its distinctive electrical, optical, and mechanical attributes. Over the past years, plenty of graphene-based metasurfaces, such as absorbers, polarization converters have been proposed as representative of THz functional devices. However, these studies often focus on investigating the amplitude, polarization state, phase, or propagation direction of EM waves independently, with limited efficiency and narrow bandwidth. Designing

C. Chang et al. (Eds.): YTHZ 2024, SPPHY 400, pp. 127–132, 2024.
https://doi.org/10.1007/978-981-97-3786-4_26

dynamically tunable multi-functional THz devices with broadband and high efficiency remains a significant challenge to enhance their widespread applications.

In this paper, we introduce a multi-functional broadband THz polarization converter, which offers the capability to switch between linear-to-linear polarization conversion and linear-to-circular polarization conversion by adjusting the Fermi level of graphene.

2 Design and Simulation Setup

The structure of the designed polarization converter is shown in Fig. 1, which comprises a gold substrate, dielectric layer 1, 7-layer graphene, dielectric layer 2, and patterned gold layer from bottom to top. The gold layer at the bottom is thick enough to prevent the transmission of EM waves. The top patterned gold layer functions as the resonance layer, composed of three different lengths of gold strips symmetrically arranged along the *u*-axis. The conductivity of gold is set as 4.56×10^7 S/m. Multilayered graphene plays a role of gated element by adjusting its Fermi level, which includes 7 layers of monolayer graphene separated by HTL resin with a thickness of 0.1 μm. Both dielectric layers 1 and 2 are HTL resin with the relative permittivity of 2.8 and loss tangent of 0.08 [4]. The optimized geometric parameters are shown in Table 1.

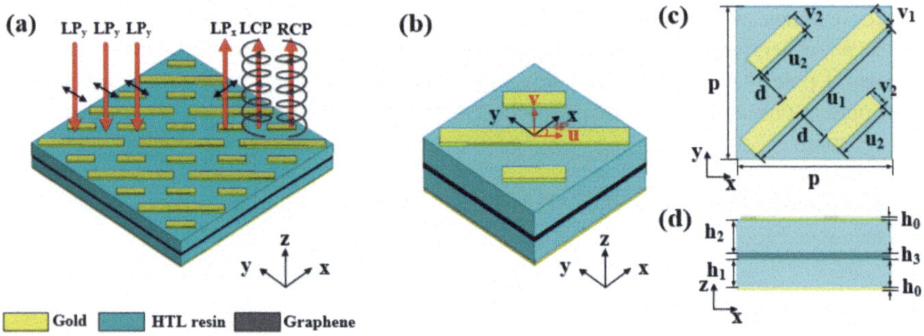

Fig. 1. Schematic diagram of the proposed THz polarization converter. (a) the 3×3-unit cells (b) the unit cell (c) top view and (d) side view.

Table 1. The optimized parameters of our proposed structure (unit: μm).

h_0	h_1	h_2	h_3	u_1	v_1	u_2	v_2	p	d
0.2	18	22	3	120	14	40	12	100	22.5

In this paper, all the simulation results are calculated by the finite-difference time-domain (FDTD) method in simulation software CST Microwave Studio. The unit cell boundary conditions are set along both the *x*-axis and *y*-axis, and open (add space) conditions are set along the *z*-axis. Graphene material for optical applications is created as 2D material, wherein T is set to 293 K, τ is set to 0.05 ps.

When y-polarized wave is perpendicular to the structure along the negative z-axis, the Polarization Conversion Ratio (*PCR*) is introduced to assess the performance of linear-to-linear polarization conversion, which is calculated as [8–11]:

$$PCR = \frac{|r_{xy}|^2}{|r_{xy}|^2 + |r_{yy}|^2} \tag{1}$$

where r_{yy} and r_{xy} represent the co-polarized and cross-polarized reflection wave, respectively. When $r_{yy} = 0$ and $r_{xy} = 1$, *PCR* equals 1, which means all y-polarized waves are converted into x-polarized waves.

For characterizing the efficacy of linear-to-circular polarization conversion, the Stokes Parameters are employed, defining the ellipticity χ as [8–11]:

$$\chi = \frac{2|r_{xy}||r_{yy}|\sin(\Delta\phi)}{|r_{xy}|^2 + |r_{yy}|^2} \tag{2}$$

where $\Delta\varphi$ represents the difference between phases φ_{yy} and φ_{xy}. When χ equals -1 or + 1, it signifies that the reflected wave is perfect right-handed circularly polarized (RCP) wave or perfect left-handed circularly polarized (LCP) wave, respectively.

The relative bandwidth (*RB*) is calculated as $f_{\max} - f_{\min}/[f_{\max} + f_{\min}/2]$.

3 Results and Discussion

Figure 2(a) illustrates the reflectance magnitudes of r_{yy} and r_{xy}, along with the PCR. Notably, when E_f is 0 eV, in the broadband region spanning from 0.52 to 1.41 THz (RB: 91%), most of the reflected waves are cross polarized, with a small portion being co-polarized. In addition, PCR is greater than 0.85,which indicates the implementation of linear-to-linear polarization conversion.

Upon transitioning E_f to 1 eV, the graphene layers assume a metallic state, and thus work as reflection mirrors. As depicted in Fig. 2(b), within the frequency range of 0.7–1.6 THz (RB: 78%), χ is less than -0.85, and $\Delta\varphi$ approaches 270°, signifying the presence of RCP waves in the reflection spectra. Conversely, in the frequency range of 1.8–3.1 THz (RB: 53%), χ exceeds 0.85, and $\Delta\varphi$ is close to 90°, indicating the prevalence of LCP waves in the reflection spectra.

A comparison between reported graphene-based metasurfaces polarization converters and our work is presented in Table 2. It can be seen that our proposed polarization converter has more diverse functions and a larger relative bandwidth, highlighting the exceptional performance in both linear-to-linear and linear-to-circular conversion within the THz range.

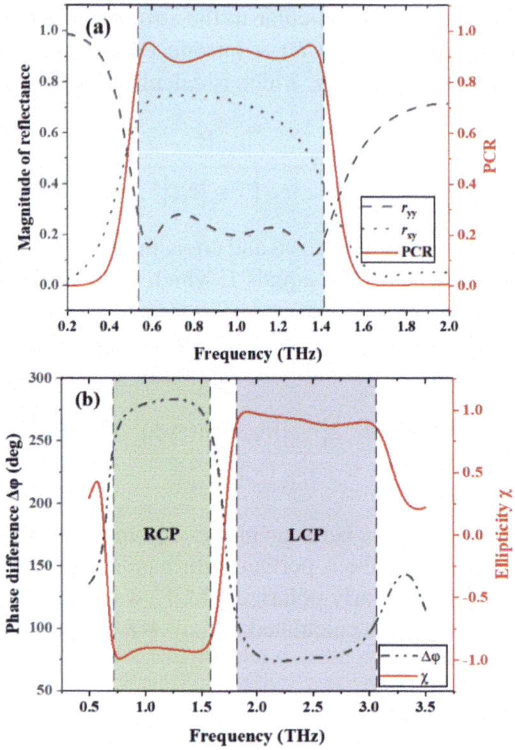

Fig. 2. (a) Magnitude of reflectance r_{yy}, r_{xy} and PCR ($E_f = 0$ eV). (b) Phase difference $\Delta\varphi$ and ellipticity χ ($E_f = 1$ eV).

4 Summary

We have introduced a multi-functional and switchable broadband THz polarization converter, capable of operating in two distinct modes: linear-to-linear polarization conversion and linear-to-circular polarization conversion. When the Fermi level of graphene is 0 eV, the polarization converter can realize a broadband linear-to-linear polarization conversion in the band of 0.53–1.41 THz (*RB*:91%) with PCR larger than 0.85. When the Fermi level of graphene is 1 eV, the polarization converter can realize a broadband linear-to-circular polarization conversion in the band of 0.7–1.6 THz (*RB*:78%) and 1.8–3.1 THz (*RB*:53%) with ellipticity less than -0.85 and larger than 0.85, respectively. Our work holds considerable promise for various applications in THz devices, including optical switches, modulators, communication systems, and sensors.

Table 2. Comparisons between reported polarization converters and our work.

Ref	Active material	Polarization conversion mode	Frequency range (THz)	Relative bandwidth	Performance
[5]	graphene	LP to LP	1.28–2.13	49.9%	PCR > 0.85
[8]	graphene	LP to RCP	9.87–11.03	11.1%	$\chi < -0.95$
		LP to LCP	13.16–14.43	9.2%	$\chi > 0.95$
[9]	graphene	LP to LP	1.38–1.85	29%	PCR > 0.9
		LP to RCP	1.12–1.72	42.3%	$\chi = -1$
[10]	graphene	LP to LP	0.78–0.97	21.7%	PCR > 0.9
		LP to RCP	0.61–0.67	9.3%	$\chi < -0.8$
		LP to LCP	0.72–0.97	29.6%	$\chi > 0.8$
[11]	graphene	LP to LP	2.92–6.26	72.8%	PCR > 0.9
		LP to RCP	4.45–6.47	37%	$\chi < -0.9$
This work	graphene	LP to LP	0.53–1.41	91%	PCR > 0.85
		LP to RCP	0.7–1.6	78%	$\chi < -0.85$
		LP to LCP	1.8–3.1	53%	$\chi > 0.85$

References

1. Cui, Y., Xu, Y., Han, D., et al.: Hidden-information extraction from layered structures through terahertz imaging down to ultralow SNR. Sci. Adv. **9**(40), eadg8435 (2023)
2. Nagatsuma, T., Horiguchi, S., Minamikata, Y., et al.: Terahertz wireless communications based on photonics technologies. Opt. Express **21**, 23736–23747 (2013)
3. Wang, R., Xu, L., Huang, L., et al.: Ultrasensitive terahertz biodetection enabled by quasi-BIC-based metasensors. Small **19**, 2301165 (2023)
4. Shen, Z., Li, S., Xu, Y., et al.: Three-dimensional printed ultrabroadband terahertz metamaterial absorber. Phys. Rev. Appl. **16**(1) (2021)
5. Han, D., Li, W., Sun, T., et al.: 2D-to-3D buckling transformability enabled reconfigurable metamaterials for tunable chirality and focusing effect. Photonics Res. **11**, 1770–1780 (2023)
6. Shrekenhamer, D., Chen, W.-C., Padilla, W.J.: Liquid crystal tunable metamaterial absorber. Phys. Rev. Lett. **110**(17), 177403 (2013)
7. Zhu, J., Li, S., Deng, L., et al.: Broadband tunable terahertz polarization converter based on a sinusoidally-slotted graphene metamaterial. Opt. Mater. Express **8**, 1164–1173 (2018)
8. Yu, F., Wang, Y., Yang, G., et al.: Tunable reflective dual-band line-to-circular polarization convertor with opposite handedness based on graphene metasurfaces. Opt. Express **29**, 13373–13387 (2021)
9. Zhang, J., Zhang, K., Cao, A., et al.: Bi-functional switchable broadband terahertz polarization converter based on a hybrid graphene-metal metasurface. Opt. Express **28**, 26102–26110 (2020)

10. Barkabian, M., Sharifi, N., Granpayeh, N.: Multi-functional high-efficiency reflective polarization converter based on an ultra-thin graphene metasurface in the THz band. Opt. Express **29**(13), 20160 (2021). https://doi.org/10.1364/OE.427583
11. Zhang, R., You, B., Wang, S., et al.: Broadband and switchable terahertz polarization converter based on graphene metasurfaces. Opt. Express **29**, 24804–24815 (2021)

Fingerprint Spectrum Recognition Based on Surface Wave Propagation

Zeyan Zhang[1,2](✉)

[1] Peking University, Beijing 100871, China
2201110344@stu.pku.edu.cn
[2] National Innovation Institute of Defense Technology, Beijing 100071, China

Abstract. Traditional methodologies aimed at enhancing sensing performance typically revolve around optimizing the near-field response of individual meta-atoms. However, when addressing fingerprint recognition for bio-substances, the need arises for multiple pixelated metasurfaces to establish a quasi-continuous spectrum. Herein, we introduce an innovative sensing approach for Terahertz (THz) fingerprint recognition, capitalizing on surface waves (SWs). This strategy facilitates the simultaneous resolution of diverse fingerprint details within a continuous spectrum based on the transmission amplitude of SWs. Significantly, our approach circumvents the complexities associated with meta-atom design and the constraints imposed by pixelated metasurfaces on device integration. Furthermore, it broadens the research landscape concerning metasurface-excited SWs.

Keywords: metasurface · plasmonic coupler · surface waves · fingerprint recognition

1 Introduction

The terahertz (THz) frequency, distinguished by attributes such as low energy, deep penetration, and precise resolution, has propelled THz technology as a formidable tool in biological works [1–12], accompanied by the evolution of THz sources [13, 14]. Recent progressions have ushered in diverse THz metasurface-based methodologies, harnessing these characteristics for exceptionally sensitive sensing [15–23]. Surface waves (SWs) are extensively utilized in the design of electromagnet part, with the often-neglected the capability of long-range propagation of SWs holding substantial promise of bio-sensing.

Within this investigation, we conceived a metasurface employing Pancharatnam–Berry (PB) meta-atoms to initiate and efficiently propagate SWs. Our discoveries highlight an elevated sensitivity in the transmission phase, facilitating smooth and uninterrupted fingerprint recognition. These results emphasize the capability of our approach to concurrently identify multiple fingerprint molecules, alongside the actual biomolecules.

C. Chang et al. (Eds.): YTHZ 2024, SPPHY 400, pp. 133–137, 2024.
https://doi.org/10.1007/978-981-97-3786-4_27

2 Results and Discussions

Illustrated in Fig. 1a and 1c are the components of the metasurface, showcasing a conventional metal-insulator-metal cavity comprising a gold resonator antenna and a quartz substrate. Defining the structure are parameters such as $p = 277.6$ μm, $h = 100$ μm, $r = 126.6$ μm, $w = 20$ μm, and $\alpha = 33°$. Adjusting the thickness of the quartz substrate plays a crucial role in achieving a reflected phase difference close to π between the x-pol and y-pol throughout the entire frequency spectrum. This adjustment is vital for the metasurface's efficacy in exciting SWs, as depicted in Fig. 1b.

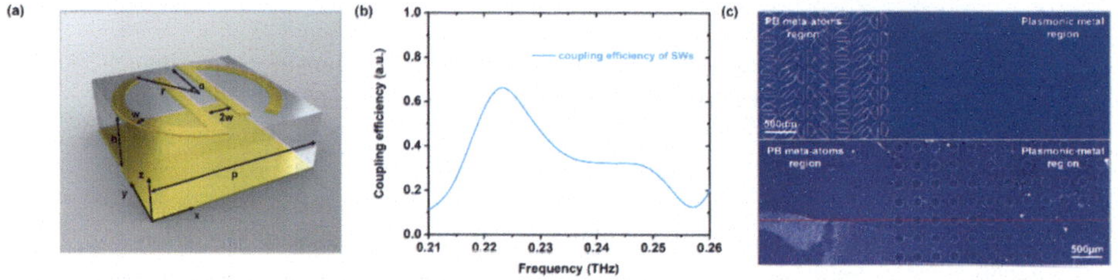

Fig. 1. (a) The schematic representation of PB-meta-atoms [17]; (b) Evaluation of the SWs coupling efficiency of the metasurface upon irradiation with THz waves. (c) Micrographs depicting both the front and back views of the metasurface structure.

In pursuit of our objective to investigate the metasurface's capacity for recognizing fingerprint spectra and to broaden its application in the realm of bio-molecular sensing, we utilize the commercial software COMSOL Multiphysics. This software, coupled with the Drude-Lorentz model, enables us to simulate the behavior of the molecular layers. Assuming a uniform distribution of molecules on the plasmonic metal surface and employing the Drude-Lorentz model, we accurately describe the dielectric constant [17]:

$$\varepsilon_m(\omega) = \varepsilon_\infty + \frac{\omega_p^2}{\omega_m^2 - \omega^2 - i\Gamma_m\omega} \tag{1}$$

Here, ε_∞ represents the background permittivity, and the vibration resonant frequency f_m of the molecular layers, as illustrated in Fig. 2a. To determine the SWs propagation length, we establish it as N_x times the cell period, serving as the representation of the transmission distance. Molecular layers, characterized by a thickness of t, are applied to the surface of the device. Illustrated in Fig. 2b are the field distributions of SWs transmitted beneath a molecular layer exhibiting a vibration mode at $f_m = 0.225$ THz, showcasing notable amplitude attenuation solely at 0.225 THz. Figure 2c exhibits the normalized reductions in transmission amplitude as molecular layers with varying fm values propagate, juxtaposed with the background dielectric constant $\varepsilon_\infty = 2.8$. As fm is modulated within the operational frequency range, the distinctive vibration modes of the molecular layer become apparent in the decline of the transmission curve. The extent of this decline reflects the interaction between SWs and the molecular layer.

In order to assess the metasurface's capability to detect multiple fingerprint spectra simultaneously, we establish two distinct cases. Throughout these cases, we maintain consistent values of $t = 10$ μm and $N_x = 20$. In both *case 1* and *case 2*, we arrange three and five molecular layers, each possessing distinct fm values, in an alternating fashion. Figure 2d presents the curve depicting the decline in transmission amplitude, effectively capturing the vibration modes of all molecular layers. It is noteworthy that different

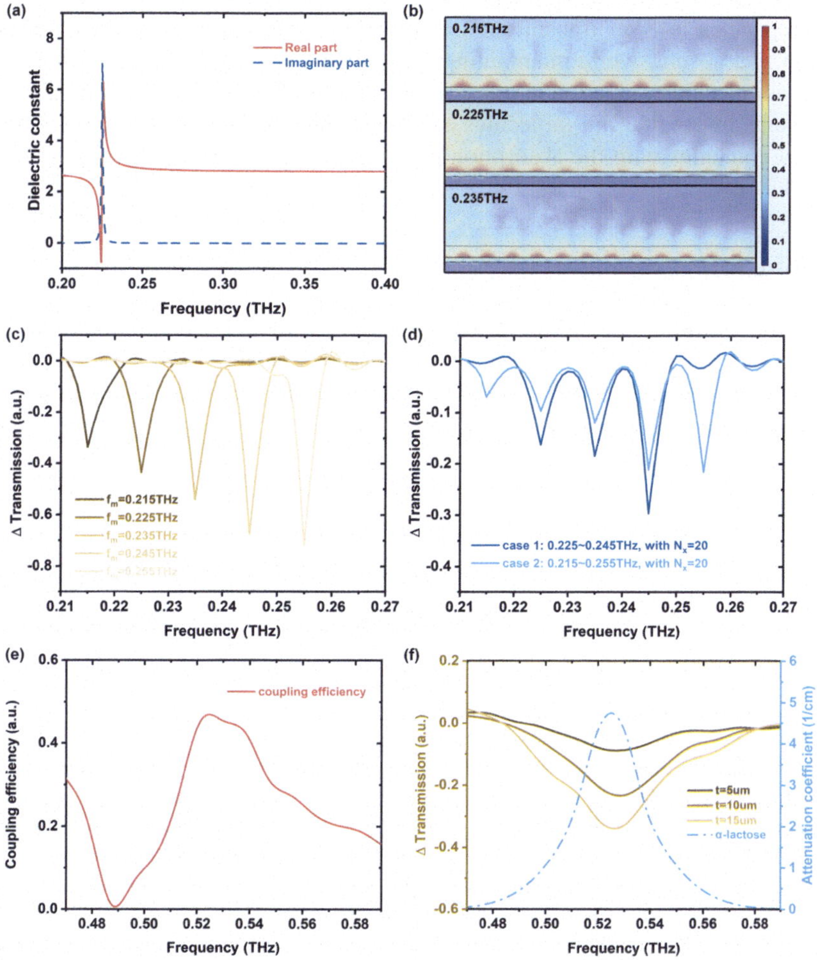

Fig. 2. (a) The layer under examination is represented using the Drude-Lorentz model [17]. (b) Field distributions of SWs, focusing on Re(Ez), vary across different frequencies when a molecular layer is present [17]; The disparities in normalized amplitude are observed in scenarios involving single (c) or multi-frequency vibration cases of the covered layer [17]; (e) By adjusting structural parameters, the coupling efficiency of another metasurface is analyzed with respect to SWs; (f) Changes in SWs transmission amplitude are scrutinized upon the application of α-lactose layers.

vibration modes only affect a restricted segment of the SWs transmission within the relevant frequency range. This characteristic permits the accumulation of diverse vibration mode information without encountering mutual interference during propagation.

To showcase the metasurface's capacity for identifying the unique fingerprint spectrum of real bio-molecules, we employ simulations for an additional metasurface. Structural parameters are adjusted to specifically target the fingerprint characteristics of α-lactose. The coupling efficiency of this adjusted metasurface is depicted in Fig. 2e, effectively capturing the absorption peak of α-lactose at 0.525 THz. Figure 2f delineates the discrepancies in SWs transmission amplitude between scenarios where α-lactose is present and absent, while maintaining $N_x = 30$ as the fixed transmission distance. Notably, a noticeable reduction in transmission amplitude is evident around 0.525 THz. These results imply that bio-molecules can induce absorption losses in SWs when the frequency coincides with the absorption peak, and these losses are directly influenced by the thickness.

3 Conlusion

In summary, we introduce an innovative approach for realizing continuous fingerprint spectrum recognition utilizing a compact meta-atoms structure, obviating the necessity for intricate configurations found in pixelated metasurfaces. Our findings extend the application scope of SWs and hold considerable potential for diverse detection scenarios.

References

1. Wang, P., et al.: Progress on cutting-edge infrared-terahertz biophysics. IEEE Trans. Microw. Theory Tech. **70**(11), 5117–5140 (2022)
2. Liu, G., et al.: Myelin sheath as a dielectric waveguide for signal propagation in the mid-infrared to terahertz spectral range. Adv. Funct. Mater. **29**(7), 1807862 (2019)
3. Xiang, Z., et al.: A primary model of THz and far-infrared signal generation and conduction in neuron systems based on the hypothesis of the ordered phase of water molecules on the neuron surface. Sci. Bull. **65**, 308 (2020)
4. Li, Y., et al.: Terahertz wave enhances permeaility of the voltage-gated calcium channel. J. Am. Chem. Soc. **143**, 9 (2021)
5. Liu, X., et al.: Nonthermal and reversible control of neuronal signaling and behavior by midinfrared stimulation. Proc. Natl. Acad. Sci. **118**(10), e2015685118 (2021)
6. Zhang, J.X., et al.: Non-invasive, opsin-free mid-infrared modulation activates cortical neurons and accelerates associative learning. Nature Comm. **12**, 2730 (2021)
7. Yu, Y., et al.: Terahertz photons improve cognitive functions in posttraumatic stress disorder. Research **6**, 0278 (2023)
8. Peng, W., et al.: High frequency terahertz waves disrupt Alzheimer's amyloid fibril formation. eLight 3, 18 (2023)
9. Tan, X., et al.: Neuromodulation of chemical sysnaptic transmission driven by THz photons. Research **2022**, 0010 (2022)
10. Li, Y., et al.: Physicochemical insights on terahertz wave diminished side effects of drugs from slow dissociation. ACS Nano **16**(5), 8419 (2022)
11. Wu, K., et al.: Terahertz wave accelerates DNA unwinding: a molecular dynamics simulation study. J. Phys. Chem. Lett. **11**(17), 7002–7008 (2020)

12. Yin, J., et al.: Terahertz wave as a non-drug and non-invasive strategy for long-term inhibition of cancer cells through telomerase activity suppression. ACS Nano (2024). https://doi.org/10.1021/acsnano.3c09216
13. Gao, M., et al.: Enhanced and frequency-tunable terahertz emission from two crossed air-plasmas. Phys. Rev. Res. **5**, 023091 (2023)
14. Chang, C., et al.: High-gain Thompson-scattering X-Ray free-electron laser by time-synchronic laterally tilted optical wave. Phys. Rev. Lett. **110**, 064802 (2013)
15. Lou, J., et al.: Optically controlled ultrafast terahertz metadevices with ultralow pump threshold. Small **17**, 2104275 (2021)
16. Wang, R., et al.: Electric Fano resonance-based terahertz metasensors. Nanoscale **13**, 18467–18472 (2021)
17. Zhang, Z., et al.: Advanced terahertz refractive sensing and fingerprint recognition through metasurface-excited surface waves. Adv. Mater., 2308453 (2024)
18. Lou, J., et al.: Calibration-free, high-precision, and robust terahertz ultrafast metasurfaces for monitoring gastric cancers. PNAS **119**(43) (2022)
19. Wang, P., et al.: An ultra-sensitive metasurface biosensor for instant cancer detection based on terahertz spectra. Nano Res. **16**, 7304–7311 (2023)
20. Jiao, Y., et al.: Photoactive terahertz metasurfaces for ultrafast switchable sensing of colorectal cells. Mater. Horiz. **9**, 2984–2992 (2022)
21. Ma, Z., et al.: Identification and quantitative detection of two pathogenic bacteria based on a terahertz metasensor. Nanoscale **15**, 515–521 (2023)
22. Wang, R., et al.: Ultrasensitive terahertz biodetection enabled by quasi-BIC-based metasensors. Small **19**, 2301165 (2023)
23. Sun, L., et al.: A pixelated frequency-agile metasurface for broadband terahertz molecular fingerprint sensing. Nanoscale **14**, 9681–9685 (2022)

Ultrasensitive Terahertz Detection via Different Novel Metasurfaces

Jiangfeng Fu[1], Anqi Tao[2], Lin Qiu[3], Fu Zheng[1], Tong Wu[1], Hongyi Tang[1], Xinyu Cui[1], Duo liu[1], and Jiuhui Jiang[1(✉)]

[1] Department of Orthodontics, Peking University School and Hospital of Stomatology; Beijing Key Laboratory of Digital Stomatology; National Clinical Research Center for Oral Diseases and National Engineering Laboratory for Digital and Material Technology of Stomatology, Haidian District, Beijing 100081, China
2311110632@bjmu.edu.cn

[2] Department of Periodontics, Peking University School and Hospital of Stomatology; Beijing Key Laboratory of Digital Stomatology; National Clinical Research Center for Oral Diseases and National Engineering Laboratory for Digital and Material Technology of Stomatology, Haidian District, Beijing 100081, China

[3] Central Laboratory, Peking University School and Hospital of Stomatology; Beijing Key Laboratory of Digital Stomatology; National Clinical Research Center for Oral Diseases and National Engineering Laboratory for Digital and Material Technology of Stomatology, Haidian District, Beijing 100081, China

Abstract. In recent decades, Terahertz (THz) bioanalytical sensors are considered as an excellent sensing platform due to the advantages such as label-free, accurate, nonionizing, high signal to noise ratio, and non-damaging [1–10]. It is able to distinguish biological sample features including structure, composition, molecular rotation modes and vibration modes rapidly. However, the lack of powerful radiation sources hinders the use of them. Also, weak light-matter interactions due to the mismatch between THz wavelengths and the absorption of bio-molecules has become the main disadvantages, restricting the sensitivity of THz-based biosensors [11–13]. To optimize the performance of THz biosensors, we proposed various advanced methods such as giant near-field enhancement around the metasurface-analytes boundary via quasi-BICs [14], shifting Lorentz-lattice coupled-resonance to electromagnetism-induced transparency (EIT) by photoexcitation of the silicon bridge [15], and widening spectral range via synchronous regulating graphene and C-shape rings [16]. Our results show that these THz metasurfaces bring light into next generation of trace biological substances sensing system.

Keywords: Terahertz · Metasurface · Terahertz biosensor · Fano Resonance · BIC

1 Electromagnetism-Induced Transparency (EIT) Based Metasensors

The EIT phenomenon refers to the extremely narrow absorption dip of the absorbance spectrum, which is caused by the coupling of different atomic energy levels [29] (Fig. 1).

C. Chang et al. (Eds.): YTHZ 2024, SPPHY 400, pp. 138–142, 2024.
https://doi.org/10.1007/978-981-97-3786-4_28

To shape THz sensors into a more practical and flexible sensing platform, Jiao et al. [16] proposed the scheme of shifting a Lorentz-lattice into EIT within an extremely short time of 32 ps via photoexciting the silicon bridge. In this situation, these two modes can be assumed as working at the same time. The strong field confinement enabled the combination of two frequency-sensing function to dig more slight useful details and improve the accuracy. As designed, the Lorentz-lattice has the frequency-sensitivity of 96.1 GHz per refractive index unit (RIU) and EIT mode is 118.4 GHz per RIU. This novel metasurface distinguished the different concentrations of normal and cancerous colorectal cells at the level of 2 * 10^5 cells per ml. The great performance of the metasensor steps solidly further towards more sensitive, intelligent, accurate and flexible THz sensors.

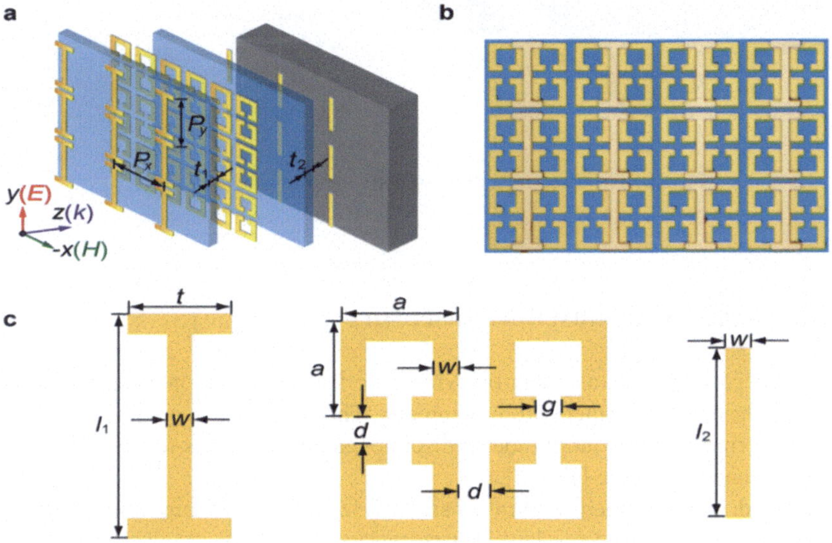

Fig. 1. Schematic of the unit cell supporting EIT [29].

2 Quasi-BIC-Based Metasensors

Theoretically BIC induces an infinite Q factor resonance, and external wave is unable to excite the metasurface directly. Whereas, they can be changed into quasi-BICs via destruction of symmetry and this enables free illuminations to directly excite them. If we kept those asymmetry factors to an extreme slight level, the quasi-BICs could also guarantee resonances with extreme high Q factor as well as tremendous near-field enhancement within the metamaterials [30, 31] (Fig. 2).

Wang et al. [11] utilized the magnetic dipole mode (MD) to structure the THz plasmonic metasensor with quasi-BICs. With the help of functional colloidal gold nanoparticles (AuNPs), the metasurface composed of split-ring resonator was able to detect trace amount of spiked solution, whose concentration is as low as 1 pM. As shown, the sensitivity of this unique quasi-BICs metasensor is 674 GHz/RIU. Then inflammatory factors C-reactive protein was detected as well as Serum Amyloid A. Comparing to the concentration of 0 pM, a shift of 17 GHz was observed after the concentration increased by 1 pM. The research illustrated the giant improvement of biosensors and lifted up the bio-detection method into superior development stage, which will diagnose and identify the people infected with bacteria and virus more sensitively and accurately.

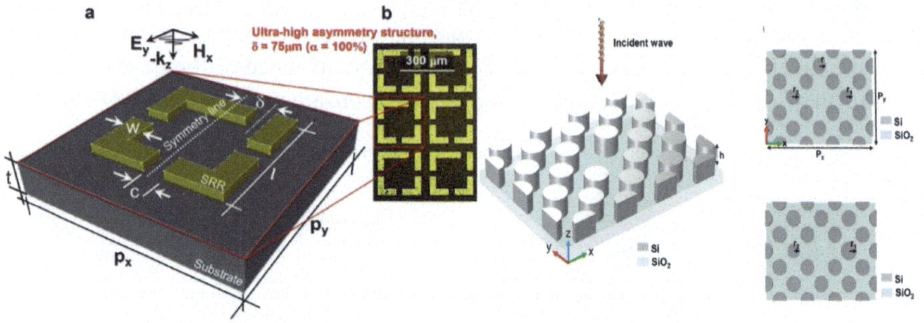

Fig. 2. The structures of quasi-BIC metasurfaces [30, 31].

3 Fano-Resonance-Based Metasensors

Fano resonance manifests as a sharp asymmetric line shape resonance, which can be designed by destroying the meta-atom-symmetry. The amplitude of continuous scattering waves interfering with the amplitude of discrete excitation waves results in the unique linearity of Fano resonance [17, 18].

Wang et al. [19] designed an ultra-sensitive THz metasensor consisting of three periodically arranged gold microrods. Limiting the asymmetry to the degree that enables leaky electric dipole resonance to couple with bound toroidal dipoles modes. The novel electric Fano resonance enhances nearfields dramatically which greatly strengthens the light–analyte interactions. The detection Interleukin-6, which is a cytokine with numerous functions, illustrated the amazing sensitivity of the novel metasensor. As the result showed, it succeeded distinguishing solutions with concentration difference as tiny as 1nM. The resonant frequency altered 23 GHz while changing the concentration from 0 nM to 1 nM. It is promising that this design would aggressively push forward the inspection of trace biological substance concentration change.

4 Toroidal-Resonance-Based Metasensors

In toroidal resonance, the surfaces consisting of a pair of split-ring (C-shape) structures suppress the loss of radiation between free space and metasurface. The coupling within the symmetric pattern is enhanced via the toroidal moment, resulting in the enhancement of quality factor response [13, 16, 20]. The design of C-shape also made the miniaturization of devices easier.

Wang et al. [15] placed the graphene in the near-field of the ring metastructure, whose electromagnetic properties has a key role in influence of C-shape metastructure. Due to the variety of Fermi level of graphene, precise tuning of ultra-wideband varying from 1.2 to 2.7 THz was realized. The detection of glucose confirmed the realization of arranging some special metastructure pixel to spectra with abundant absorption features, establishing a corresponding relationship between one tiny spatial metasensor unit and multiple spectral information of the detected object. It overcomes the 'false positive' drawback caused by traditional single characteristic spectrum detection. It is worth noticing that this design can easily extend the spectral region with change of geometric scaling parameter of C-shape ring or the Fermi level of graphene.

5 Conclusion

Biosensing based on THz metastructures shows immense potential that can't be underestimated for various applications in medicine and chemistry [21–24]. Overall, these works open a new avenue for ultra-fast THz sensors quantitively identifying low-concentration solutions more versatile, flexible and broadband miniaturized safely.

References

1. Jiaming, L., et al.: Frequency selective fingerprint sensor: the Terahertz unity platform for broadband chiral enantiomers multiplexed signals and narrowband molecular AIT enhancement. PhotoniX (2023). https://doi.org/10.1186/s43074-023-00108-1
2. Xiao, S., et al.: Active modulation of electromagnetically induced transparency analogue in terahertz hybrid metal-graphene metamaterials. Carbon **126**, 271–278 (2018)
3. Zhang, J., et al.: Highly sensitive detection of malignant glioma cells using metamaterial-inspired THz biosensor based on electromagnetically induced transparency. Biosens. Bioelectron. **185**, 113241 (2021)
4. Li, Y., et al.: Terahertz wave enhances permeaility of the voltage-gated calcium channel. J. Am. Chem. Soc. **143**, 9 (2021)
5. Liu, G., et al.: Myelin sheath as a dielectric waveguide for signal propagation in the mid-infrared to terahertz spectral range. Adv. Funct. Mater. **29**(7), 1807862 (2019)
6. Zhang, J.X., et al.: Non-invasive, opsin-free mid-infrared modulation activates cortical neurons and accelerates associative learning. Nat. Commun. **12**, 2730 (2021)
7. Liu, X., et al.: Nonthermal and reversible control of neuronal signaling and behavior by midinfrared stimulation. Proc. Natl. Acad. Sci. **118**(10), e2015685118 (2021)
8. Xiang, Z., et al.: A primary model of THz and far-infrared signal generation and conduction in neuron systems based on the hypothesis of the ordered phase of water molecules on the neuron surface. Sci. Bullet. **65**, 308 (2020)
9. Wu, K., et al.: Terahertz wave accelerates DNA unwinding: a molecular dynamics simulation study. J. Phys. Chem. Lett. **11**(17), 7002–7008 (2020)
10. Ahmadivand, A., et al.: Terahertz plasmonics: the rise of toroidal metadevices towards immunobiosensings. Mater. Today **32**, 108–130 (2020)
11. Zhang, Z., et al.: Advanced terahertz refractive sensing and fingerprint recognition through metasurface-excited surface waves. Adv. Mater. **36**(14), 2308453 (2024)
12. Lou, J., et al.: Calibration-free, high-precision, and robust terahertz ultrafast metasurfaces for monitoring gastric cancers. PNAS **119**, 43 (2022)

13. Wang, R., et al.: Ultrasensitive terahertz biodetection enabled by quasi-BIC-based metasensors. Small **19**(35), 2301165 (2023)
14. Jiao, Y., et al.: Photoactive terahertz metasurfaces for ultrafast switchable sensing of colorectal cells. Mater. Horiz. **9**(12), 2984 (2022). https://doi.org/10.1039/D2MH00787H
15. Sun, L., et al.: A pixelated frequency-agile metasurface for broadband terahertz molecular fingerprint sensing. Nanoscale **14**(27), 9681–9685 (2022). https://doi.org/10.1039/D2NR01561G
16. Tan, T.C., et al.: Active control of nanodielectric-induced THz Quasi-BIC in flexible metasurfaces: a platform for modulation and sensing. Adv. Mater. **33**, 2100836 (2021)
17. Kang, M., Liu, T., Chan, C.T., Xiao, M.: Applications of bound states in the continuum in photonics. Nat. Rev. Phys. **5**(11), 659–678 (2023)
18. Melik-Gaykazyan, E., et al.: From Fano to quasi-BIC resonances in individual dielectric nanoantennas. Nano Lett. **21**(4), 1765–1771 (2021)
19. Wang, R., et al.: Electric Fano resonance-based terahertz metasensors. Nanoscale **13**(44), 18467–18472 (2021). https://doi.org/10.1039/D1NR04477J
20. Zhang, W., et al.: Terahertz metamaterials for biosensing applications: a review. Biosensors **14**(1), 3 (2024)
21. Wang, P., et al.: Progress on cutting-edge infrared-terahertz biophysics. IEEE Trans. Microw. Theory Tech. **70**(11), 5117–5140 (2022)
22. Wang, P., et al.: An ultra-sensitive metasurface biosensor for instant cancer detection based on terahertz spectra. Nano Res. **16**, 7304–7311 (2023)
23. Yu, Y., et al.: Terahertz photons improve cognitive functions in posttraumatic stress disorder. Research **6**, 0278 (2023)
24. Yin, J., et al.: Terahertz photons inhibit cancer cells long term by suppressing nano telomerase activity. ACS Nano **18**, 4796 (2024)
25. Zhu, Z., et al.: Tunable surface wettability via terahertz electrowave controlled vicinal subnanoscale water layer. Nano Lett. **24**(10), 3243–3248 (2024)
26. Zhang, Q.L., Zhou, T., Chang, C., et al.: Ultrahigh-flux water nanopumps generated by asymmetric terahertz absorption. Phys. Rev. Lett. **132**(18) (2024)
27. Zhu, Z., et al.: Characteristic fingerprint spectrum of neurotransmitter norepinephrine with broadband terahertz time-domain spectroscopy. Analyst **144**, 2504–2510 (2019)
28. Tang, C., et al.: Integrating terahertz metamaterial and water nanodroplets for ultrasensitive detection of amyloid β aggregates in liquids. Sens. Actuat. B. Chem. **329**, 129113 (2021)
29. Zhang, X., et al.: Electromagnetically induced absorption in a three-resonator metasurface system. Sci. Rep. **5**(1), 10737 (2015). https://doi.org/10.1038/srep10737
30. Kyaw, C., et al.: Polarization-selective modulation of supercavity resonances originating from bound states in the continuum. Commun. Phys. **3**(1), 212 (2020). https://doi.org/10.1038/s42005-020-00453-8
31. Saadatmand, S.B., et al.: Quasi-BIC based all-dielectric metasurfaces for ultra-sensitive refractive index and temperature sensing. Sci. Rep. **13**(1), 20625 (2023). https://doi.org/10.1038/s41598-023-48051-2

Detection of Sepsis Inflammatory Factors Based on the Terahertz-Metasurface Biosensor

Bin Zhang and Xiaohui Du(✉)

First Medical Center, Chinese PLA General Hospital, Beijing, China
15063320960@163.com

Abstract. Sepsis is a highly dynamic and complex condition with the potential to progress rapidly to critical organ dysfunction and death. Timely detection and management of sepsis is critical as it has the potential to significantly reduce mortality and improve overall clinical outcomes. However, current clinical detection techniques are limited by plenty of factors that make it difficult to meet the requirements, and therefore a rapid and highly sensitive detection method is urgently needed. Here, we propose a plasma terahertz metasurface biosensor based on the principle of surface plasmon resonance. Meanwhile, we integrated an antibody-functionalized metasurface for real-time binding detection of inflammatory factors calcitoninogen (PCT) and interleukin-10 (IL-10) solutions with detection limits as low as 1 μg/mL. In summary, the designed terahertz metal surface biosensor not only has the advantages of low detection limit and high degree of automation, but also has a broad application prospect in biomedical field.

Keyword: Sepsis · Terahertz · Metasurface

1 Introduction

Since the vibration and rotation energy levels of many biological macromolecules are mostly in the THz band [1–4], and different molecular classes have different absorption and dispersion characteristics [5–10], the structure and properties of substances can be analyzed and identified by means of the characteristic frequency[11–21]. Meanwhile, metasurface is an artificially-constructed, patterning, periodically-arranged, ultrathin structure consisting of sub-wavelength optical resonators [22–24]. By combining or tailoring meta-atoms, metasurfaces have the characteristic ability to manipulate light scattering in the subwavelength range and can produce multifarious optical phenomena [25–27]. Hence, the utilization of terahertz in conjunction with metasurfaces technology holds great promise in the field of biosensing and biomedical application.

In this work, we demonstrate a novel metasurface biosensor to implement quantitative detection. The periodic nanostructure on the surface of the metal array elements can produce strong electric and toroidal dipole coupling, which gives the sensor significant physical characteristics such as high sensitivity and strong resonance strength. In order to verify the sensing performance of the biosensor, we placed PCT and IL-10 in the detection area to achieve quantitative detection of multiple concentration gradients, with

C. Chang et al. (Eds.): YTHZ 2024, SPPHY 400, pp. 143–146, 2024.
https://doi.org/10.1007/978-981-97-3786-4_29

the lowest detection concentration up to 1 ug/mL. In short, the biosensor developed by us has excellent performance, overcomes many limitations of terahertz sensing technology in practical clinical applications, and has broad application prospects in the biomedical field.

2 Materials and Methods

2.1 Metasurface Framework

The detailed structural design of the elementary atom consisting of an isolated metal rod and a pair of resonant rings. The length of the metal rod is L = 40 μm and the pair of resonant rings is located symmetrically on either side of it with a spacing of b = 1.5 μm and 3 μm. The length of the resonant rings is a = 30 μm, the radius of the semicircle is r = 5 μm, and the magnitude of the splitting gap is c = 1.5 μm. The width of the arms of the metal structure is w = 3 μm.

2.2 Detection Procedures

It principally involves three aspects: surface activation, antibody immobilization and target capture, including hydrophilization treatment, APTES solution, glutaraldehyde solution, anti-inflammatory factor antibodies, BSA solution, inflammatory factor. The detection mechanism is based on layer-by-layer or step-by-step surface modification through chemical processes making it optimized for capturing and quantifying biomolecules.

In this study, all experiments were conducted at room temperature (25 °C) and in a dry environment (1.5% humidity). During the test, 50 μL samples were uniformly dropped on the surface of the cell using a skin test gun, then allowed to incubate at room temperature for 1 h, then rinsed with PBS buffers to remove unbound impurities, and finally dried in nitrogen. After drying, the metasurface was placed on the detection platform of the THz-TDS system, and the metasurface is detected when both reach the required environmental conditions.

3 Results and Discussion

Firstly, the two inflammatory factors were diluted into three concentration gradients of 1 μg/mL, 10 μg/mL, 100 μg/mL using PBS solution and flowed onto the surface of the microsensors in increasing order of concentration for incubation. We found that with the increase of the concentration, the resonance curves of the two inflammatory factors showed a tendency to move to the lower frequency. The PCT resonances have redshifts of 12.45 GHz, 20.79 GHz and 36.77 GHz respectively, while IL-10 resonances have redshifts of 12.36 GHz, 22.68 GHz and 32.53 GHz, as shown in Fig. 1(a) and Fig. 1(b).

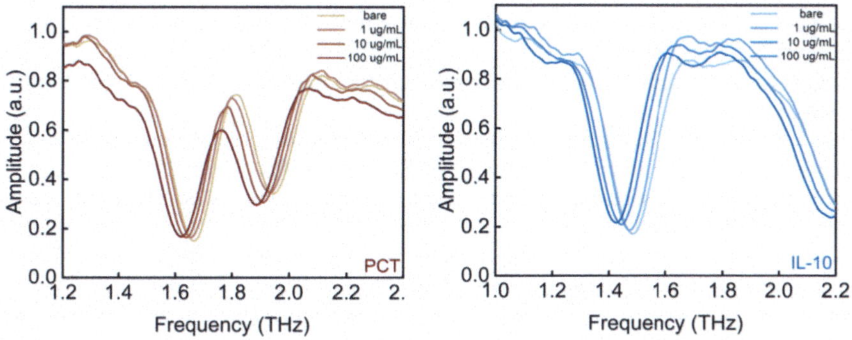

Fig. 1. Experimental transmission spectra of PCT (a) and IL-10 (b) at different concentrations. The resonance curves of both were red-shifted with increasing concentration and shifted by different amounts.

4 Conclusion

Our proposed biosensor has excellent performance. This is not only reflected in the high sensitivity and real-time detection based on terahertz (THz) and the high Q value of the biosensor built on the metasurface, which can provide ultra-high local field intensity terahertz reflection signal when detecting different inflammatory factors, so as to achieve rapid and high-precision detection of inflammatory factors in liquid samples, but also reflected in the combination of antigen and antibody. The high specificity of the biosensor ensures that the target inflammatory factors can be accurately captured in the solution, and the target inflammatory factors can be directly and quickly detected without pretreatment of the sample. In summary, we prove the feasibility of the biosensor in clinical application, which opens up a new way for the detection of other clinical indicators, and has broad application prospects.

References

1. Wang, P., et al.: Progress on cutting-edge infrared-terahertz biophysics. IEEE Trans. Microw. Theory Tech. **70**(11), 5117–5140 (2022)
2. Liu, G., et al.: Myelin sheath as a dielectric waveguide for signal propagation in the mid-infrared to terahertz spectral range. Adv. Funct. Mater. **29**(7), 1807862 (2019)
3. Wu, K., et al.: Terahertz wave accelerates DNA unwinding: a molecular dynamics simulation study. J. Phys. Chem. Lett. **11**(17), 7002–7008 (2020)
4. Xiang, Z., et al.: A primary model of THz and far-infrared signal generation and conduction in neuron systems based on the hypothesis of the ordered phase of water molecules on the neuron surface I. Sci. Bullet. **65**(4), 308 (2020)
5. Li, Y., et al.: Terahertz wave enhances permeability of the voltage-gated calcium channel. J. Am. Chem. Soc. **143**, 9 (2021)
6. Liu, X., et al.: Nonthermal and reversible control of neuronal signaling and behavior by midinfrared stimulation, Proc. Natl. Acad. Sci. **118**(10), e2015685118 (2021)
7. Zhang, J.X., et al.: Non-invasive, opsin-free mid-infrared modulation activates cortical neurons and accelerates associative learning. Nat. Commun. **12**, 2730 (2021)

8. Yu, Y., et al.: Terahertz photons improve cognitive functions in posttraumatic stress disorder. Research **6**, 0278 (2023)
9. Peng, W., et al.: High frequency terahertz waves disrupt Alzheimer's amyloid fibril formation. eLight **3**, 18 (2023)
10. Chang, C., et al.: High-gain Thompson-scattering X-ray free-electron laser by time-synchronic laterally tilted optical wave. Phys. Rev. Lett. **110**, 064802 (2013)
11. Wang, R., et al.: Ultrasensitive terahertz biodetection enabled by Quasi-BIC-Based metasensors. Small **19**(35) (2023)
12. Zhang, Z., et al.: Advanced terahertz refractive sensing and fingerprint recognition through metasurface-excited surface waves. Adv. Mater. **36**(14), 2308453 (2024)
13. Lou, J., et al.: Optically controlled ultrafast terahertz metadevices with ultralow pump threshold. Small **17**(44) (2021)
14. Zhu, Z., et al.: Characteristic fingerprint spectrum of neurotransmitter norepinephrine with broadband terahertz time-domain spectroscopy. Analyst **144**, 2504–2510 (2019)
15. Tang, C., et al.: Integrating terahertz metamaterial and water nanodroplets for ultrasensitive detection of amyloid β aggregates in liquids. Sens. Actuat. B Chem. **329**, 129113 (2021)
16. Tan, X., et al.: Neuromodulation of chemical synaptic transmission driven by THz photons. Research **6**, 1–16 (2022)
17. Zhang, Q.L., Zhou, T., Chang, C., et al.: Ultrahigh-flux water nanopumps generated by asymmetric terahertz absorption. Phys. Rev. Lett. **132**(18) (2024)
18. Zhu, Z., et al.: Tunable surface wettability via terahertz electrowave controlled vicinal subnanoscale water layer. Nano Lett. **24**(10), 3243–3248 (2024)
19. Yin, J., et al.: Terahertz photons inhibit cancer cells long term by suppressing nano telomerase activity. ACS Nano **18**(6), 4796–4910 (2024)
20. Sun, L., et al.: Inhibition of cancer cell migration and glycolysis by terahertz wave modulation via altered chromatin accessibility. Research 9860679 (2022)
21. Qiao, Z., et al.: Infrared nanoimaging of nanoscale sliding dislocation of collagen fibrils. Nano Res. **15**(3), 2355–2361 (2022)
22. Wang, R., et al.: Electric Fano resonance-based terahertz metasensors. Nanoscale **13**(44), 18467–18472 (2021)
23. Ma, Z., et al.: Identification and quantitative detection of two pathogenic bacteria based on a terahertz metasensor. Nanoscale **15**(2), 515–521 (2023)
24. Sun, L., et al.: A pixelated frequency-agile metasurface for broadband terahertz molecular fingerprint sensing. Nanoscale **14**, 9681–9685 (2022)
25. Wang, P., et al.: An ultra-sensitive metasurface biosensor for instant cancer detection based on terahertz spectra. Nano Res. **16**, 7304–7311 (2023)
26. Lou, J., et al.: Calibration-free, high-precision, and robust terahertz ultrafast metasurfaces for monitoring gastric cancers. PNAS **119**(43) (2022)
27. Jiao, Y., et al.: Photoactive terahertz metasurfaces for ultrafast switchable sensing of colorectal cells. Mater. Horiz. **9**, 2984–2992 (2022)

Research on Infrared and Terahertz Spectral Characterization Systems

Jiaying Zhu(✉), Jialiang Chuang, and Miaoyan Wang

The 11th Research Institute of China Electronics Technology Group Corporation, Beijing 100015, China
zhujiaying17@163.com

Abstract. Electromagnetic waves in the infrared (3–11 μm) and terahertz (30–3000 μm) bands resonate absorption reactions with chemical bonds in biological tissues. Combined with spectral characterization systems, relevant pathological tissues can be distinguished. This paper provides an overview of infrared and terahertz spectral characterization systems. Each system has its own advantages in terms of application requirements, and is complementary. Based on different types of infrared radiation sources, the nonlinear crystal tuning spectral characterization system, broadband supercontinuum radiation spectral characterization system, and quantum cascade spectral characterization system are elaborated. Based on the radiation detection mechanism of photoconductive antennas, fiber type and free space optical field modulation terahertz time-domain spectroscopy systems are introduced. By summarizing the system compositions and principles of infrared and terahertz spectral characterization systems, this paper provides reference for scientific researchers to select the corresponding band spectral characterization system.

Keywords: infrared · spectral characterization system · terahertz time-domain spectroscopy system

1 Introduction

The electromagnetic waves in the infrared (3–11 μm) and terahertz (30–3000 μm) bands have unique spectral characteristics [1–5]. The fingerprint spectral characteristics of these frequencies band can be used to detect, identify, and quantify different substance molecules [6–10]. Therefore, infrared and terahertz spectral imaging characterization systems have important application value in disease diagnosis, biomedical research, and other fields.

Infrared and terahertz spectral characterization systems obtain spectral information of tissue samples through point by point scanning imaging. Specific tissue structural components can be qualitatively identified without fluorescence labeling, and imaging characterizes the biochemical components and lesion areas of biological tissues. This is the common feature of infrared and terahertz spectral characterization systems. The spectral characterization technology in these bands have been widely applied in the biomedical field.

C. Chang et al. (Eds.): YTHZ 2024, SPPHY 400, pp. 147–152, 2024.
https://doi.org/10.1007/978-981-97-3786-4_30

2 Overview of Infrared Spectral Characterization Systems

Infrared spectral characterization systems, as an analytical tool to identify the composition of tissue components by acquiring spatial and spectral information of biological tissues, have found wide application in biomedicine such as cancer detection.

In infrared spectral characterization systems, the most important difference is the generation mechanism of the infrared radiation source, which is mainly classified into three types: nonlinear crystal tuning spectral characterization system, broadband supercontinuum radiation spectral characterization system, and quantum cascade spectral characterization system.

2.1 Nonlinear Crystal Tuning Spectral Characterization System

Infrared radiation sources based on nonlinear crystals, such as optical parametric oscillators (OPOs) or differential frequency generation (DFG), have relatively narrow linewidths and wide tuning ranges [19]. Figure 1 shows a spectral characterization system based on an Optical Parametric Oscillator, but this technique is usually limited by low optical conversion efficiencies and requires a complex optical system as well as sophisticated mechanical parts for tuning.

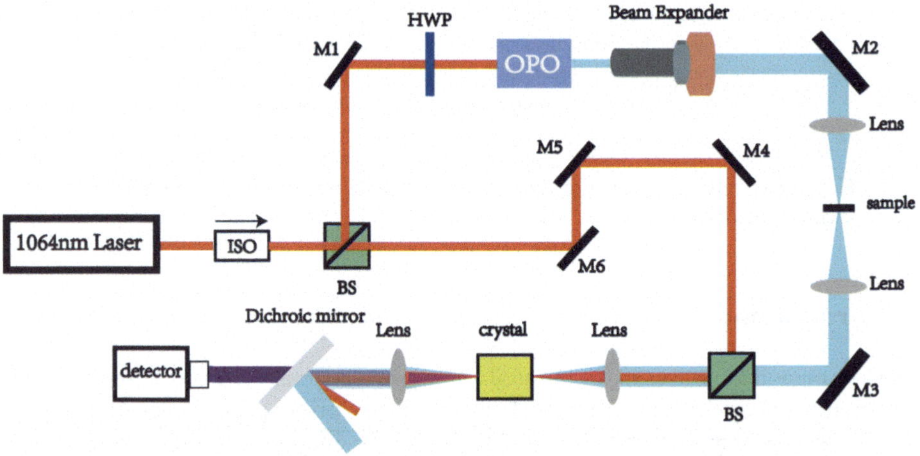

Fig. 1. Infrared spectral characterization system based on nonlinear crystal tuning

2.2 Broadband Supercontinuum Radiation Spectral Characterization System

The supercontinuum radiation source is generated based on the nonlinear effect and provides a wider spectral width, up to 10μm at long wavelengths, which ensures the output of a higher quality beam. The infrared spectral characterization system based on a sulfide photonic crystal fiber (PCF) to generate a supercontinuum radiation source [2], which has a radiation source that can be output in a single mode to achieve diffraction-limited imaging performance. This radiation source has the characteristics of low cost, high brightness, and miniaturization.

2.3 Quantum Cascade Spectral Characterization System

Quantum cascade lasers based on frequency tunable have narrower bandwidths and higher brightness in specific wavelength bands. In addition, quantum cascade lasers combined with corresponding gratings can achieve narrowband tuning, providing more accurate wavelength selection and higher signal-to-noise ratio. Typically, several tunable quantum cascade lasers are integrated together to achieve imaging and characterization within the infrared broadband range [20]. Figure 2 represents an infrared spectral imaging system based on Quantum Cascade Laser (QCL), which can perform fast spatial scanning imaging and discrete spectral profile measurements. By studying machine learning algorithms on tissue imaging results, good classification results were obtained.

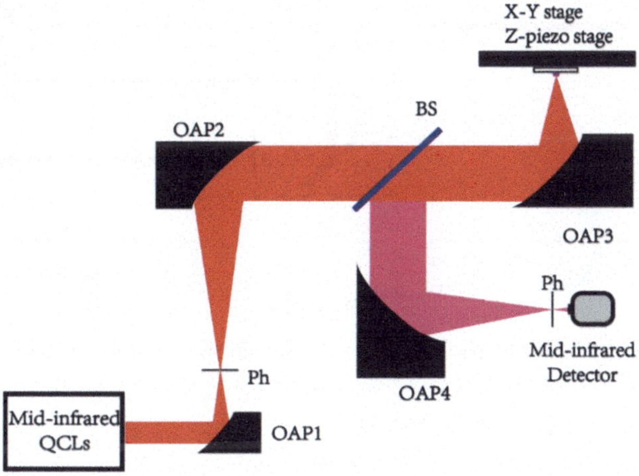

Fig. 2. Infrared spectral characterization system based on quantum cascade lasers

3 Overview of Terahertz Spectral Characterization Systems

Terahertz time-domain spectroscopy system is a typical representative of terahertz spectral characterization systems, as a correlation detection device, obtaining the information of the phase and amplitude, and calculating the absorption coefficient and the refractive index coefficient, which can be used for the qualitative analysis of the measured sample tissue. It has become a potential research device in the fields of biomedical and materials science. The terahertz time-domain spectroscopy system can be divided into two categories: fiber type and free space light. This paper focuses on the terahertz time-domain spectroscopy characterization system based on photoconductive antennas.

3.1 Fiber Type Terahertz Time-Domain Spectral Characterization System

Fiber type terahertz time-domain spectroscopy system has the advantages of simple structure, practicality and compact size [21]. Figure 3 shows the fiber type terahertz time-domain spectroscopy system, grating stretcher is placed in front of the beam-splitting device to compensate for the nonlinear effects generated by the optical fiber. An all-fiber optic emitter and detector are used for terahertz transmission or reflection measurements on the sample.

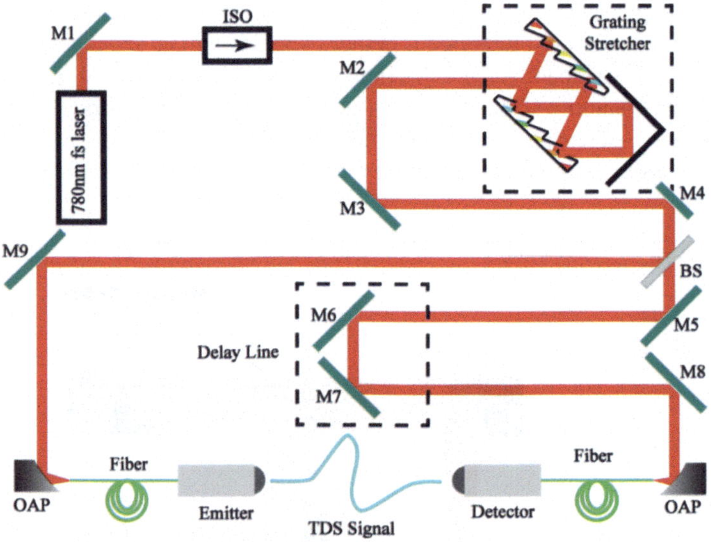

Fig. 3. Fiber type terahertz time-domain spectroscopy system

3.2 Free Space Terahertz Time-Domain Spectral Characterization System

In order to improve the pump radiation utilization and emission power of the photoconductive array antenna, Fig. 4 shows a terahertz time-domain spectroscopy system with pump light field modulation. The 780 nm spatial light modulator is used to modulate the laser, and the modulated structured light matched the cross finger array antenna to pump terahertz, which improved the signal-to-noise ratio of the terahertz time-domain signal [11]. This is helpful for the detection of substances with severe absorption of terahertz.

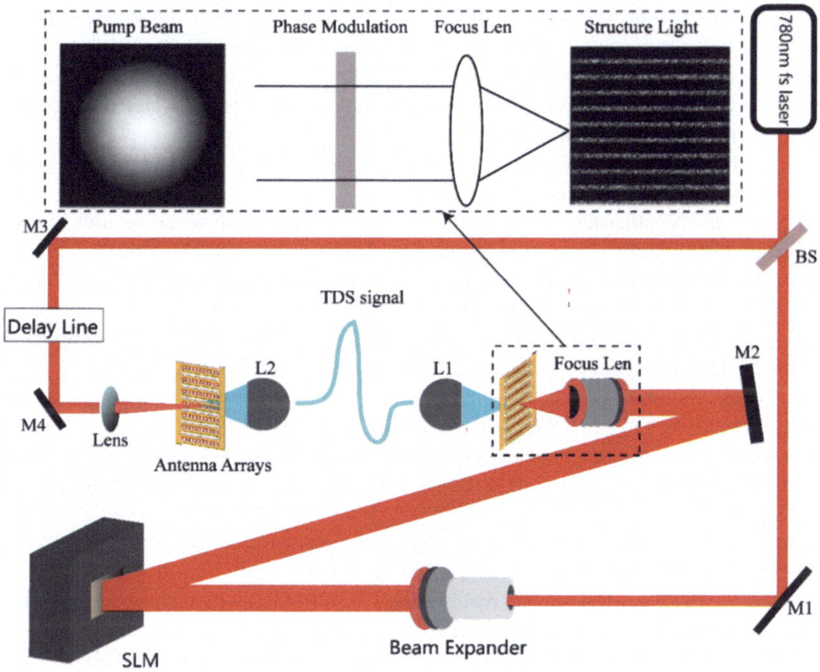

Fig. 4. Terahertz time-domain spectroscopy system with optical field modulated pumping

4 Conclusion

The infrared and terahertz spectral characterization systems outlined in this paper are beneficial for distinguishing relevant biological tissues, especially for distinguishing cancer tissues [12, 13], and provide reference for studying the pathogenesis of diseased tissues [14–18]. The infrared spectral characterization system outlined provides higher imaging resolution of samples due to shorter wavelengths. The fiber type terahertz time-domain spectroscopy system is beneficial for system integration and enables flexible imaging of samples. Finally, the terahertz time-domain spectroscopy system based on optical field modulation array synthesis improves the signal-to-noise ratio of time-domain signals, providing a solution for detecting substances with severe absorption of terahertz.

References

1. Kawase, K., Ogawa, Y., Watanabe, Y., et al.: Non-destructive terahertz imaging of illicit drugs using spectral fingerprints. Opt. Express **11**(20), 2549–2554 (2003)
2. Petersen, C.R., Prtljaga, N., Farries, M., et al.: Mid-infrared multispectral tissue imaging using a chalcogenide fiber supercontinuum source. Opt. Lett. **43**(5), 999–1002 (2018)
3. Zhang, J., et al.: Non-invasive, opsin-free mid-infrared modulation activates cortical neurons and accelerates associative learning. Nat. Commun. **12**(1), 2730 (2021)

4. Liu, G., et al.: Myelin sheath as a dielectric waveguide for signal propagation in the mid-infrared to terahertz spectral range. Adv. Funct. Mater. **29**(7), 1807862 (2019)
5. Xiang, Z., et al.: A primary model of THz and far-infrared signal generation and conduction in neuron systems based on the hypothesis of the ordered phase of water molecules on the neuron surface. Sci. Bullet. **65**, 308 (2020)
6. Li, Y., et al.: Terahertz wave enhances permeability of the voltage-gated calcium channel. J. Am. Chem. Soc. **143**(11), 4311–4318 (2021)
7. Liu, X., et al.: Nonthermal and reversible control of neuronal signaling and behavior by midinfrared stimulation. Proc. Natl. Acad. Sci **118**(10), e2015685118 (2021)
8. Wu, K., et al.: Terahertz wave accelerates DNA unwinding: a molecular dynamics simulation study. J. Phys. Chem. Lett. **11**(17), 7002–7008 (2020)
9. Chang, C., et al.: High-gain Thompson-scattering X-ray free-electron laser by time-synchronic laterally tilted optical wave. Phys. Rev. Lett. **110**, 064802 (2013)
10. Lou, J., et al.: Calibration-free, high-precision, and robust terahertz ultrafast metasurfaces for monitoring gastric cancers. Proc. Natl. Acad. Sci. **119**, 43 (2022)
11. Wang, R., et al.: Interdigitated photoconductive antenna pumped by reconfigurable structured light for terahertz emission and modulation. IEEE Trans. Microw. Theo. Tech. **71**(8), 3661–3667 (2023)
12. Wang, P., et al.: An ultra-sensitive metasurface biosensor for instant cancer detection based on terahertz spectra. Nano Res. **16**, 7304–7311 (2023)
13. Jiao, Y., et al.: Photoactive terahertz metasurfaces for ultrafast switchable sensing of colorectal cells. Mater. Horiz. **9**, 2984–2992 (2022)
14. Wang, R., et al.: Ultrasensitive terahertz biodetection enabled by quasi-BIC-based metasensors. Small **19**, 2301165 (2023)
15. Zhang, Z., et al.: Advanced terahertz refractive sensing and fingerprint recognition through metasurface-excited surface waves. Adv. Mater. 2308453 (2024)
16. Zhu, Z., et al.: Characteristic fingerprint spectrum of neurotransmitter norepinephrine with broadband terahertz time-domain spectroscopy. Analyst **144**(8), 2504–2510 (2019)
17. Tang, C., et al.: Integrating terahertz metamaterial and water nanodroplets for ultrasensitive detection of amyloid β aggregates in liquids. Sens. Actuat. B Chem. **329**, 129113 (2021)
18. Zhu, Z., et al.: Broadband terahertz signatures and vibrations of dopamine. Analyst **145**(18), 6006–6013 (2020)
19. Junaid, S., Chaitanya Kumar, S., Mathez, M., et al.: Video-rate, mid-infrared hyperspectral upconversion imaging. Optica **6**(6), 702–708 (2019)
20. Mittal, S., et al.: Simultaneous cancer and tumor microenvironment subtyping using confocal infrared microscopy for all-digital molecular histopathology. PNAS **115**(25), E5651–E5660 (2018)
21. Ellrich, F., Weinland, T., Molter, D., et al.: Compact fiber-coupled terahertz spectroscopy system pumped at 800 nm wavelength. Rev. Sci. Instrum. **82**(5), 053102 (2011)

A Method to Improve the Detection Capability of Millimeter-Wave/Terahertz-Wave Personnel Security Screening Systems

Yingkang Jin[1](✉), Xiaoyi He[1], Handan Jing[1], Lingbo Qiao[2], and Zhimin Zheng[1]

[1] SHENMUTEK Company Limited, Beijing 100098, China
jinyingkang@shenmutek.com

[2] Department of Engineering Physics, Tsinghua University, Beijing 10084, China

Abstract. This paper introduces a method to improve the detection capabilities of existing millimeter-wave/terahertz-wave personnel security scanners. For the shortcomings of current millimeter-wave/terahertz-wave holographic imaging devices in detection, a new working mode is added in addition to the holographic imaging mode. By obtaining terahertz transmission images, the blind zones in holographic images can be effectively compensated, and the detection capability of the security screening devices can be comprehensively improved.

Keywords: Personnel security screening · Holographic Imaging · Transmission Imaging

1 Technical Background and Challenge

Nowadays, security screening, especially personnel security screening, is an important field of terahertz imaging technology applications [1–5]. Now that millimeter-wave/terahertz-wave signals have strong capability to penetrate normal clothing and they are non-ionizing, these signals can be safely used for human body imaging with millimeter level image resolution. Analysis of these images can reveal whether anomalies are carried on the human body [1].

Millimeter-wave/terahertz-wave personnel security scanners are based on the holographic imaging principle [2]. The transmitting antenna transmits wide-band electromagnetic waves to a human body, and the adjacent receiving antenna receives the reflection waves, obtaining the amplitude and phase information of the reflection signals. When a coverage scan of the whole human body is completed by transmitting/receiving antennas (mechanical scan, electronic scan or the combination of both), an image of the human body refection coefficients can be obtained by reconstruction of all amplitude and phase information [6].

Figure 1 shows a part of a holographic terahertz image of a human body. In the image, the human body area has a high gray scale value (the human body has a relatively higher terahertz reflection rate), whereas the background area is close to 0 (There is no object in the background). In the human body area, since the human body can be

C. Chang et al. (Eds.): YTHZ 2024, SPPHY 400, pp. 153–158, 2024.
https://doi.org/10.1007/978-981-97-3786-4_31

seen as an intermediate background, objects concealed under the clothes can be clearly distinguished from the human body, no matter its reflection rate is higher (higher gray scale value) or lower (lower gray scale value). However, in the area outside the human body, objects with very low refection rates (materials with wave-absorbing character, for example, explosive powder) may not be distinguished because its low gray scale value may be integrated into the background, which results in blind zones where such objects may not be detected. In real world applications, such blind zones may be the sides of a human body, on shoulders and in the pockets of a loose coat, which brings potential safety issues [7].

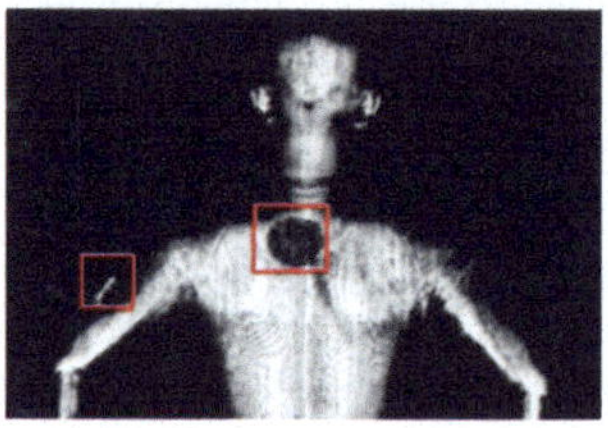

Fig. 1. Terahertz holographic image of a human body (part)

2 Method

2.1 Concept

In order to detect absorption materials placed in the background area (blind zone), it is necessary to find the difference of terahertz wave signals with/without absorption materials in such area. If an absorption material is placed, the terahertz wave signal transmitted from the transmitting antenna will be greatly attenuated by the absorption material, therefore the reflection and transmission signal are both very low. On the contrary, if no object is placed, the transmitting terahertz wave signal will not be reflected and will then continue traveling forward, which can be considered as a high transmission signal with very little reflection attenuation. Hence, it is possible to analyze whether there is an absorption material in such area by detecting the transmission signal.

At present, personnel security scanners are normally composed of two imaging units, each of which contains multiple terahertz transmitting and receiving antennas, and implements holographic imaging to one side (front or back) of a person independently. In order to obtain the transmission signal, an additional work mode is created based on the normal separate imaging of the two imaging units. As shown in Fig. 2, to get the transmission signal at a specific position, a transmitting antenna at one side must work in coordination with the receiving antenna at the corresponding position on the other side. When the two-dimensional scan (mechanical scan or electronic scan) is completed, a two-dimensional transmission image of the region where the person exists is generated. On this image, if no object is placed, the transmission signal is relatively higher (r1 in Fig. 2). With the absorption material blocking, the transmission signal is close to 0 (r2 in Fig. 2).

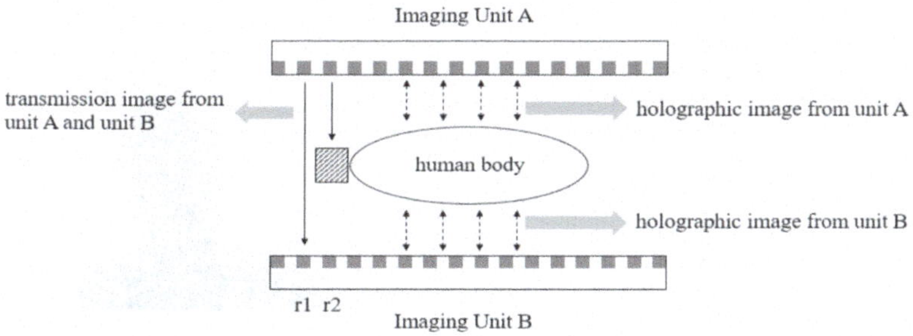

Fig. 2. Diagram of the work mode of both hologram and transmission images by using two imaging units. As shown in the figure, there can be significant differences in the transmission signals with/without an absorption material in the background area.

2.2 Experimental Result

An experiment is implemented based on the above concept. Two one-dimensional transceiver arrays are set horizontally towards each other and moved synchronously in the vertical direction. When a certain transmitting antenna on one transceiver array works, a receiving antenna at the corresponding position on the other array works. After a scan, a two-dimensional human body transmission image is obtained and compared with a one side holographic image, as shown in Fig. 3.

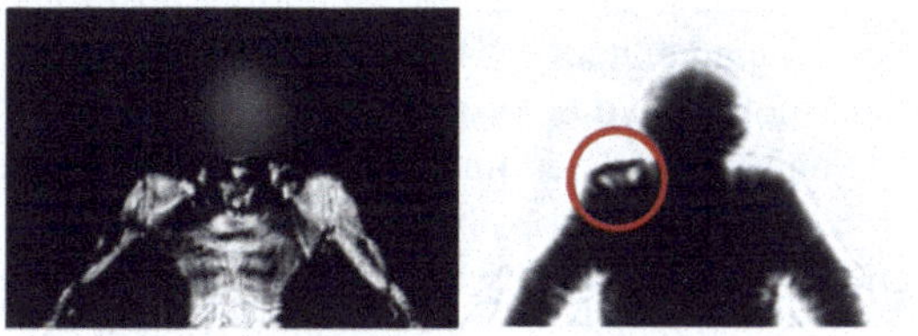

Fig. 3. Holographic image vs. transmission image

In this experiment, an object of absorption material is placed on the right shoulder of a person. In the holographic image (left of Fig. 3), there is no significant difference between the right shoulder and the left shoulder, and the background gray scale values are both close to 0. In the transmission image (right of Fig. 3), it is obvious that something exists in the background area on the right shoulder. This solution is proved to be able to effectively detect the absorption material in such background area and therefore eliminate the blind zone of holographic images.

3 Artifact Analysis

In the transmission imaging experiments, it was found that there are multiple layers of striped artifacts at the edges of the human body. As shown in Fig. 4b, the artifacts happen to be located at the edge of the human body and within the back-ground area, which may interfere with the determination of whether there is absorbing material in the background area.

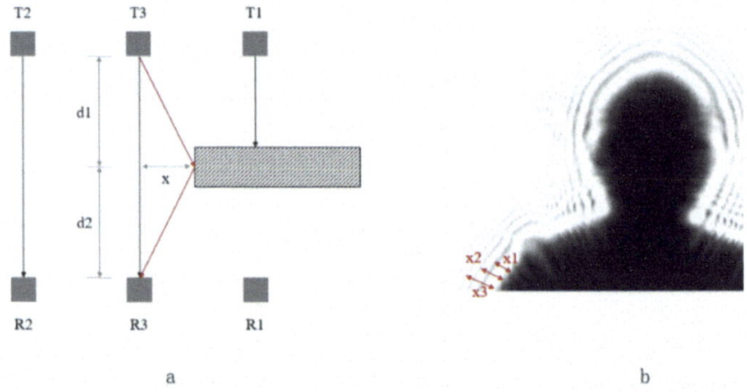

Fig. 4. Schematic diagram of edge artifacts in transmission images and their causes. a. Schematic diagram of the optical path generated by interference fringes; b. Interference fringes appear as artifacts on the edges of the human body in transmission image.

To reduce the artifacts, the cause of artifacts was analyzed. As shown in Fig. 4a, there is an object on the optical path connecting the transmitting antenna T1 and the corresponding receiving antenna R1, resulting in a weak transmission signal and a lower gray scale value of the pixel in the transmission image. There is no object on the optical path connecting the transmitting antenna T2 and the corresponding receiving antenna R2, resulting in a strong transmission signal and a higher gray scale value of the pixel in the transmission image. Artifacts appear at the corresponding positions of the transmitting antenna T3 and the receiving antenna R3. There is no object present on the optical path connecting T3 and R3, which should be displayed as a high gray scale value, however, R3 receives not only terahertz waves emitted by T3, but also terahertz waves emitted from T3 and reflected at the edges of the object. If the optical path difference between these two electromagnetic waves is exactly an odd multiple of the half wavelength, the coherent result is that the signal received at R3 becomes weaker. As a result, a low gray scale value is formed at that pixel.

The locations of the interference fringes were calculated based on the wavelength and the spatial distance between the human body and two arrays, which is consistent with the location of the artifacts in the transmitted image (see Table 1), proving that the artifacts are interference fringes generated by electromagnetic wave coherence.

Table 1. Positions of interference fringes

d1 (mm)	d2 (mm)	Artifact position (mm)	Optical path difference (mm)
400	400	$\times 1 = 24.5$	1.5 (0.5 times wavelength)
400	400	$\times 2 = 42.5$	4.5 (1.5 times wavelength)
400	400	$\times 3 = 54.9$	7.5 (2.5 times wavelength)

Since the cause of the artifacts have been confirmed, they can be corrected based on the wavelength and the distance between the human body and the arrays. In engineering applications, a simpler solution can be adopted. Due to the correlation between the position of interference fringes and wavelength, several electromagnetic waves of different frequencies can be used to generate transmission images for averaging. As shown in Fig. 5, the artifacts in the average image are significantly reduced, and their impact on subsequent image analysis can be ignored.

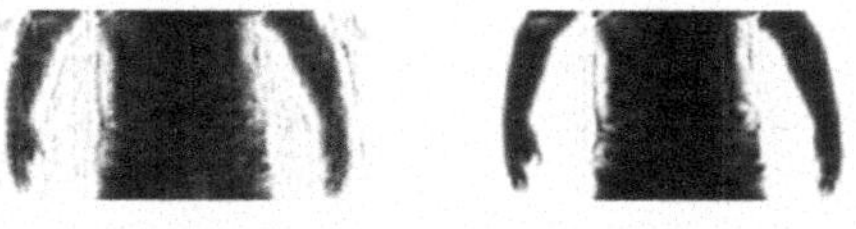

Fig. 5. The average effect of transmission images at different frequencies. a. Single frequency transmission image; b. The average effect of multi frequency transmission images.

4 Application and Prospect

The conventional millimeter-wave/terahertz-wave holographic imaging personnel security screening scanners implement target recognition algorithms to analyze the holographic image and provide location information of suspicious items after generating the holographic image. Based on the results, we can get the holographic image and transmission image of the tested person at the same time in one scan, and then use different target detection algorithms to analyze the holographic image and transmission image respectively, to comprehensively obtain the information of suspicious objects carried by the tested person. This method has been applied to the personnel security screening product developed by SHENMUTEK company and Tsinghua University, which significantly improved the detection performance. The product passed the highest standard certification of ECAC personnel security screening in 2023, which proved the effectiveness and engineering feasibility of this method [8].

References

1. Ahmed, S.: Microwave imaging in security—two decades of innovation. IEEE J. Microwaves **1**(1), 191–201 (2021)
2. Sheen, D., et al.: Three-dimensional millimeter-wave imaging for concealed weapon detection. IEEE Trans. Microwave Theory Techn. **49**(9), 1581–1592 (2001)
3. Liu, G., et al.: Myelin sheath as a dielectric waveguide for signal propagation in the mid-infrared to terahertz spectral range. Adv. Funct. Mater. **29**(7), 1807862 (2019)
4. Li, Y., et al.: Terahertz wave enhances permeability of the voltage-gated calcium channel. J. Am. Chem. Soc. **143**(11), 4311–4318 (2021)
5. Zhang, J., et al.: Non-invasive, opsin-free mid-infrared modulation activates cortical neurons and accelerates associative learning. Nat. Commun. **12**, 2730 (2021)
6. Lingbo, Q., et al.: Exact reconstruction for near-field three-dimensional planar millimeter-wave holographic imaging. J. Infrared Millimter Terahertz Waves **36**, 1221–1236 (2015)

7. Gumbmann, F., et al.: Walk through screening with multistatic mmW technology. In: Proceedings of SPIE 9993, Millimeter Wave and Terahertz Sensors and Technology IX, 999306 (2016)
8. ECAC CEP public equipment list–Security scanners (SSc). https://www.ecac-ceac.org/images/activities/security/ECAC-CEP-SSc-Public_equipment_list_update_20230818-rev1

Terahertz Anapole-Mode Metasurface for Highly Sensitive Refractive Index Sensing

Chenglin Yang, Zehao Liang, Shulin Xue, and Liuyang Zhang(✉)

School of Mechanical Engineering, Xi'an Jiaotong University, Xi'an 710049, China
liuyangzhang@xjtu.edu.cn

Abstract. Terahertz metasurface sensing has attracted tremendous attention in biomedical applications due to its advantages of label-free, non-invasive, and non-destructive detection. However, the sensing sensitivity needs to be further improved to achieve the sensing of trace samples. Here, we proposed a non-radiating anapole mode metasurface to enhance light-matter interaction and achieve highly sensitive refractive index sensing properties. The proposed metasurface consists of two E-shaped structures which can induce electric dipole and toroidal dipole simultaneously and their destructive interference caused the anapole mode. The electromagnetic field distribution and the multipole scattering theory indicated the excitation of the anapole mode. The refractive index sensing properties of the designed metasurface is tested which shows a high sensitivity of 416 GHz/GIU (refractive index unit). Our work provides a new design strategy for metasurface sensing devices and demonstrates its effectiveness in refractive index sensing applications.

Keywords: terahertz metasurface · anapole mode · refractive index sensing

1 Introduction

Terahertz (THz) waves are electromagnetic waves between microwaves and infrared, locating in the transition band from electronics to photonics. It has demonstrated unique advantages and broad application prospects in information perception due to its attractive characteristics, including non-ionization, non-destruction, timesaving, and capacity to reflect molecular fingerprint spectrum [1–7]. Due to the low-energy property, THz wave leads to little damage to biological substances and thus has attracted much attention in biomedical sensing applications [8–12]. However, some critical and common problems, such as weak interaction between THz wave and samples and complicated manufacturing of THz devices, are limiting the sensing performance improvement and applications of this technology.

Metasurfaces are artificially designed subwavelength periodic structures. Through carefully design the structure of the unit cell and its arrangements, metasurfaces can exhibit exotic physical properties, such as negative refractive index [1], broadband absorption [2, 13], electromagnetically induced transparency (EIT) [14, 15], etc. THz metasurfaces can generate local electric field hotspots and thus enhance the light-matter

C. Chang et al. (Eds.): YTHZ 2024, SPPHY 400, pp. 159–163, 2024.
https://doi.org/10.1007/978-981-97-3786-4_32

interaction, which exhibits a highly sensitivity to minimal environmental changes. However, the most designed electrical dipole resonances show low sensing sensitivity due to the strong free-space coupling with high radiative loss. Anapole mode, a non-radiating charge-current configuration can realize the efficient localization of the electromagnetic field, which gives it greater potential for sensing applications.

Here, an anapole-mode metasurface is proposed to achieve highly sensitive refractive index sensing. The unit cell of the designed metasurface consists of two E-shaped structures. By adjusting the size of the intermediate gap, the excitation of the anapole mode can be excited when the electric dipole and toroidal dipole meet the perfect destructive interference condition. Subsequently, the mechanism of the anapole-mode resonant is analyzed through the multiple scattering theory. Finally, the refractive index sensing property of the proposed anapole-mode metasurface is tested, demonstrating a high sensitivity of 416 GHz/RIU (refractive index unit), which shows the potential in label-free biomedical sensing.

2 Results and Discussion

2.1 Structure and Resonance

Figure 1 (a) shows the schematic diagram of the proposed anapole-mode metasurface. The structure is metal–dielectric–metal based sandwiched configuration in which its top and the bottom are copper layer and are separated by a dielectric layer which can be fabricated by a simple three-steps 3D printing process [16, 17]. The thickness of the bottom copper layer is thick enough to prohibit the transmission of terahertz waves. With normal *y*-polarized plane wave illumination, the anapole mode will be excited. The unit cell consists of two E-shaped structures. The structural parameters are as follows: p_x = 160 μm, p_y = 140 μm, l_x = 140 μm, l_y = 120 μm, h = 20 μm, w = 20 μm, g_1 = 36 μm, g_2 = 20 μm. The finite element software CST Microwave Studio is utilized to calculate the resonance of the proposed anapole-mode metasurface. In the simulation, the conductivity of Cu is set as 5.96×10^7 S/m, and the relative permittivity of the BIO resin is set as 2.667(1–0.046j) at 1 THz. The top and bottom Cu layer is set as 200 nm to prohibit the THz transmission.

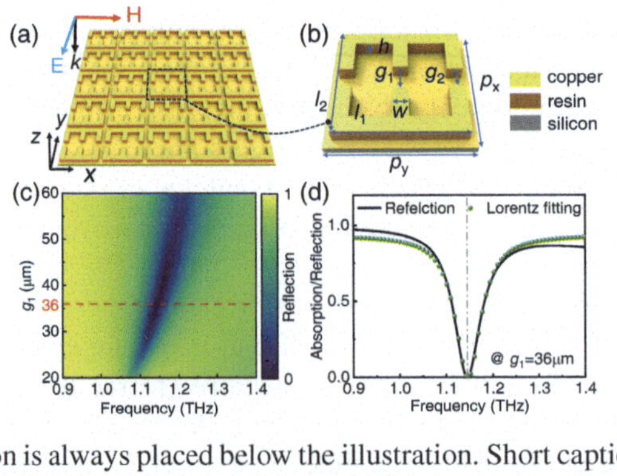

Fig. 1. A figure caption is always placed below the illustration. Short captions are centered, while long ones are justified. The macro button chooses the correct format automatically.

Figure 1 (c) shows the simulated reflection spectra. With the structure parameter g_1 varies from 20 μm to 60 μm. It is clearly observed that the reflection dip decreases first and then increases with the resonance frequency experiences a blue-shift at the same time. Meanwhile, a near-zero reflection can be observed when $g_1 = 36$ μm. As shown in Fig. 1 (d), a strong Lorentz-like resonance located around 1.145 THz and its peak reflection value is close to 1.0, which shows a perfect absorption.

2.2 Anapole Mode Analysis

To better understand the mechanism of the resonance, the distributions of the electric field (**E**) and magnetic field (**H**) in the *x-y* and *x-z* planes of the metasurface at the resonance frequency of 1.145 THz are illustrated in Figs. 2(a) and (b). We can see from Fig. 2(a) that a pair of ring currents with opposite spins are first excited on top of the copper layer, which excites two magnetic fields with opposite directions at both sides. As shown in Fig. 2(b), the induced magnetic fields produce a circular head-to-tail magnetic dipole ($\boldsymbol{M}$), that is, toroidal dipole ($\boldsymbol{T}$) along the *y*-axis. In addition, due to the charge oscillations on the middle bar, an electric field along the *y*-axis is generated in the middle gap, as shown in Fig. 2(a), which results in the formation of an electric dipole $\boldsymbol{P}$ in the direction opposite to that of the electric field. When the intensity and phase of $\boldsymbol{P}$ and $\boldsymbol{T}$ meet the case of $\boldsymbol{P} = \mathrm{ik}\boldsymbol{T}$, the destructive interference between the two will result in a critical non-radiative anapole mode.

To further reveal the excitation mechanism of the anapole mode, the scattering power of electromagnetic multipole is calculated according to multiple scattering theory. Here, electric dipole, magnetic dipole, toroidal dipole, electric quadrupole (Q_e), and magnetic quadrupole (Q_m) are considered as the five leading multipoles. As shown in Fig. 2(c), $\boldsymbol{P}$ and $\boldsymbol{T}$ dominates the resonance with the same scattering power, while all of the other three multipoles are effectively suppressed, indicating that the resonance is mainly excited by the electrical part of the incident electromagnetic wave. Figure 2(d) shows the phase of $\boldsymbol{P}$ and ik$\boldsymbol{T}$. The difference of the two shows zero at 1.145 THz. Both the intensity and

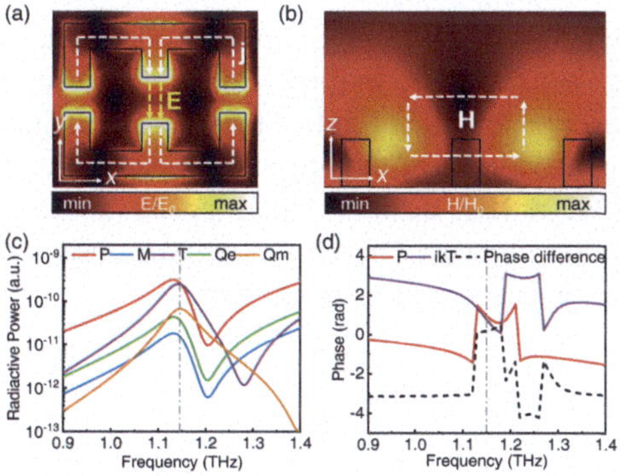

Fig. 2. (a) Electric field in the x-y plane, (b) magnetic field in the x-z plane at the resonance of 1.145 THz of the metasurface. (c) Scattered powers of multipole decomposition. (d) Phase of Cartesian $\boldsymbol{P}$, ik$\boldsymbol{T}$ and their difference.

the phase of the multipole implies that the critical excitation condition of anapole mode is satisfied.

2.3 Sensing Performance

We test the potential maximum sensing sensitivity of the proposed anapole metasurface by placing the analyte with a thickness to be semi-infinite but varying the refractive index from 1.0 to 2.0 above the metasurface. In Fig. 3(a), the reflection spectra show a redshift with the refractive index increasing from 1.0 to 2.0, and the sensitivity can be calculated as 416 GHz/RIU (refractive index unit), as shown in Fig. 3(b), which indicate that the proposed metasurface can effectively enhance terahertz sensing. This might be due to the non-radiative characteristic of the anapole mode, which would produce strong field confinement and thus enhance the interaction between the incident THz wave and the analyte on the metasurface. In addition, the designed out-of-plane relief-type structure contributes to the sensing performance as well.

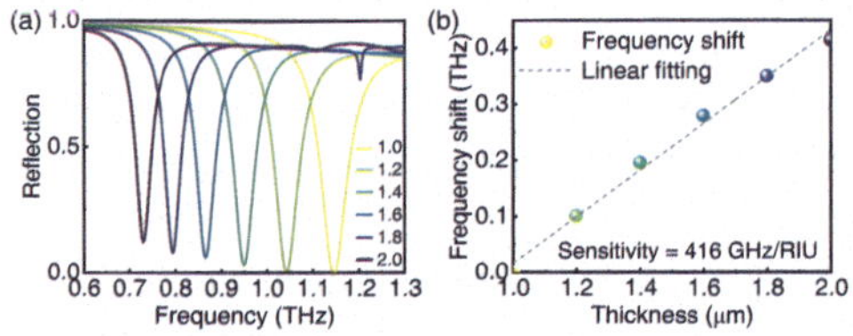

Fig. 3. (a) Simulated reflection spectra with an analyte of semi-infinite thickness but the refractive index n varied from 1.0 to 2.0 placed on the metasurface. (b) The frequency shift of the resonant dips in (a) and their linear fitting.

3 Conclusion

In this paper, we demonstrated a terahertz anapole-mode metasurface used for highly sensitive refractive index sensing. The multipole expansion theory is utilized to reveal the mechanism of the resonance which implies the critical excitation of anapole mode. Furthermore, the sensing properties of the proposed metasurface is tested which shows a high sensitivity of 416 GHz/RIU. Besides, the proposed metasurface can be fabricated by a three-steps 3D printing process. Therefore, our work might provide an alternative method for highly sensitive THz metasensor devices.

References

1. Valentine, J., et al.: Three-dimensional optical metamaterial with a negative refractive index. Nature **455**(7211), 376–379 (2008)
2. Shen, Z., Li, S., Xu, Y., Yin, W., Zhang, L., Chen, X.: Three-dimensional printed ultrabroadband terahertz metamaterial absorbers. Phys. Rev. Appl. **16**(1), 014066 (2021)

3. Liu, G., et al.: Myelin sheath as a dielectric waveguide for signal propagation in the mid-infrared to terahertz spectral range. Adv. Funct. Mater. **29**(7), 1807862 (2019)
4. Li, Y., et al.: Terahertz wave enhances permeability of the voltage-gated calcium channel. J. Am. Chem. Soc. **143**, 9 (2021)
5. Zhang, J.X., et al.: Non-invasive, opsin-free mid-infrared modulation activates cortical neurons and accelerates associative learning. Nature Comm. **12**, 2730 (2021)
6. Liu, X., et al.: Nonthermal and reversible control of neuronal signaling and behavior by midinfrared stimulation. Proc. Natl. Acad. Sci. **118**(10), e2015685118 (2021)
7. Wu, K., et al.: Terahertz wave accelerates DNA unwinding: a molecular dynamics simulation study. J. Phys. Chem. Lett. **11**(17), 7002–7008 (2020)
8. Zhang, Z., et al.: Advanced terahertz refractive sensing and fingerprint recognition through metasurface-excited surface waves. Adv. Mater **36**(14), 2308453 (2024)
9. Ma, Z., et al.: Identification and quantitative detection of two pathogenic bacteria based on a terahertz Metasensor. Nanoscale **15**(2), 515–521 (2023)
10. Lou, J., et al.: Calibration-free, high-precision, and robust terahertz ultrafast metasurfaces for monitoring gastric cancers. Proc. Natl. Acad. Sci. **119**, e2209218119 (2022)
11. Wang, P., et al.: An ultra-sensitive metasurface biosensor for instant cancer detection based on terahertz spectra. Nano Res. **16**(5), 7304–7311 (2023)
12. Yin, W., et al.: Broadband and multiband terahertz metamaterials based on 3-D-printed liquid metal-filled microchannel. IEEE Trans. Microwave Theory Techn. **17**, 3333 (2023)
13. Li, S., et al.: 3D-printed terahertz metamaterial for electromagnetically induced reflection analogue. J. Phys. D Appl. Phys. **55**(32), 325003 (2022)
14. Jiao, Y., et al.: Photoactive terahertz Metasurfaces for ultrafast switchable sensing of colorectal cells. Mater. Horiz. **9**(12), 2984–2992 (2022)
15. Shen, Z., et al.: Chiral metasurfaces with maximum circular dichroism enabled by out-of-plane Plasmonic system. Laser Photonics Rev. **16**(12), 2200370 (2022)
16. Yin, W., et al.: Highly sensitive terahertz sensing with 3D-printed metasurfaces empowered by a toroidal dipole. Opt. Lett. **47**(21), 5513–5516 (2022)
17. Shen, Z., et al.: Terahertz spin-selective perfect absorption enabled by quasi-bound states in the continuum. Opt. Lett. **47**(3), 505 (2022)

Realization of Quasi-Optical Output and Beam Deflection Based on Bifunctional Metasurface

Di Guo, Quansheng Zhang, Changsheng Shen, Zhaofu Chen, and Ningfeng Bai(✉)

Southeast University, Nanjing 210096, Jiangsu, China
{guodi,bnfeng}@seu.edu.cn

Abstract. On the demand of vacuum electronic devices, it is very important to realize high power beam quasi-optical and phase control. In this paper, a W-band bifunctional metasurface array is proposed as a mode convertor for gyrotron, which is based on the principle of propagation phase. The quasi-optical output of the metasurface is realized based on the rotation of the metasurface unit cell structure, which is based on the inverse process of the generation of radial vector field. On the other hand, the beam deflection can be realized based on the transmission phase principle and generalized Fresnel law. The beam deflection function of the metasurface array is verified by two beam deflection angles. The designed metasurface array has a good transmittance based on the simulation results, where the reflection S-parameter of the overall structure is less than −13 dB, and it has good performance with the bandwidth of 10 GHz. The proposed metasurface structure can realize quasi-optical, dual-function, and has the characteristics of miniaturization. The metasurface is expected to be used in microwave radar and satellite communication and this design method can also be applied to the THz band.

Keywords: Metasurface · Bifunctional · Quasi-optical · Deflection

1 Introduction

With the maturity of high-power vacuum electronic devices, especially gyrotrons, it is particularly important to realize quasi-optical based on their working modes and flexibly control their phases and amplitudes. However, its transmission mode is a high-order mode, which cannot be effectively utilized in space, so it needs to be regulated to meet the application requirements. As an artificial electromagnetic metamaterial, metasurface can flexibly control the phase and amplitude of electric field [1–6]. Zhao et al. realized the transformation from plane wave to the required special mode based on metasurface. Therefore, the quasi-optical output process based on TE01 mode can be considered as the inverse process of this process [1]. Based on this principle, Fu et al. realized the quasi-optical output by controlling the phase and amplitude TE01 mode [2, 3].

In this paper, based on the work of fu et al., a dual-function metasurface based on TE10 mode is proposed, which is the working mode of gyrotron. The metasurface modulates the phase while realizing quasi-optical, so that the TE10 mode can transform the

C. Chang et al. (Eds.): YTHZ 2024, SPPHY 400, pp. 164–168, 2024.
https://doi.org/10.1007/978-981-97-3786-4_33

Gaussian beam and realize the beam deflection. The use of this method is based on the gyrotron with mature technology, and only the geometric parameters and rotation angle of the unit cell structure need to be considered. The designed metasurface tends to be miniaturized and integrated, which meets the current demand for device miniaturization. This design method is expected to be applied to microwave radar and satellite communication.

2 The Design of the Unit Cell

The design principle of the metasurface is based on the Jones matrix. The functions implemented in this part require line-to-line polarization conversion. Therefore, the complex amplitude in the Jones matrix needs to meet the following conditions [7, 8]:

$$|T_{xx}| = |T_{yy}| = 1, \quad |T_{xy}| = |T_{yx}| = 0, \quad \delta_{xy} = \varphi_{xx} - \varphi_{yy} = \pm\pi \tag{1}$$

where T_{xx} and T_{xy} are the complex amplitude of x-polarized and y-polarized outgoing respectively when x-polarized incidence; T_{yx} and T_{yy} are the complex amplitude of x-polarized and y-polarized outgoing respectively when y-polarized incidence; φ_{xx} represents the phase change of x-polarized wave after passing through the unit cell structure, and φ_{yy} represents the phase change of y-polarized wave after passing through the unit cell structure.

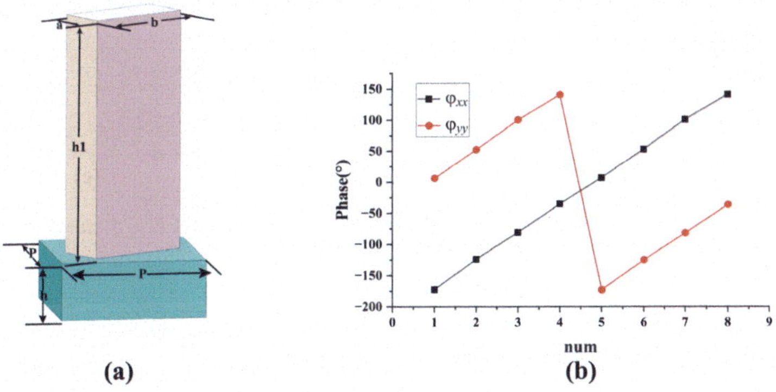

Fig. 1. (a) Schematic illustration of the designed alumina unit cell on the glass substrate. (b) The change phase of φ_{xx} and φ_{yy} of eight metasurface unit cell structures.

By adjusting the rotation angle of the metasurface unit cell structure, the polarization direction of each position of the incident light field can be adjusted regularly to achieve polarization in the same direction finally. The unit cell structure of the designed metasurface is shown in Fig. 1(a). The substrate material is SiO2 and thickness h = 0.4 mm. The period of the unit cell structure is p = 1.5 mm and the thickness is h1 = 2.5 mm. Its length and width are "a" and "b" respectively. The phase modulation of 2π can be achieved by changing the values of "a" and "b", which is shown in Fig. 1(b). The phase distribution of 2π and the requirements of beam deflection function can be achieved by the elaborately selected eight structures.

3 System Design and Simulation Results Analysis

The standard W band circular waveguide is used as the excitation source, as shown in the Fig. 2(a), the radius of the incoming waveguide port is 2.39 mm, and the radius of the outgoing waveguide port is 5 mm. The designed metasurface consists of 256 unit cell structures. The distance between the metasurface base and the waveguide exit port is 8 mm.

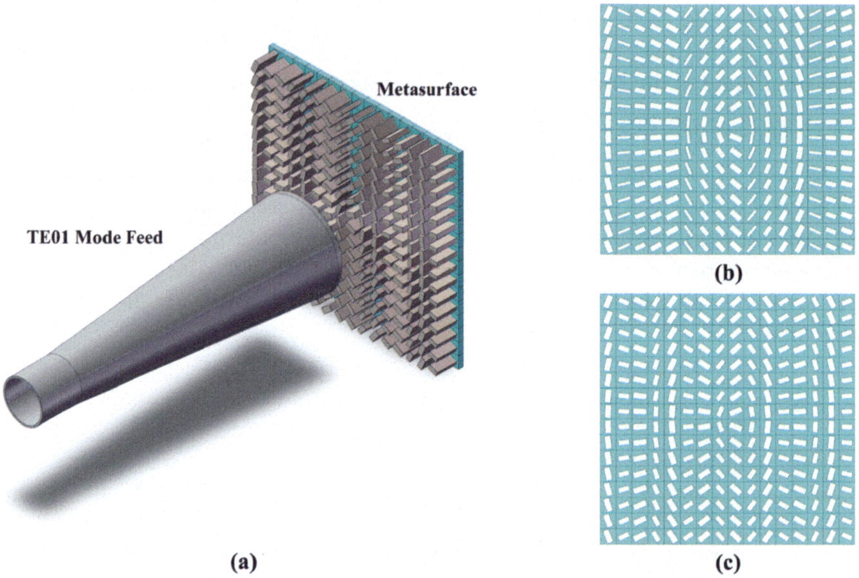

Fig. 2. (a) Schematic illustration of the overall simulation structure. (b) The metasurface topology when $d\Phi$ is $\pi/4$. . (c) The metasurface topology when $d\Phi$ is $\pi/2$.

The generalized Fresnel's law can be expressed as [9]:

$$sin(\theta_t)n_t - sin(\theta)n_i = \frac{\lambda_0}{2\pi}\frac{d\Phi}{dx} \tag{2}$$

where θ_i is the incidence angle, n_i is the refractive index in the incident light space, θ_t is the exit angle, and n_t is the refractive index in the exit light space; λ_0 is the working wavelength; dx takes the period p of the unit cell structure.

In this article, $\theta_i = 0°$, $n_i = n_t = 1$, $d\Phi$ is set as $\pi/4$ and $\pi/2$ respectively. Therefore, the theoretical values of θ_t are calculated to 14.48°and 30° respectively. The plane topology of the metasurface structure with a small deflection angle is shown in Fig. 2(b), and the plane topology of the metasurface structure with a large deflection angle is shown in Fig. 2(c).

Figure 3(a) and Fig. 3(b) show the far-field diagrams of the above two cases respectively. The beam deflection angle generated by the simulation structure is equal to the theoretical calculation, and it can be seen from the illustrations of the two pictures that the polarization directions of each position of the outgoing radio field are the same, and the electric field energy is concentrated in the central part, indicating that we have realized the quasi-optical function. Figure 3(c) and Fig. 3(d) show that the reflection S- parameters in both cases are less than −13 dB, which fully demonstrates the good transmittance of the designed structure.

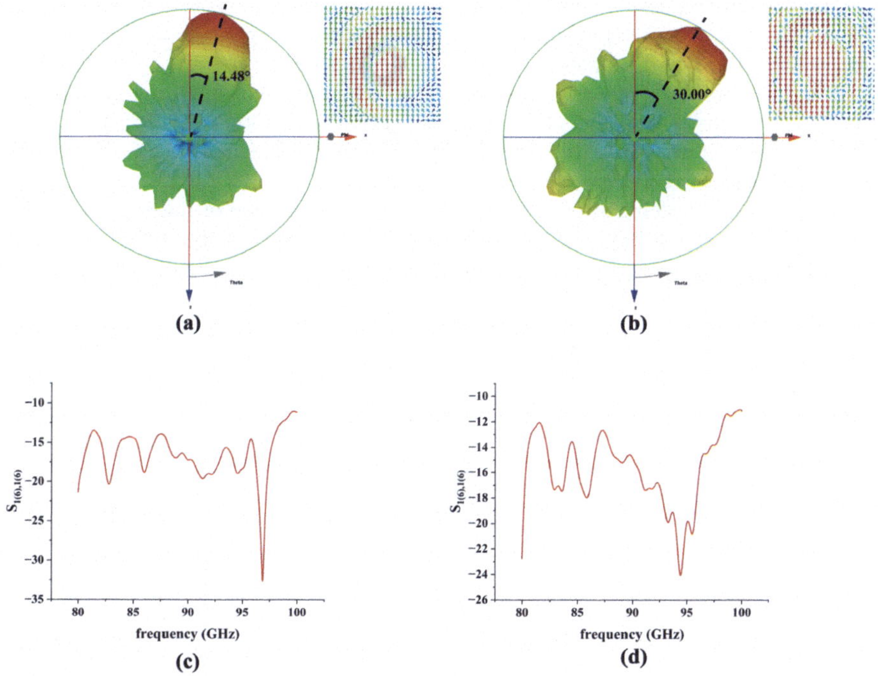

Fig. 3. (a) Three-dimensional far-field diagram when $d\Phi$ is $\pi/4$. (b) Three-dimensional far-field diagram when $d\Phi$ is $\pi/2$. (c) The reflection S-parameter when $d\Phi$ is $\pi/4$. (d) The reflection S-parameter when $d\Phi$ is $\pi/2$.

4 Conclusions

In summary, a metasurface based on the transmission phase theory is designed. Under the premise of achieving of the quasi-optical output, the beam deflection is realized based on the generalized Fresnel law. The simulation results show that the design is consistent with the theoretical calculation value and has a small reflection coefficient. This method can effectively control the beam exit angle while simplifying the quasi-optical system, and can achieve miniaturization. This structure is expected to be applied to microwave radar and satellite communication, and this design concept can also be applied to the THz band.

References

1. Zhao, A.R.Z., et al.: Nanoscale polarization manipulation and encryption based on dielectric metasurfaces. Adv. Opt. Mater. **6**(19), 1800490 (2018)
2. Fu, W., Hu, S., Zhang, C., Guan, X., Yan, Y.: Compact quasi-optical mode converter based on anisotropic metasurfaces. Opt. Express **29**(11), 16205–16213 (2021)
3. Han, M., et al.: All-dielectric metalens for quasi-optical mode and polarization conversion. Opt. Express **30**(19), 34797–34808 (2022)
4. Zhang, Z., et al.: Advanced terahertz refractive sensing and fingerprint recognition through metasurface-excited surface waves. Adv. Mater **36**(14), 2308453 (2024)
5. Lou, J., et al.: Calibration-free, high-precision, and robust terahertz ultrafast metasurfaces for monitoring gastric cancers. Proc. Natl. Acad. Sci. **119**(43), e2209218119 (2022)
6. Wang, R., et al.: Ultrasensitive terahertz biodetection enabled by quasi-BIC-based metasensors. Small **19**, 2301165 (2023)
7. Huang, C., et al.: Asymmetric electromagnetic wave transmission of linear polarization via polarization conversion through chiral metamaterial structures. Phys. Rev. B **85**,195131 (2012)
8. Zhao, R.Z., et al.: Nanoscale polarization manipulation and encryption based on dielectric metasurfaces. Adv. Opt. Mater. **6**(19), 1800490 (2018)
9. Yu, N.F., et al.: Light propagation with phase discontinuities: generalized laws of reflection and refraction. Science **334**(6054), 333–337 (2011)

Study of a Double-Open-Ring Terahertz Metamaterial Sensor

Yue Zhang, Lixia Yang, Shaomeng Wang(✉), and Yubin Gong

School of Electronic Science and Engineering, University of Electronic Science and Technology of China, Chengdu 611731, China
wangsm@uestc.edu.cn

Abstract. A double-opening ring-type terahertz metamaterial sensor that can be used to detect the refractive index of a material has been designed, which consists of two parts: a metamaterial structure nested with double-opening circular and square rings and a polyimide substrate. When terahertz is vertically incident on the metamaterial surface, the sensor structure forms three high-Q resonance peaks (center frequencies $f_1 = 0.74\text{THz}, f_2 = 1.39\text{THz}, f_3 = 1.70\text{THz}$.) in the range of 0.4–1.8 THz. By analyzing the current distribution on the surface of the metamaterial structure, the mechanism of different resonance generation can be obtained. In addition, the application of the sensor in refractive index sensing is also explored. Under a certain thickness of the object to be measured, the sensor has superior sensing characteristics with a sensing sensitivity of up to 161.6GHz/RIU, 279.3 GHz/RIU and 309.2 GHz/RIU at the resonant frequencies respectively. At the same time, the sensor is also insensitive to the polarization angle in the range of 0–30 degrees, which makes the performance more stable. Its multiple resonance peaks can be utilized for high sensitivity refractive index sensing.

Keywords: Terahertz · metamaterial · high sensitivity · biological detection

1 Introduction

Terahertz waves are characterized by low photon energy and high resolution, which will not damage the composition of biological samples, and many biological macromolecules show strong resonance absorption in the terahertz band, so that terahertz waves can be well applied in biological detection and pharmaceutical research [1]. Through the precise design of the structural units and combinations of metamaterials, it is possible to make them possess electromagnetic properties that are not found in natural materials, but are independent of the atomic or molecular structure of the material [2]. In 2023, Jiang et al. proposed a dual-band asymmetric knife-cut resonant ring sensor capable of rapidly and accurately detecting low concentrations of hyaluronic acid, providing a new idea for

This work was supported in part by the National Natural Science Foundation of China (under Grants T2241002, 32000914, and 61921002), and in part by Medico-Engineering Cooperation Funds for University of Electronic Science and Technology of China (Nos. ZYGX2021YGLH004 and ZYGX2021YGLH018).

C. Chang et al. (Eds.): YTHZ 2024, SPPHY 400, pp. 169–174, 2024.
https://doi.org/10.1007/978-981-97-3786-4_34

terahertz sensor design [3]. Zhang, et al. reported a mid-infrared modulation (MIM) stimulation technique that can activate neurons in the brain without introducing any exogenous genes, providing a new application scenario for terahertz sensor [4–9].

2 Design and Simulation

The terahertz metamaterial structural unit consists of two double-open resonant rings and a dielectric substrate. The resonant rings are made of metallic gold with a conductivity $\sigma = 4.561 \times 10^7$ S/m and a thickness of 50 nm; the substrate is made of a flexible polyimide material with a relative dielectric constant $\varepsilon_r = 3.5(1 + i0.0025)$ and thickness of 30 μm (Fig. 1).

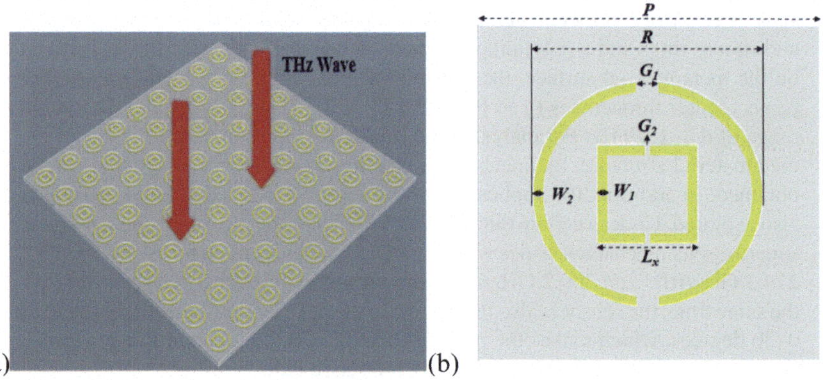

(a) (b)

Fig. 1. (a) Structure diagram of the designed sensor. (b) Schematic diagram of the three-dimensional metamaterial unit cell.

Table 1. Optimized parameters.

Parameters	Values(μm)
P	160
R	54
G_1	10
G_2	2
W_1	5
W_2	5
L_c	47

The structure is arranged periodically along the x- and y-axis directions, and electromagnetic waves in the x-polarization direction are incident perpendicularly on the surface of the structure along the z-axis direction. The unit structure parameters of the sensor are shown in Table 1. The electromagnetic properties of the designed metamaterial structure are modeled and simulated using the frequency domain solver of CST Microwave Studio.

3 Result and Discussion

3.1 Mechanistic Analysis

Figure 2 illustrates the transmission spectrum of the sensor unit structure, and it can be seen that the structure has three resonance peaks in the range of 0.4–1.8 THz, with center frequencies of $f_1 = 0.74\text{THz}, f_2 = 1.39\text{THz}, f_3 = 1.70\text{THz}$.

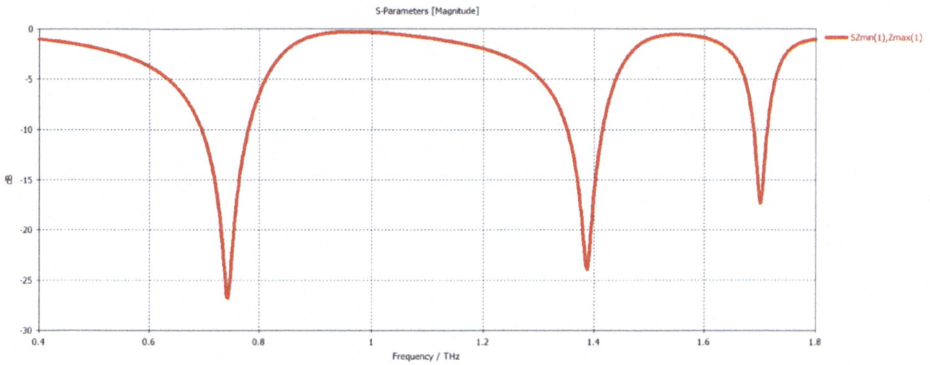

Fig. 2. Transmission spectrum of the designed sensor.

Figuring out the electromagnetic response of a metamaterial structure to an applied electromagnetic field is key to designing metamaterials. The electromagnetic resonance mechanism of the designed metamaterial structural unit can be analyzed by observing the surface current distribution of the structural unit at the resonance frequency through the field monitor.

The polarization direction of the incident electromagnetic wave is consistent with the direction of the metal ring structure. From Fig. 3(a), it can be seen that at the resonance point f_1, the surface current mainly oscillates in the left and right arms of the outer ring, which is generated by the dipole resonance of the outer ring. From Fig. 3(b), it can be seen that the resonant frequency f_2 is generated by the dipoles in the same direction on the outer and inner square rings, while there is some coupling between the inner and outer rings. From Fig. 3(c), it can be seen that the surface current at the resonant frequency f_3 oscillates mainly on the inner square ring, producing a typical dipole resonance. It can be seen that different resonance modes form different surface current distributions, and the performance indexes for substance detection will be different.

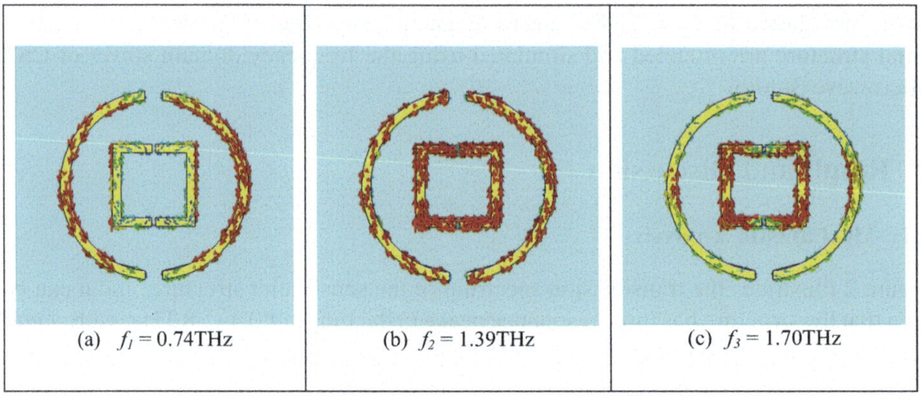

(a) $f_1 = 0.74$THz (b) $f_2 = 1.39$THz (c) $f_3 = 1.70$THz

Fig. 3. Surface current distribution in metallic resonant layers for different resonant modes.

3.2 Performance Analysis

The essence of terahertz sensors is to convert the difference in dielectric constant into the difference in electromagnetic signal (i.e., frequency shift) to achieve the detection. The sensing performance of metamaterial terahertz sensors is usually evaluated in terms of sensitivity S and quality factor Q, which can be expressed as follows:

$$Q = f/FWHM \tag{1}$$

$$S = \Delta f/\Delta n \tag{2}$$

The quality factor and sensitivity at three different resonant frequencies can be calculated as shown in Table 2 (Fig. 4).

Table 2. Q and S at three frequency points

Frequency(THz)	$f_1 = 0.74$	$f_2 = 1.39$	$f_3 = 1.70$
Q	62.8	112	149
S(GHz/RIU)	161.6	279.3	309.2

The resonant frequency is almost constant over the range of polarization angle variations from 0 to 30°, but the Q value decreases as the polarization angle deviation increases. At the same time, the performance of the designed sensor remains stable within a line width processing error of 1um (Fig. 5).

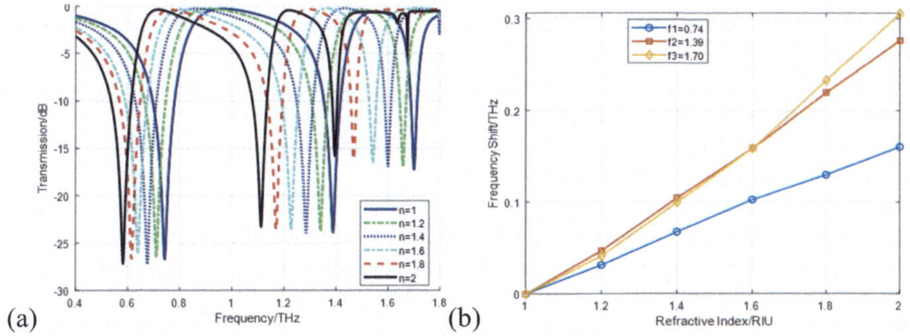

Fig. 4. (a) Spectrum of different refractive index. (b) Relationship between frequency shift and refractive index variation.

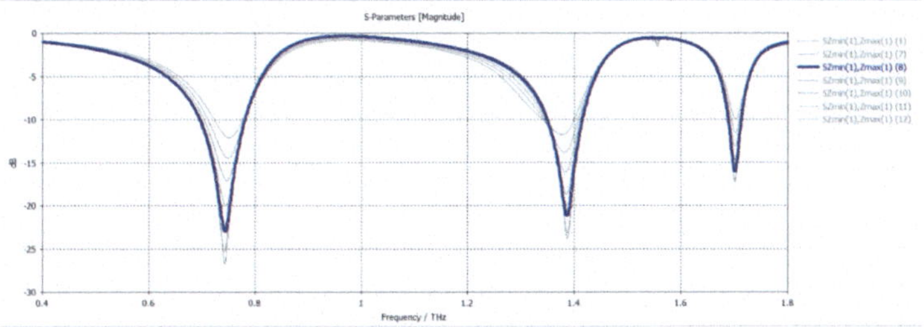

Fig. 5. Spectra at different polarization angles

4 Conclusion

In this paper, a terahertz metamaterial sensor based on a double-open ring structure with high sensitivity is designed. The formation mechanism of the three resonant peaks and the sensing performance are analyzed by analog simulation. When the thickness of the analyte is 15μm, the sensitivity of the refractive index corresponding to the three resonance frequencies is obtained to be 161.6, 279.3 and 309.2 GHz/RIU. Meanwhile, the designed sensor has a certain degree of polarization stability, which reduces the requirement of incident light in the actual detection, and the processing error has less influence on the performance of the device. Considering its multi-resonance and high sensitivity characteristics, the sensor has greater practical significance.

References

1. Liu, D.Y., Zhou, L., Huang, L.H., et al.: Microfluidic integrated capacitive biosensor for C-reactive protein label-free and real-time detection. Analyst **146**(17), 5380–5388 (2021)
2. Li, L., et al.: A four-band terahertz metamaterial sensor based on symmetric E-shaped structure. Coatings **12**(11), 1694 (2022)
3. Q. H. et al.: High-sensitivity terahertz dual-resonance Fabry–Perot sensor for detecting hyaluronic acid. Acta Optica Sinica **43**(9), 0928001–9 (2023)

4. Zhang, J., et al.: Non-invasive, opsin-free mid-infrared modulation activates cortical neurons and accelerates associative learning. Nat. Commun. **12**, 2730 (2021)
5. Liu, G., et al.: Myelin sheath as a dielectric waveguide for signal propagation in the mid-infrared to terahertz spectral range. Adv. Funct. Mater. **29**(7), 1807862 (2019)
6. Li, Y., et al.: Terahertz wave enhances permeability of the voltage-gated calcium channel. J. Am. Chem. Soc. **143**(11), 4311–4318 (2021)
7. Zhang, Z., et al.: Advanced terahertz refractive sensing and fingerprint recognition through metasurface-excited surface waves. Adv. Mater. **36**(14), 2308453 (2024)
8. Lou, J., et al.: Calibration-free, high-precision, and robust terahertz ultrafast metasurfaces for monitoring gastric cancers. Proc. Natl. Acad. Sci. **119**(43), e2209218119 (2022)
9. Wang, R., et al.: Ultrasensitive terahertz biodetection enabled by quasi-BIC-based metasensors. Small **19**, 2301165 (2023)

The Transmission Characteristics of Micrometer-Scale Materials in the Terahertz Frequency Range

Zihao Liu, Liangsheng Li(✉), Hongcheng Yin, He Cai, Quanchun Yu, and Xianli Zhu

National Key Laboratory of Scattering and Radiation, Beijing 100854, China
liuzihaoht207@163.com, liliangshengbititp@163.com

Abstract. The transmission characteristics of 4mm chopped carbon fibers (diameter 7μm), 4μm alumina powder, and 4μm copper powder were investigated using terahertz time-domain spectroscopy and Fourier transform spectroscopy. To mitigate the absorption of the samples on terahertz radiation, the samples were mixed with polyethylene at certain proportions and compressed into pellets. The terahertz time-domain Spectroscopy covered the frequency range of 0.1 to 2 THz, while the Fourier transform spectroscopy covered 2 to 10 THz. By concatenating the two spectral segments, the transmittance and mass extinction coefficients of the three types of mix-powder pellets at different concentrations were obtained over the entire terahertz frequency range (0.1 to 10 THz).

Keywords: Terahertz time-domain spectroscopy · Fourier transform spectroscopy · Transmission characteristics

1 Introduction

Terahertz radiation refers to electromagnetic waves with frequencies ranging from 0.1 to 10 THz and wavelengths between 3 mm and 30 μm. It occupies the spectral region between infrared and microwave radiation, serving as a transitional band from electronics to optics. Unlike microwave and infrared technologies, the development of terahertz technology is not yet mature. Issues such as miniaturization, high power, and portability of terahertz radiation sources and detectors remain unresolved, limiting the full utilization of the terahertz spectrum. Therefore, the terahertz frequency band can be regarded as a gap between the microwave and infrared bands, often termed the "terahertz gap", which remains to be explored and developed [1].

In recent years, significant breakthroughs have been achieved in terahertz source and detection technologies, propelled by advancements in photoconductive antennas and femtosecond laser techniques. Consequently, terahertz-related technologies have found widespread applications in various fields including imaging [2], communication [3], non-invasive detection [4], medicine [5–9] and secure detection [10].

We employed terahertz time-domain spectroscopy and Fourier transform spectroscopy to test the transmission characteristics of 4mm chopped carbon fibers

C. Chang et al. (Eds.): YTHZ 2024, SPPHY 400, pp. 175–179, 2024.
https://doi.org/10.1007/978-981-97-3786-4_35

(diameter 7μm), 4μm alumina powder, and 4μm copper powder. In order to mitigate the absorption of the samples on terahertz radiation, the samples were mixed with polyethylene (PE) at certain proportions and compressed into pellets. The T-ray5000 fiber terahertz time-domain spectroscopy produced by API covered the frequency range of 0.1 to 2 THz, while the VERTEX 80v Fourier transform spectroscopy manufactured by Bruker covered the frequency range of 2 to 10 THz. By concatenating the two spectral segments, the transmittance and mass extinction coefficients of the mix-powder pellets at different concentrations were obtained over the entire terahertz frequency range (0.1 to 10 THz).

2 Experiment Setup

2.1 Sample Preparation

To measure the terahertz frequency range transmission characteristics of powdered objects at different concentrations, the commonly used method of pellet pressing is selected. This method compresses the powdered or granular sample into a solid pellet under high pressure, facilitating experimental measurements. It allows for effective control of powder concentration, enabling the analysis of the relationship between sample transmittance and concentration. Pelletized samples exhibit durability for repeated measurements. PE is widely used as a diluent material in terahertz frequency range pellet testing due to its low absorption coefficient and ease of forming stable pellets under low pressure. PE powder is uniformly mixed with the tested particles (carbon fibers, alumina, copper powder) in the desired proportions. The mixture is then placed into a pellet mold and pressed in a pellet press machine at 20 MPa pressure for 5 min. This process yields pelletized samples with a diameter of 30 mm comprising the tested particles mixed with PE. The weight of the samples is measured using a precision electronic balance with a repeatability of 0.1 mg, ensuring accurate sample ratios. The types of samples measured are listed in Table 1.

Table 1. The types of samples and their composition ratios for pellet preparation.

Samples	The composition ratios
4mm chopped carbon fibers (diameter 7μm)	5/10 mg of carbon fibers + 400 mg of PE powder
4μm alumina	20/60/100/140 mg of alumina + 400 mg of PE powder
4μm copper	20/40/60/80/100 mg of copper + 400 mg of PE powder

2.2 Terahertz Systems

The two terahertz systems used in the experiment are illustrated in the diagrams below. Figure 1. (a) depicts the experimental setup for measuring the transmission characteristics of sample particles using terahertz time-domain spectroscopy. The system utilizes laser-excited photoconductive antennas for terahertz emission and reception, allowing direct acquisition of spectra from 0.1 to 2 THz. Figure 1. (b) shows the Fourier transform spectroscopy system, with a SiC heater rod as the emission source and a deuterated triglycine sulfate (DTGS) crystal as the detector. This system covers the high-frequency range of 2 to 10 THz.

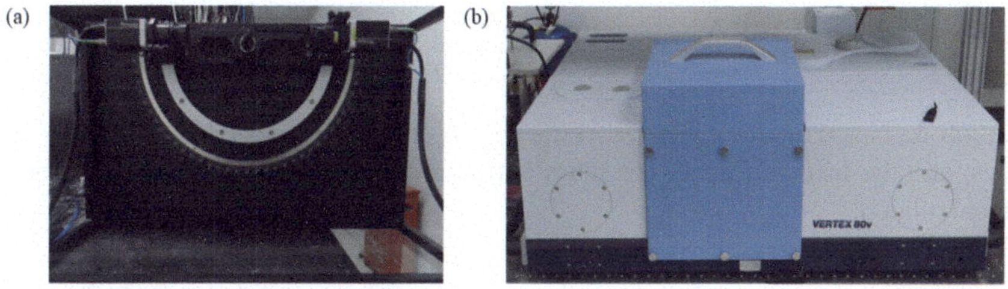

Fig. 1. (a) terahertz time-domain spectroscopy. (b) Fourier transform spectroscopy.

2.3 Data Processing Steps

a. When no sample is present, the terahertz waves emitted from the transmission end directly enter the detector, and the time-domain waveform or interference pattern $E_{ref}(t)$ collected by the detector is recorded. For terahertz time-domain spectroscopy tests, the system is filled with dry air to reduce the humidity to below 4%. For Fourier transform spectroscopy tests, the sample chamber is evacuated to remove moisture and carbon dioxide, minimizing their absorption of terahertz waves.
b. The pelletized samples are placed in the sample area, and the terahertz transmission through the samples is measured to obtain the time-domain waveform or interference pattern $E_{sam}(t)$. The testing conditions remain the same as in the previous step.
c. Both the reference and sample time-domain waveform or interference pattern are subjected to windowing (using a Hanning window) before Fourier transformation. This process yields the energy spectrum of both the reference $P_{ref}(\omega)$ and sample $P_{sam}(\omega)$.
d. Calculate the transmittance

$$T(\omega) = \frac{P_{sam}(\omega)}{P_{ref}(\omega)} \tag{1}$$

e. Calculate the mass extinction coefficients

The extinction performance of sample particles is commonly described using the mass extinction coefficient $\alpha\left(\mathrm{m}^2/\mathrm{g}\right)$. Under certain conditions, the attenuation of sample particles to terahertz waves follows Lambert-Beer's law.

$$T = \frac{I}{I_0} = I_0 \exp(-\alpha C_m L) \tag{2}$$

where: I_0 is the intensity of the incident terahertz wave, I is the intensity after passing through the sample, T is the transmittance, C_m is the mass density of the sample particles, and L is the distance the light travels through the pellet. The above equation indicates that, under constant sample concentration and thickness, for a given intensity of incident terahertz wave, the transmittance is determined by the extinction coefficient α. Therefore, α serves as the fundamental basis for evaluating the extinction performance of sample particles. According to the above equation, the formula for calculating the extinction coefficient is:

$$\alpha = \frac{1}{C_m L} \ln \frac{1}{T} \tag{3}$$

Therefore, based on the thickness of the pellet sample, the mass density of the sample particles, and the transmittance, the mass extinction coefficient of the sample particles can be calculated.

3 Result and Analysis

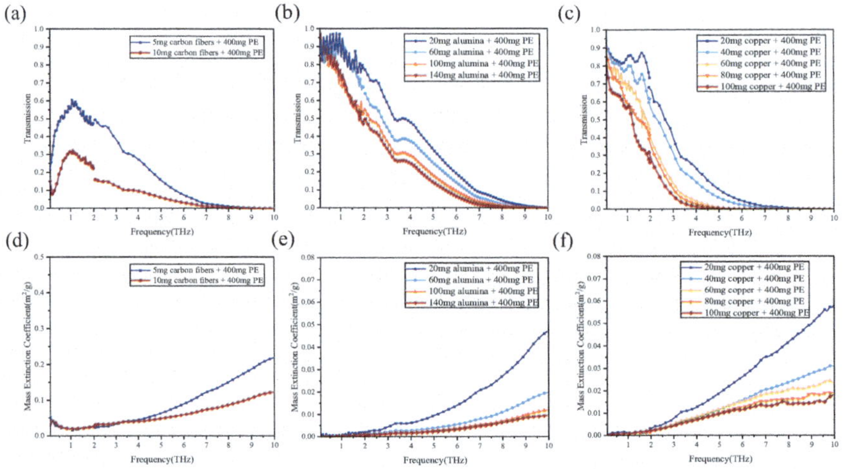

Fig. 2. Terahertz (a) – (c) transmission and (d) – (f) mass extinction coefficient spectra of different concentrations of carbon fibers, alumina, and copper.

The transmission of different concentrations of carbon fibers, alumina, and copper have been measured by terahertz time-domain spectroscopy and Fourier transform spectroscopy, shown in Fig. 2. For different concentrations of carbon fibers, the transmittance decreases overall with increasing concentration at various frequency points, while the mass extinction coefficient shows a similar trend at low frequencies. However, for the same concentration of carbon fibers, the transmittance exhibits an initial increase followed by a decrease trend as the frequency increases, reaching peak around 1 THz, while the mass extinction coefficient shows a trend of initial decrease followed by an increase, reaching its lowest point around 1 THz. At high frequencies, the lower the concentration, the higher the mass extinction coefficient.

For different concentrations of alumina, the transmittance decreases overall with increasing concentration at various frequency points. For the same concentration of alumina, the transmittance exhibits a general decreasing trend as the frequency increases. For different concentrations of alumina, the mass extinction coefficient shows a similar trend at low frequencies. However, at high frequencies, the lower the concentration of alumina, the higher the mass extinction coefficient. For the same concentration of alumina, the mass extinction coefficient increases with increasing frequency. The trend of copper powder is similar to that of alumina.

4 Conclusion

The results indicate that for samples with different concentrations, the transmittance decreases overall with increasing concentration at various frequency points. For samples with the same concentration, the transmittance generally decreases as the frequency increases. Regarding the mass extinction coefficient, samples with different concentrations exhibit similar trends at low frequencies. However, at high frequencies, lower concentration samples tend to have higher mass extinction coefficients. For samples with the same concentration, the mass extinction coefficient generally increases with frequency. The mass extinction coefficient of carbon fiber is significantly higher than that of alumina and copper, with copper's mass extinction coefficient slightly higher than that of alumina. We believe our study will contribute to application of micrometer-scale materials in the terahertz frequency range.

References

1. Huang, Y., Shen, Y.C., Wang, J.Y.: From terahertz imaging to terahertz wireless communications. Engineering **22**, 106–124 (2023)
2. Lyu, N.F., Zuo, J., Zhao, Y.M.: Terahertz synthetic aperture imaging with a light field imaging system. Electronics **9**(5), 830 (2020)
3. Feng, W., Wei, S.T., Cao, J.C.: 6G technology development vision and terahertz communication. Acta Physica Sinica **70**(24), 244303 (2021)
4. Liu, Y., et al.: The medical application of terahertz technology in non-invasive detection of cells and tissues: opportunities and challenges. RSC Adv. **9**(17), 9354–9363 (2019)
5. Liu, G., et al.: Myelin sheath as a dielectric waveguide for signal propagation in the mid-infrared to terahertz spectral range. Adv. Funct. Mater. **29**(7), 1807862 (2019)
6. Li, Y., et al.: Terahertz wave enhances permeability of the voltage-gated calcium channel. J. Am. Chem. Soc. **143**(11), 4311–4318 (2021)
7. Taday, et al.: Selected applications of terahertz pulses in medicine and industry. Appl. Sci. **12**(12), 6169 (2022)
8. Zhang, J., et al.: Non-invasive, opsin-free mid-infrared modulation activates cortical neurons and accelerates associative learning. Nat. Commun. **12**, 2730 (2021)
9. Wu, K., et al.: Terahertz wave accelerates DNA unwinding: a molecular dynamics simulation study. J. Phys. Chem. Lett. **11**(17), 7002–7008 (2020)
10. Chandel, et al.: Saliency and superpixel improved detection and segmentation of concealed objects for passive terahertz images. Opt. Eng. **62**(2), 023101 (2023)

Design of an Ultra-wide Band Rectangular Microstrip Patch Antenna for Early Breast Cancer Detection

Xuanxuan Zhang[1,2], Lixia Yang[1], Haiqing Liu[2,3](✉), Yuan Yao[2,3], Zhiyong Zou[3], Cuizhen Wang[3,4], and Yan Wang[3,4]

[1] School of Electronic and Information Engineering, Anhui University, Hefei 230039, China
lixiayang@yeah.net
[2] Insitute of Energy, Hefei Comprehensive National Science Center, Hefei 230031, Anhui, China
hqliu@ipp.ac.cn
[3] Institute of Plasma Physics, Hefei Institutes of Physics Science, Chinese Academy of Sciences, Hefei 230031, China
[4] University of Science and Technology of China, Hefei 230031, China

Abstract. In recent years, breast cancer has become the most diagnosed cancer in the world. Regular screening and early treatment can not only reduce the incidence of breast cancer, but also improve the treatment effect and prolong patients' survival significantly. Compared with the current main clinical screening methods for breast cancer, microwave imaging technology is safe and cost-effective, uitable for regular daily screenings for breast cancer. The microwave antenna serves as the front-end system for transmitting and receiving signals, crucially influencing the detection depth and imaging resolution of the system. In this paper, a rectangular microstrip patch antenna for breast cancer monitoring is designed, and its feasibility of is verified through simulation. The antenna is designed at dimensions of 30 mm × 30 mm, with rectangular radiation patches etched onto the front of the dielectric substrate, featuring grooves, while a defect ground structure is employed at the bottom of the antenna. The antenna simulation results show that the best return loss (S_{11}) at the operating frequency of 8.5 GHz is around −23.1 dB, and the maximum bandwidth is 11.7 GHz relative to the return loss which is less than −10 dB. The proposed antenna can identify tumors with a depth of 5.5 cm and a radius of 2 mm through simulation. It provides a technical basis for the further early realization of breast tumor screening in the later stage.

Keywords: Reflection coefficient · Transmission coefficient · Voltage Standing Wave Ratio · Breast phantom model

1 Introduction

According to the data from the World Health Organization [1], breast cancer has become the most common cancer among women It is estimated that there are as many as 2.26 million new cases of cancer, and 24.5% of newly diagnosed women with cancer.

C. Chang et al. (Eds.): YTHZ 2024, SPPHY 400, pp. 180–185, 2024.
https://doi.org/10.1007/978-981-97-3786-4_36

Numerous studies have been conducted to monitor tumors based on electromagnetic technology [1–7]. In 2020, Abdul Aziz *et al.* designed a circular microstrip patch antenna in the ISM band, and two types of breast phantom models cuboid and hemispherical were created to analyze the results [2]. The breast phantom models with tumors and without tumors were simulated separately, and the presence of breast tumors can be detected according to the return loss of the antenna. The antenna can detect tumors with a radius of 6 mm. In 2021, DN Elsherif *et al.* designed a low-frequency microstrip patch antenna with a working frequency of 2.45 GHz, using a model similar to real human breasts for simulation [8]. The tumor is placed at different locations in the breast phantom, and it is concluded that the change of the tumor location changes the electrical properties of the antenna, so that the tumor can be localized according to the change of the electrical properties. The antenna can detect tumors with a radius of 5 mm. In 2022, M. Firoz Ahmed *et al.* designed a new multi-trough rectangular microstrip patch antenna with a working frequency of 6.85 GHz, using a simple tapered breast model. Simulated the current density, electric field, and magnetic field parameters of the antenna in the breast phantom model with and without tumor, and obtained that changes in electrical characteristics indicate the presence of tumors. This antenna can detect 2 mm tumors [9].

This paper presents a small planar UWB antenna measuring 30 × 30 × 1.6 mm^3. The antenna has a simple structure, few geometric parameters, and a large bandwidth, and the microband antenna is ideal for Ultra-wide band systems, especially for tumor detection using microwave imaging technology.

2 Methods and Models

2.1 Antenna Design

Table 1. Parameter values of the proposed antenna.

parameter	size(mm)	parameter	size(mm)
ws	30	Ls	30
wf	3.06	Lf	9.75
lp	14.5	wp_1	2.5
wp_2	2.25	lp_2	2.5
lp_3	3.5	wg_1	3.06
h	1.6	Wp	19.2
lp_1	2.25	wp_3	12
lg_1	2.5		

The antenna was designed on an FR4 dielectric substrate with a thickness of h = 1.6 mm a relative dielectric constant of 4.4 F/m, and with a size of 30 mm × 30 mm (Ls × Ws). The feeder is provided by a microstrip line with a normalized impedance of

50 Ω. The design results are shown in Fig. 1, and the proposed antenna dimensions are shown in Table 1.

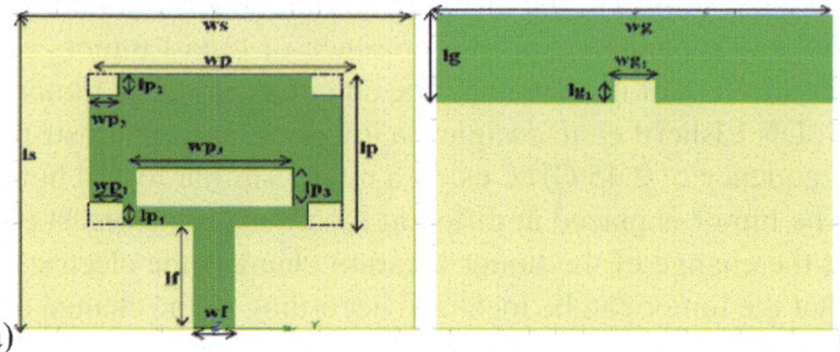

(a) (b)

Fig. 1. Proposed antenna geometry (a) front view, (b) rear view.

2.2 Breast Phantom Model Design

In order to study the antenna of breast tumor detection ability, this paper adopts the hemispherical three-layer model simulation human breast phantom model, with a semicircle radius of 55 mm, which consists of skin tissue layer thickness of 2 mm, fat tissue layer thickness of 15 mm, glandular tissue layer thickness of 38 mm [10]. Tumor tissue was simulated using a sphere with a radius of r (2–6 mm). The initial value of r was set to 2 mm. During the simulation, different breast tissues were simulated based on conductivity, dielectric constant, and density. The electromagnetic parameters of breast tissue are in Table 2, and the simulation diagram is shown in Fig. 2 (a) [11]. This paper conducted two scenario simulations: the first scenario used a single antenna placed near the breast phantom model, as shown in Fig. 2 (b). The second scenario uses two antennas on both sides of the breast phantom model, as shown in Fig. 2 (c).

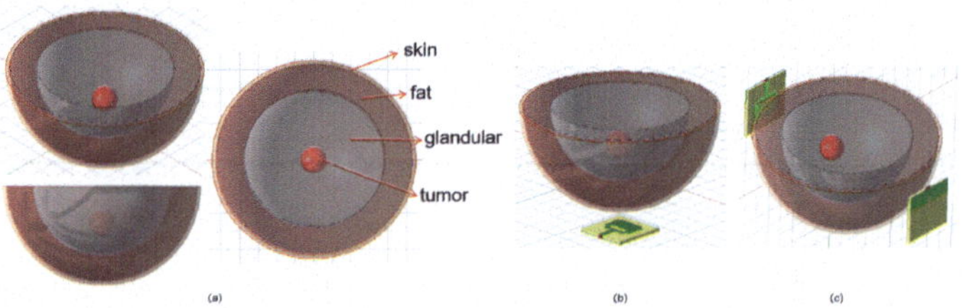

Fig. 2. (a) Breast phantom model design, uses (b) one antenna and (c) two antennas to detect breast phantom model with or without breast tumors.

Table 2. Electromagnetic parameters of the breast tissue.

Material	Relative permittivity ε_r (F/m)	Conductivity σ (S/m)	Mass Density ρ (Kg/m^3)	Thickness(mm)
Skin	36	4	1100	2
Fat	9	0.4	700	15
glandular	11	0.5	1020	35
tumor	50	4	1050	2r

3 Results and Discussion

3.1 Analysis of Antenna Parameters in Free Space

The S-parameters describe the input and output relationship between the antenna ports, and whether the antenna has radiation can be inferred from the S-parameters diagram. S_{11} is the input reflection coefficient, which describes the reflection efficiency of the antenna. The smaller the S_{11} value, the better, the smaller the energy reflected back by the antenna itself, and the higher the efficiency of the antenna. S_{21} is the forward transmission coefficient, also known as insertion loss, which describes how much energy is transmitted to the destination end. The higher the value, the better. The ideal value is 0 dB, and the higher the S_{21}, the higher the transmission efficiency. The VSWR is used to measure whether the antenna matches well with the system. The VSWR value of the antenna is usually less than 2, and when the standing wave ratio is equal to 1, both the feeder and the antenna are perfectly matched. Figure 3 shows the relationship between S_{11} amplitude, phase value, and VSWR of the antenna in free space with frequency. It is observed from Fig. 3 (a-b) that when the resonance frequency is 8.5 GHz, the echo loss is -23.1 dB. When $S_{11} < -10$ dB, the maximum bandwidth obtained is 11.7 GHz, which has ultra-wide band characteristics. Figure 3 (c) shows the change of voltage standing wave ratio with frequency at VSWR < 2, indicating a good impedance match over the entire operating frequency range. Thus, the designed ultra-wide band patch antenna can be used in systems for microwave detection of breast cancer.

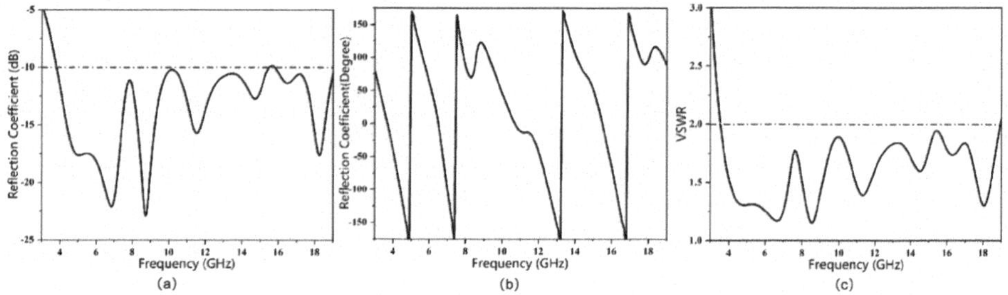

Fig. 3. Proposed antenna (in free space): S_{11} (a) magnitude in dB, (b) phase in degree, and (c) VSWR values.

3.2 Analysis of Antenna Parameters in the Breast Phantom Model

In order to detect the propagation path and characteristics of signals in breast tissue, this paper configures two antennas on both sides of the breast to further test the transmission of signals through the breast. Place the antenna above the breast phantom model, with a distance of 12 mm between the antenna and the breast. For these two scenarios, S parameters with and without tumors were recorded separately. As shown in Figs. 4, 5. Figure 4 shows the amplitude and phase values of the proposed antenna S_{11} in the first scenario, without and with different tumor sizes and with different depths.

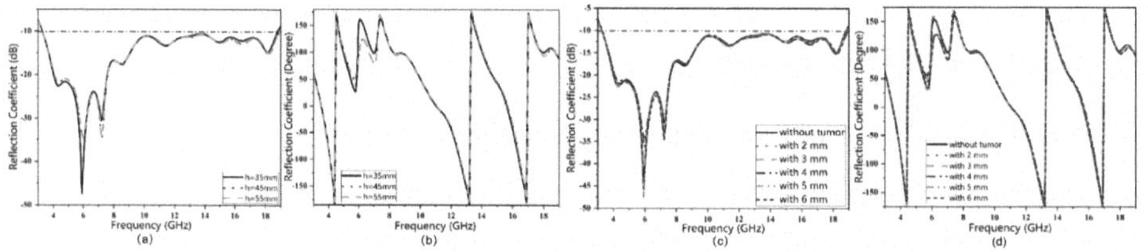

Fig. 4. With different depths/sizes of tumors, the S_{11} of the proposed antenna (as shown in Fig. 3 (a) for the first scenario): (a, c) amplitude in dB and (b, d) phase value in degrees.

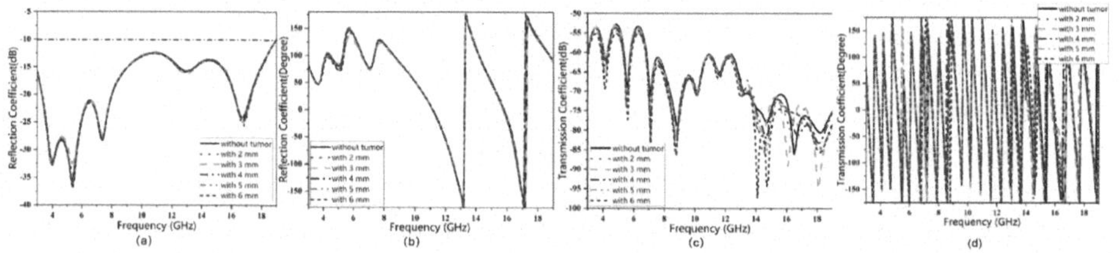

Fig. 5. Under different tumor sizes, the proposed antenna (second scenario, as shown in Fig. 2 (c)): S_{11} (a) amplitude in dB and (b) phase in degrees; S_{21} (c) amplitude in dB and (d) phase in degrees.

Figure 5 shows the magnitude and phase values of the second scenario (shown in Fig. 2 (c)) S_{11} and S_{21}. As observed in Figs. 4, 5, the reflection coefficient of each frequency point is under −10 dB in the 4–18 GHz frequency band range. The results indicate that as the depth of the tumor increases, the value of s11 changes less significantly; As the tumor size increases, the value of S_{11} changes significantly in the frequency range of 3–7 GHz. From the S_{21} results of Fig. 5 (c), it is concluded that the larger tumor communicates with a higher insertion loss. Furthermore, the S_{21} results follow smaller insertion losses at lower frequencies and larger ones at higher frequencies, with a worst-case insertion loss of approximately 97 dB.

4 Conclusion

In this paper, we designed a microband rectangular patch antenna for breast tumor detection. The antenna was designed on an FR 4 dielectric substrate with a relative dielectric constant of = 4.4 F/m and a thickness of h = 1.6 mm. The designed antenna has a maximum bandwidth of 11.7 GHz in free space and identifies breast tumors with a radius of 2mm. The simulation results show that the optimal echo loss at 8.5 GHz is −23.1 dB; and meets VSWR < 2 in the whole bandwidth range, indicating that the designed antenna is perfectly matched. Meanwhile, the antenna S parameters value of different tumor sizes is analyzed. The larger the tumor size, the shallower the tumor depth, the value of S parameters becomes more sensitive, and the change in S parameters value becomes obvious. Therefore, the design of the antenna provides a promising capability for early detection of breast cancer.

Acknowledgements. This work was supported by the Institute of Energy, Hefei Comprehensive National Science Center (Grant No. 21KZZ517, 22KZZ514, 24JYJB01, 23KHH126); the University Synergy Innovation Program of Anhui Province (No. GXXT-2022-014, GXXT-2021-028).

References

1. Siegel, R.L., et al:. Cancer Statistics, 2021. CA Cancer J. Clin. **71**(1), 7–33 (2021)
2. Aziz, A., et al.: On-body circular patch antenna for breast cancer detection. IEEE International Electromagnetics and Antenna Conference (IEMANTENNA), pp. 029–034. Canada (2019)
3. Lou, J., et al.: Calibration-free, high-precision, and robust terahertz ultrafast metasurfaces for monitoring gastric cancers. Proc. Natl. Acad. Sci. **119**(43), e2209218119 (2022)
4. Wang, P., et al.: An ultra-sensitive metasurface biosensor for instant cancer detection based on terahertz spectra. Nano Res. **16**, 7304–7311 (2023)
5. Jiao, Y., et al.: Photoactive terahertz metasurfaces for ultrafast switchable sensing of colorectal cells. Mater. Horiz. **9**, 2984–2992 (2022)
6. Yin, J., et al.: Terahertz photons inhibit cancer cells long term by suppressing nano telomerase. ACS Nano **18**(6), 4796–4910 (2024)
7. Sun, L., et al.: Inhibition of cancer cell migration and glycolysis by terahertz wave modulation via altered chromatin accessibility. Research **2022**, 9860679 (2022)
8. Elsherif, D.N., Makkey, M.Y.: Early Detection of Breast Cancer using Microstrip Patch Antenna. International Conference on Electronic Engineering (ICEEM), pp. 1–6. IEEE, Menouf, Egypt (2021)
9. Rakib, et al.: A meandered line patch antenna at low frequency range for early stage breast cancer detection. IJEEI **9**, 662–674 (2021). https://doi.org/10.52549/.v9i3.2824
10. Sreelekshmi, S., Sankar, S.P.: Design and analysis of novel miniaturised triple band antenna for breast cancer detection. Microsyst. Technol. **28**, 1715–1726 (2022)
11. Elsheikh, D., Eldamak, A.R.: Microwave textile sensors for breast cancer detection. In: 38th National Radio Science Conference (NRSC), pp. 288–294. Egypt (2021)

THz Bioeffect and Biophysics

High-Frequency Terahertz Waves Regulate the Dynamics of Mitochondrial Network in Human Pluripotent Stem Cells, Induced Neuron Stem Cells and Motor Neurons

Yuchen Tian[1] and Xinyu Chen[2(✉)]

[1] The College of Life Sciences, Northwest University, Xi'an 710032, China
[2] Department of Biochemistry and Molecular Biology, Air Force Medical University, Xi'an 710032, China
Xinyu20052@163.com

Abstract. With the advent of novel Terahertz sources and devices, Terahertz application has already been implemented to a variety scenarios of life sciences. However, only a few reports focused on the biological effects of Terahertz on human pluripotent stem cells (hPSC) and its neuronal derivatives, namely induced neuron stem cells (iNSC), neurons, etc. We applied Terahertz radiation to hPSC, iNSC and motor neurons, and examined possible effects of Terahertz on the dynamics of mitochondrial networks. Real time quantitative polymerase chain reaction (RT-qPCR) and Western Blot assay results suggested fusion and fission processes of the mitochondrial membranes were perturbed, as well as the contents of mitochondrial oxidative phosphorylation (OXPHOS) complexes, of which II&V showed contrasting changes in hPSC and iNSC. In light of OXPHOS complex corresponding mRNA expression level changes, specific frequency of Terahertz may play a role in neural induction of hPSC and iNSC, providing additional clues to regenerative medicine.

Keywords: Terahertz · Mitochondria · human pluripotent Stem cell · neuron stem cell

1 Introduction

Terahertz waves refer to electromagnetic waves with frequencies ranging from 0.1 to 10 THz, corresponding to wavelengths from 30 μm to 3 mm [1]. The average energy output of Terahertz waves is as low as several millivolts, and as to the biological effects, Terahertz radiation has a unique action of mechanism compared with high energy radiation with damaging potential, such as X and γ ray [2]. Recent years, terahertz waves were applied to biomedical area [3–9]. Peng et al. discovered Amyloid-βprotein structures was altered by high frequency terahertz waves, providing alternative treating approach for Alzhmeimer's Disease [3]. In early and quick diagnostics of cancer and traumatic brain injury, Terahertz waves could also be a valuable tool for complementing current molecular testing or imaging technology [4, 5].

C. Chang et al. (Eds.): YTHZ 2024, SPPHY 400, pp. 189–194, 2024.
https://doi.org/10.1007/978-981-97-3786-4_37

Owing to the unlimited self-renewal and pluripotent differentiation potential, human pluripotent stem cells (embryonic stem cells and induced pluripotent stem cells, hESC and hiPSC) are drawing constant attention in disease modeling and regenerative medicine since 1998. The neuronal derivatives of hPSC, namely induced neural stem cells(iNSC), are multipotent, capable of differentiating to nearly all neural lineages, is fundamental to brain development and cognition [10]. In recent years, there has been a number of findings providing compelling evidence that mitochondrial play a pivotal role in stem cell function instead of non-existent or negligible role in traditional view. Mitochondrial fusion and fission are implicated in cellular ageing, and mitochondrial dynamics is critical for the full pluripotency and embryonic developmental potential of pluripotent stem cells. Emerging reports now indicates mitochondrial dynamics influences NSC fate decisions, of which MFN1, MFN2, OPA1 function in fusion, DRP1, FIS1 in fission. Inherited mutations and accumulated damage to mitochondrial over the course of neurodevelopment and adult neurogenesis serve as key factors underlying neurodevelopment disorders and neurodegenerative diseases, respectively [11]. Thereby, depleting of damaged mitochondrial depletion and replenishing of healthy mitochondrial may serve as valuable route for Terahertz based therapies and interventions for cognitive defects.

In this report, we aimed to investigate high frequency Terahertz radiation on hPSC and its neural derivatives, with a focus on mitochondrial dynamics and OXPHOS content, which may provide links between Terahertz intervention and its biological effects on hPSC/iNSC.

2 Results and Discussion

To investigate the effect of terahertz waves on mitochondrial function of cultured hPSC (hESC, hiPSC), hiNSC, and motor neuron cells, we tested terahertz exposure on each type of cells by setting up the non-irradiation control group and the terahertz irradiation group. The cells in the THz group were irradiated for 1h under the condition of 700 mA current and 100kHz frequency (power of about 15mW), while the cells in the control group were left untreated in the same environment.

2.1 Effects of High Frequency Terahertz Waves on Mitochondrial Fusion and Fission Factors

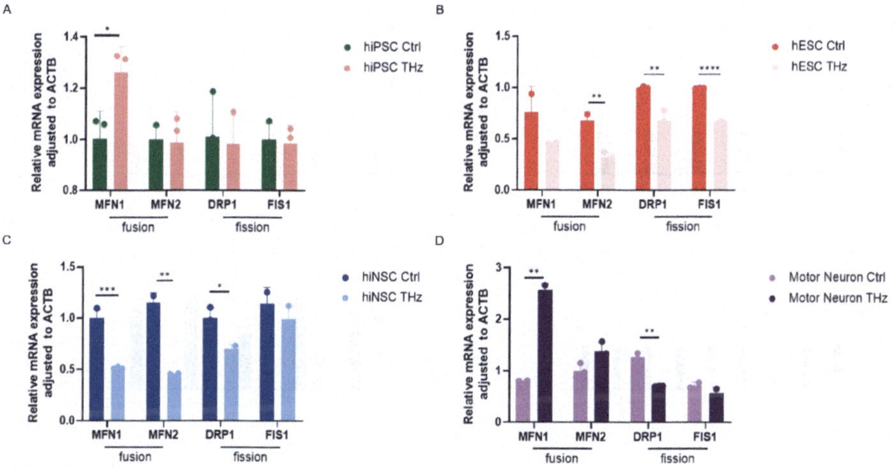

Fig. 1. Terahertz irradiation influenced mitochondrial fusion and fission related genes in A. hiPSC, B. hESC, C. hiNSC, D. Motor Neuron. qPCR was used to analyze the transcription level of genes related to mitochondrial fusion and fission.

2.2 Effect of High Frequency Terahertz Wave on OXPHOS Content

In general, we found that terahertz irradiation group can perturb mitochondrial fusion and fission related genes in all the cells examined, showing constant dropping of fission related gene DRP1 expression in hESC, iNSC and motor neuron (Fig. 1). The immunoblot of complex II and IV showed contrasting upregulation and downregulation in hESC and hiNSC groups, respectively (Fig. 2).

With the application of terahertz technology in the biomedical field, it is of paramount importance to study its biological effects The rational exposure and beneficial effects of terahertz have become a topic of increasing con- cern. Bogomazova AN et al. enriched differentially expressed genes (DEGs) using RNA sequencing data from terahertz irradiated human embryonic stem cells, and found that many DEGs were related to mitochondrial function [12]. Our experimental studies also found that terahertz radiation can perturb mitochondrial fusion and fission related genes in iNSC, ESC, and motor neuron cells as well as complex II and IV in OXPHOS con- tent. Zhong, X., et al. found that excess mitochondrial fission is associated with an impaired embryonic developmental potential [11], they suggested balanced mitochon- drial dynamics is beneficial to maintain full pluripotency of hPSC. These findings lend credence to that of iNSC which share similar energy metabolism with hPSC. There- fore, high-frequency terahertz waves may warrant further investigation to assess its role in maintain developmental potential of hPSC and hNSC under either control or stress conditions.

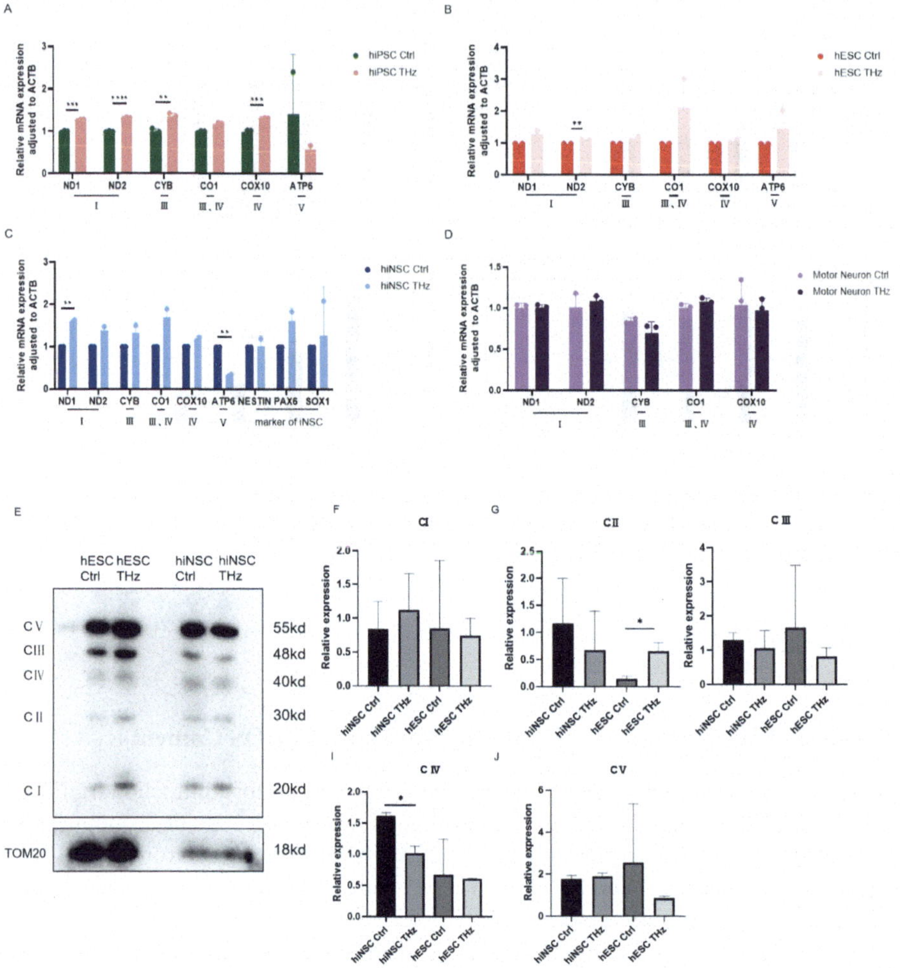

Fig. 2. Terahertz irradiation can affect OXPHOS content. A-D, qPCR was used to analyze the transcription levels of genes related to OXPHOS content. E-J, Western blotting was used to analyze the translation of genes related to OXPHOS content. The results of Western blot showed that terahertz irradiation increased the expression level of mitochondrial complex II in hESC(G) and decreased the expression level of complex IV in iNSC(I).The results of RT-qPCR showed that terahertz irradiation increased the expression level of mitochondrial complex I and decreased the expression level of mitochondrial complex V in iNSC(C), but had no effect on the mitochondrial complex content of motor neuron(D).

3 Methods

Cell line. hiPSC was generated from hair follicle derived keratinocytes by episomal vector transfection [13]. iNSC was generated by direct differentiation method from donor's keratinocyte during reprogramming [14]. H9 human Embryonic stem cell was a kind gift from Chinese Academy of Sciences. Induced motor neuron (iMN) was generated from both iNSC and H9 cells [15]. Three days before Terahertz radiation, hPSCs, iNSCs or MN were plated to achieve optimal cell confluency and state.

qPCR. Total RNA was extracted from the cells using FastPure Cell/Tissue Total RNA Isolation Kit (Vazyme, Cat.No.RC101–01) and was reverse transcribed into cDNA using a One-Step RT-PCR kit (Takara, Cat.No.RR036A).Quantification of gene expression was performed using an Applied Biosystems 7500 Fast Real-Time PCR System with SYBR® Green (Takara,Cat.No.RR820A).The relative mRNA expression levels were normalized to that of β-actin using the $2^{-\Delta\Delta CT}$ method.

Western blotting. Refer to the literature for specific methods [16].

Terahertz Radiation. The QCL was provided by the Institute of semiconductors of Chinese Academy of Sciences. The terahertz waves from the terahertz source are emitted at a certain divergence Angle. After adjusting the distance between the terahertz lens and the transmitting port of the terahertz source, the terahertz waves passing through the terahertz lens are parallel terahertz beams, and the cells attached to the dish radiate from bottom to up. The frequency and power density are 100 kHz and 4.28 mw/m^2. The irradiation time was 60 min. The cells in the control group were treated the same as those in the terahertz exposure group, but were not exposed to radiation.

References

1. Zhang, J., Li, S., Le, W.: Advances of terahertz technology in neuroscience: current status and a future perspective. iScience **24**(12), 103548 (2021)
2. Zhang, J., et al.: Terahertz Irradiation Improves Cognitive Impairments and Attenuates Alzheimer's Neuropathology in the APP(SWE)/PS1(DE9) Mouse: A Novel Therapeutic Intervention for Alzheimer's Disease. Neurosci Bull (2023). https://doi.org/10.1007/s12264-023-01145-3
3. Peng, W., et al.: High-frequency terahertz waves disrupt Alzheimer's β-amyloid fibril formation. eLight **3**(1), 18 (2023)
4. Wang, Y., et al.: Terahertz spectroscopic diagnosis of early blast-induced traumatic brain injury in rats. Biomed. Opt. Express **11**(8), 4085–4098 (2020)
5. Rahman, A., Rahman, A.K., Rao, B.: Early detection of skin cancer via terahertz spectral profiling and 3D imaging. Biosens. Bioelectron. **82**, 64–70 (2016)
6. Liu, G., et al.: Myelin sheath as a dielectric waveguide for signal propagation in the mid-infrared to terahertz spectral range. Adv. Funct. Mater. **29**(7), 1807862 (2019)
7. Zhang, J., et al.: Non-invasive, opsin-free mid-infrared modulation activates cortical neurons and accelerates associative learning. Nat. Commun. **12**, 2730 (2021)
8. Li, Y., et al.: Terahertz wave enhances permeability of the voltage-gated calcium channel. J. Am. Chem. Soc. **143**(11), 4311–4318 (2021)
9. Wu, K., et al.: Terahertz wave accelerates DNA unwinding: a molecular dynamics simulation study. J. Phys. Chem. Lett. **11**(17), 7002–7008 (2020)
10. Khacho, M., Harris, R., Slack, R.S.: Mitochondria as central regulators of neural stem cell fate and cognitive function. Nat. Rev. Neurosci. **20**(1), 34–48 (2018)
11. Zhong, X., et al.: Mitochondrial dynamics is critical for the full pluripotency and embryonic developmental potential of pluripotent stem cells. Cell Metab. **29**(4), 979-992.e4 (2019)

12. Bogomazova, A.N., et al.: No DNA damage response and negligible genome-wide transcriptional changes in human embryonic stem cells exposed to terahertz radiation. Sci. Rep. **5**, 7749 (2015)
13. Klingenstein, S., et al.: From hair to iPSCs-a guide on how to reprogram keratinocytes and why. Curr. Protoc. Stem Cell Biol. **55**(1), e121 (2020)
14. Cheng, L., et al.: Generation of urine cell-derived non-integrative human iPSCs and iNSCs: a step-by-step optimized protocol. Front. Mol. Neurosci. **10**, 348 (2017)
15. Ting, H.C., et al.: Robust generation of ready-to-use cryopreserved motor neurons from human pluripotent stem cells for disease modeling. Int. J. Mol. Sci. **23**(21), 13462 (2022)
16. Zhang, Y., et al.: Melatonin attenuates myocardial ischemia-reperfusion injury via improving mitochondrial fusion/mitophagy and activating the AMPK-OPA1 signaling pathways. J. Pineal Res. **66**(2), e12542 (2019)

Effect About Different External Terahertz Field on Permeability of Sodium Channel

Yanjiang Liu, Hongguang Wang(✉), Xiaofei Zhao, Yize Wang, Chunliang Liu, and Wen Ding

Key Laboratory for Physical Electronics and Devices of the Ministry of Education, School of Electronic Science and Engineering, Xi'an Jiaotong University, Xi'an 710049, Shaanxi, China
wanghg@xjtu.edu.cn

Abstract. Sodium channel is an important unit that regulates intracellular ion concentration and maintains the balance of sodium ions between intracellular and intracellular environments. Different functional groups of ion channels have different absorption peaks of frequency. The permeability of ion channel is related to the activity of specific functional groups in selectivity filter. These functional groups have their specific resonant frequencies, causing the affection to the permeability of ion channels when they vibrate under the action of applied external electric field with corresponding frequencies. In this paper, changes of permeability about NavMs channel under different field intensity are analyzed by applying external terahertz field at the frequency of 51.67THz as the main condition. With the Influence of external terahertz field, the permeability of channel has increased about 146.26% when field intensity is 0.6V/nm.

Keywords: Sodium channel · Terahertz wave · Carbonyl group · Ionic permeability

1 Introduction

NavMs sodium channel is a type of tetramer channel with four chains, whose structure is shown in Fig. 1 [1]. The selectivity filter (SF) region of the channel has 5 binding sites, and each chain in SF region has 1 carboxyl group, 5 carbonyl groups and 2 hydroxyl groups [2]. Each functional group has its own peak absorption frequency which is generally in the THz frequency band, with special effect about the properties of ion channels on neurons [3–13]. These functional groups are closely related to the binding sites of selectivity filters and have obvious effects on the ion permeability of channels.

It has been found that the permeability of ion channels is affected by applied external electric field [14]. Applying a kind of electric field at the peak frequency of a specific functional group can cause the resonance of corresponding functional group, which affects the permeability, structure and other characteristics of ion channels to a certain extent [14, 15].

C. Chang et al. (Eds.): YTHZ 2024, SPPHY 400, pp. 195–199, 2024.
https://doi.org/10.1007/978-981-97-3786-4_38

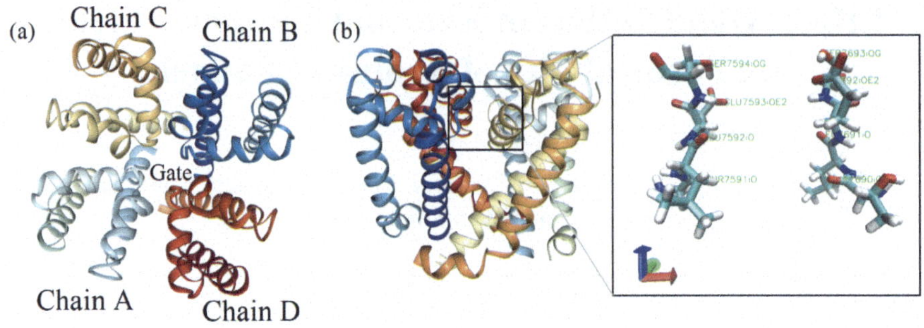

Fig. 1. The overall structure of the NavMs channel (a) and the selectivity filter (b), the right image only shows the structure of two chains for clarity

2 Methods and Results

The main part of the binding site on sodium ions is oxygen atom near the site, and the vibration of the site is able to affect the range of attraction to sodium ions, modifying their trajectories and changing the permeability of the whole channel [16]. The locations of binding sites do not strictly correspond to that of functional groups one by one, but properties of sites are obviously affected by the activity of these groups [3].

In order to explore the effect of peak frequency on ion channels, the simulation with an applied field of 51.67THz was used as the experimental group, and that without electric field was used as the control group. In this paper, the constant electric field-ion imbalance method was adopted, with NavMs channel (pdb identifier: 5HVX) as the core, and a double-layer structure simulation system was constructed with the help of GROMACS, as shown in Fig. 2(a) [17, 18]. It is known from previous studies that ions interact with multiple binding sites as they pass through the selectivity filter of the channel [1]. We have found that sodium ion has similar transmembrane process without complete dehydration, as shown in Fig. 2(b).

The GROMACS has used to calculate the spectrum of functional groups in the SF region of the channel. As shown in Fig. 3(a), the peak frequency of the carbonyl group is near 51.67 THz, and the peak frequency of the carboxyl group is near 49.32 THz.

In the simulation system, the charge difference between the two sides of the double-layer membrane was maintained to be 4e, and the MD simulation with a single duration of 400ns was carried out by applying different electric fields to NavMs channel at the frequency of 51.67 THz. In order to avoid the influence of extreme values, each group was simulated for 8 times, and the change of average ion flux over time in the whole simulation process was shown in Fig. 3(b).

In this paper, the ion flux under each field intensity was simulated and calculated, and the average value of each group of data was calculated respectively, as shown in Table 1 and Fig. 4. It can be seen that with the change of intensity and frequency of applied external field, the numbers of average flux of the control group and the experimental group present a different relationship.

Compared with the condition without electric field, it can be found that the field with a frequency of 51.67 THz can promote the permeability of ion channel, indicating that

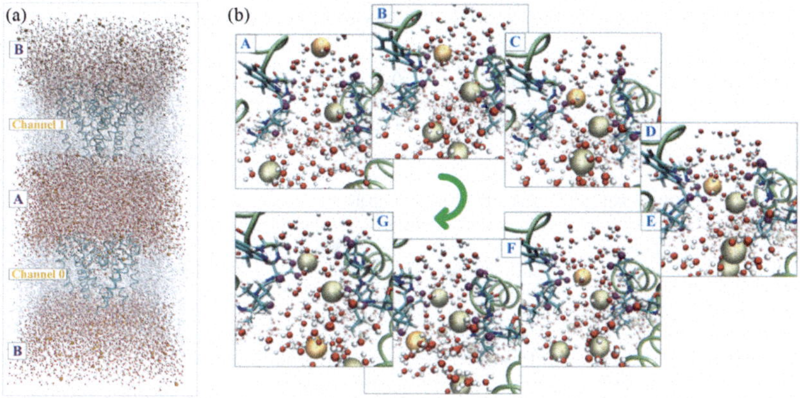

Fig. 2. (a) Two-layer simulation system with the constant electric field-ion imbalance method; (b) the ionic transmembrane process in sodium channel observed by VMD

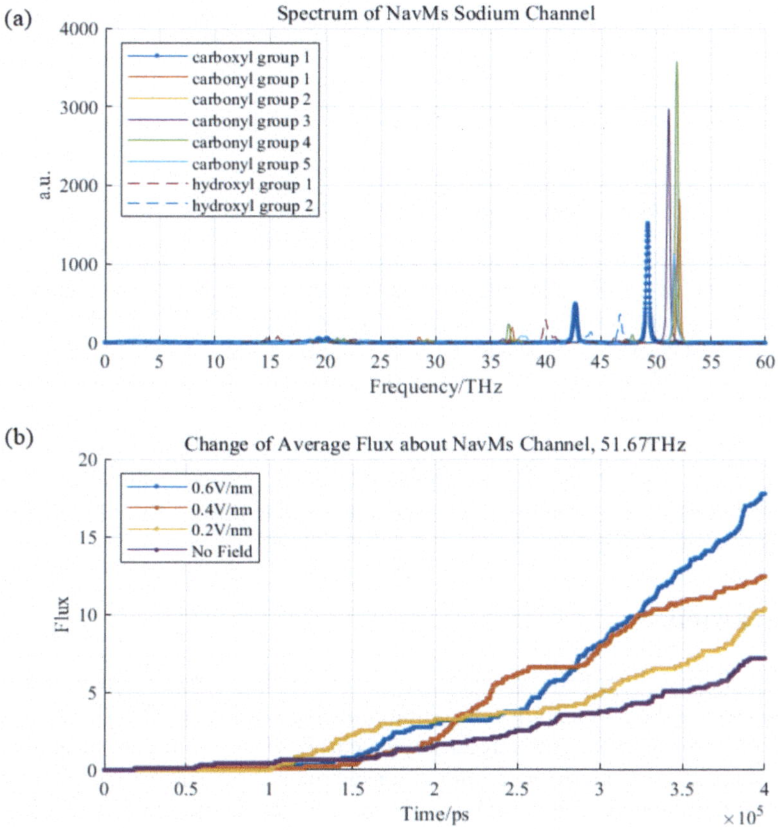

Fig. 3. (a) The absorption spectrum of NavMs channel; (b) change of average ionic flux

Table 1. Changes of average ionic flux about field intensity (51.67 THz, 4e)

Intensity/V·nm^{-1}	0	0.2	0.4	0.6
Average	7.22	10.43	12.5	17.78
Increment Rate	0	44.46%	73.13%	146.26%
Average value without a set of extreme values	6.28	10.00	10.17	14.43
Increment Rate	0	59.24%	61.94%	129.78%

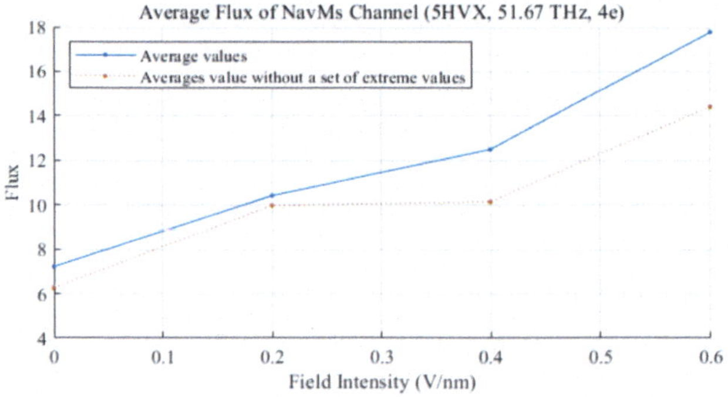

Fig. 4. Changes of average ionic flux with field intensity

the field causes the carbonyl group resonance in the SF region of ion channel, which enhances the ion permeability of the channel. In the case of 51.67 THz electric field, the average flux of ion channel increases with the augment of electric field intensity.

At a lower field intensity, the electric field has less effect on the permeability of ion channel. Fields with intensity of 0.2 V/nm and 0.4 V/nm have relatively similar effects on the permeability of ion channel. At the higher field intensity, 0.6 V/nm, the promoting effect of electric field on permeability is significantly enhanced. This result is less affected by extreme values, which remains similar when a pair of extreme values are removed from each group.

3 Conclusions

When the intensity of applied external field is relatively high, the field plays a significant role in promoting the permeability of sodium ion channels. As the frequency of electric field is consistent with the peak frequency of carbonyl group, the group in SF region resonates and enhances the permeability of ion channels. This enhancement effect is also affected by the intensity of external electric field. With the effect of external terahertz field, the permeability of channel has increased about 146.26% when field intensity is 0.6 V/nm.

Meanwhile, the effect of carbonyl group on the permeability of the channel is obvious. When the frequency of external field matches the resonance frequency of carbonyl group, the ion permeability of channel changes significantly. Further analytical verification of the underlying mechanisms and their comparison with other frequencies and functional groups is still needed.

References

1. Ke, S., et al.: A. Different inward and outward conduction mechanisms in NaVMs suggested by molecular dynamics simulations. PLoS Comput Biol. **10**(7), e1003746 (2014)
2. McCusker, et al.: Structure of a bacterial voltage-gated sodium channel pore reveals mechanisms of opening and closing. Nature Commun. **3**, 1102 (2012)
3. Ulmschneider, M., et al.: Molecular dynamics of ion transport through the open conformation of a bacterial voltage-gated sodium channel. Proc. Natl. Acad. Sci. **110**, 6364 (2013)
4. Xiao, T., et al.: Sensory input-dependent gain modulation of the optokinetic nystagmus by mid-infrared stimulation in pigeons. eLife **12**, e78729 (2023)
5. Liu, X., et al.: Nonthermal and reversible control of neuronal signaling and behavior by midinfrared stimulation. Proc. Natl. Acad. Sci. USA **118**, e2015685118 (2021)
6. Xiang, Z., et al.: A primary model of THz and far-infrared signal generation and conduction in neuron systems based on the hypothesis of the ordered phase of water molecules on the neuron surface. Sci. Bull. **65**, 308 (2020)
7. Liu, G., et al.: Myelin sheath as a dielectric waveguide for signal propagation in the mid-infrared to terahertz spectral range. Adv. Funct. Mater. **29**(7), 1807862 (2019)
8. Zhang, J., et al.: Non-invasive, opsin-free mid-infrared modulation activates cortical neurons and accelerates associative learning. Nat. Commun. **12**, 2730 (2021)
9. Wu, K., et al.: Terahertz wave accelerates DNA unwinding: a molecular dynamics simulation study. J. Phys. Chem. Lett. **11**(17), 7002–7008 (2020)
10. Chang, C., et al.: High-gain thompson-scattering X-ray free-electron laser by time-synchronic laterally tilted optical wave. Phys. Rev. Lett. **110**, 064802 (2013)
11. Yu, Y., et al.: Terahertz photons improve cognitive functions in posttraumatic stress disorder. Research **6**, 0278 (2023)
12. Yin, J., et al.: Terahertz photons inhibit cancer cells long term by suppressing nano telomerase activity. ACS Nano **18**(6), 4796–4910 (2024)
13. Zhang, Q.L., Zhou, T., Chang, C., et al.: Ultrahigh-flux water nanopumps generated by asymmetric terahertz absorption. Phys. Rev. Lett. 2024 (accepted)
14. Li, Y., et al.: Terahertz wave enhances permeability of the voltage-gated calcium channel. J. Am. Chem. Soc. **143**, 4311–4318 (2021)
15. Tan, X., et al.: Minimal-invasive enhancement of auditory perception by terahertz wave modulation. Nano Res. **15**, 5235–5244 (2022)
16. Hu, Z., et al.: Permeability enhancement of the KcsA channel under radiation of a terahertz wave. Phys. Rev. E **105**(2–1), 024104 (2022)
17. Zhao, X., et al.: Permeability enhancement of Kv1.2 potassium channel by a terahertz electromagnetic field. J. Chem. Phys. **159**(4), 045101 (2023)
18. Wang, Y., Wang, H., Ding, W., Zhao, X., Li, Y., Liu, C.: Regulation of Ion permeation of the KcsA channel by applied midinfrared field. Int. J. Mol. Sci. **24**(1), 556 (2022)

Non-invasive High-Frequency Terahertz Wave Activates Cortical Astrocytes *in Vivo*

He Yong[1], Du Chaohai[1](✉), and Kuan Zhang[2](✉)

[1] School of Electronics, Peking University, Beijing 100081, China
duchaohai@pku.edu.cn
[2] Brain Research Center and State Key Laboratory of Trauma, Burns, and Combined Injury, Third Military Medical University, Chongqing 400038, China
zk_crane@163.com

Abstract. Compared with the biological damage led to by thehigh-energy electromagnetic waves (X-ray, gamma ray, ultraviolet, etc.), high-frequency terahertz (THz) stimulation can induce unique non-thermal and non-invasive neuromodulation. In thc present study, using *in vivo* two-photon calcium imaging in the auditory cortex of mice, we find that high-frequency terahertz could stably excite astrocytes, and the induced calcium signals from astrocytes take place immediately at the end of high-frequency terahertz stimulation.

Keywords: High-frequency terahertz · Astrocytes · Neuromodulation

1 Introduction

Terahertz waves have broad applications in neuromodulation [1–8], molecular-control [9–11], and biosensor [12–18] based on the development of THz sources [19, 20]. 53.53 THz could modulate the startle responses of larval zebrafish [21] and increase the learning rate of mice by activating cortical neurons [22]. 34.88THz could Significantly delay the fibrosis process of amyloid-β (Aβ) oligomers, providing a potential neuromodulation method for the treatment of the related diseases [23]. However, most of the studies on the neuromodulation of high-frequency terahertz waves focus on neurons. Glial cells are another important part of the nervous system, and it is still unknown whether high-frequency terahertz waves have significant effects on glial cells and whether they bring about significant neuromodulation through glial cells.

Astrocytes are the most abundant and functionally complex glial cells in the brain. Astrocytes can respond to neuronal activity through intracellular calcium transients, at the same time, astrocytes can also modulate neurons by releasing gliotransmitters [24]. In addition, recent studies showed that astrocytes play an important role in memory persistence [25]. The amount and function of astrocytes in the brain make them suitable subjects for studying the biolofical effect of Terahertz waves besides neurons.

C. Chang et al. (Eds.): YTHZ 2024, SPPHY 400, pp. 200–203, 2024.
https://doi.org/10.1007/978-981-97-3786-4_39

2 Results

High-Frequency Terahertz Induces Calcium Signals of Cortical Astrocytes. We performed craniotomy on the auditory cortex of mice, and high-frequency terahertz (53.53 THz) was transmitted through optical fiber to the dura of the auditory cortex (Fig. 1a, left). At the same time, we recorded the calcium activity of astrocytic soma by two-photon fluorescence imaging (Fig. 1a, right). Rhod-2 acetoxymethyl ester (Rhod-2 AM), an astrocyte-specific calcium indicator, was used to demonstrate the changes of calcium activity (Fig. 1b). About half of astrocytes produced an induced calcium signal immediately at the end of high-frequency terahertz application (Fig. 1c).

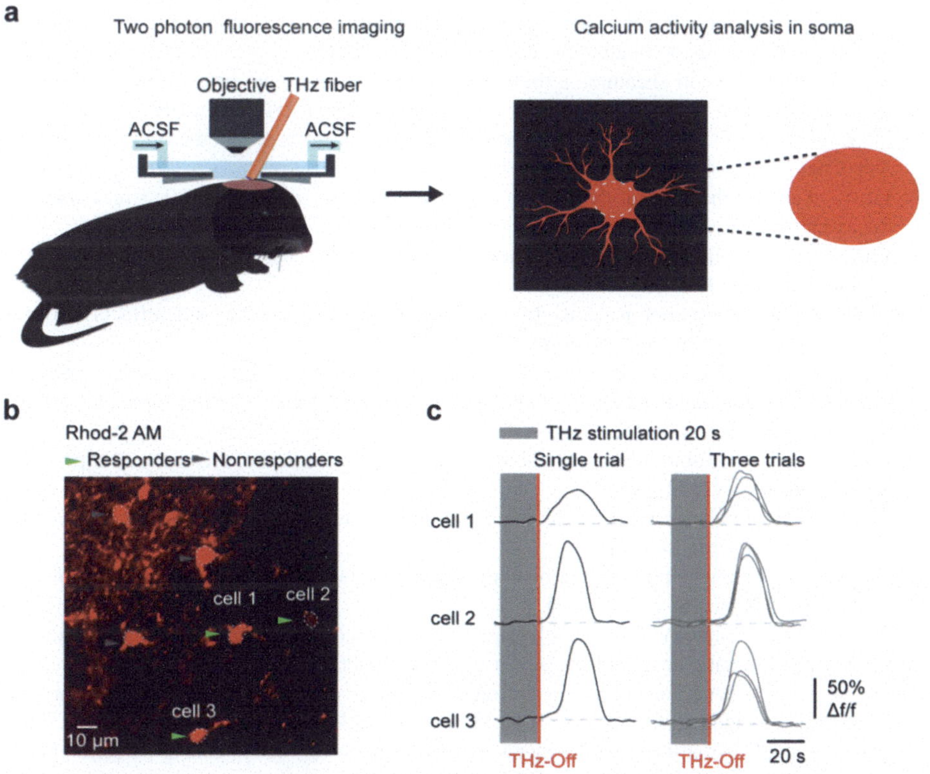

Fig. 1. Two-photon imaging records THz-induced calcium signals from astrocytes. (a) left: cartoon showing the *in vivo* two-photon imaging experiments with THz fiber. Right: cartoon showing the analysis area of calcium activity in astrocytes. (b).Two-photon image of Rhod-2 AM-stained area in the auditory cortex (imaging depth, −196 μm). (c). Calcium signals in response to THz stimulation during and after conditioning.

3 Discussion

We found that astrocytes respond to high-frequency terahertz immediately after high-frequency terahertz stimulation. We will further investigate whether the THz-induced calcium signal is from intracellular calcium release or from extracellular calcium influx. At the same time, the upstream mechanism of intracellular calcium store release or extracellular calcium influx will be explored. Finally, we will explore the behavioral significance of high-frequency terahertz stimulation on astrocytes.

References

1. Xiao, T., et al.: Sensory input dependent gain modulation of the optokinetic nystagmus by mid-infrared stimulation in pigeons. Elife **12**, e78729 (2023)
2. Liu, G., et al.: Myelin sheath as a dielectric waveguide for signal propagation in the mid-infrared to terahertz spectral range. Adv. Funct. Mater. **29**(7), 1807862 (2019)
3. Li, Y., et al.: Terahertz wave enhances permeability of the voltage-gated calcium channel. J. Am. Chem. Soc. **143**(9), 4311 (2021)
4. Yu, Y., et al.: Terahertz photons improve cognitive functions in posttraumatic stress disorder. Research **6**, 0278 (2023)
5. Tan, X., et al.: Minimal-invasive enhancement of auditory perception by terhaertz wave modulation. Nano Res. **15**(6), 4127 (2022)
6. Li, Y., et al.: Physicochemical insights on terahertz wave diminished side effects of drugs from slow dissociation. ACS Nano **16**(5), 8419 (2022)
7. Xiang, Z., et al.: A primary model of THz and far-infrared signal generation and conduction in neuron systems based on the hypothesis of the ordered phase of water molecules on the neuron surface. Sci. Bull. **65**, 308 (2020)
8. Tan, X., et al.: Neuromodulation of chemical sysnaptic transmission driven by THz photons. Research **6**, 1–16 (2022)
9. Wu, K., et al.: Terahertz wave accelerates DNA unwinding: a molecular dynamics simulation study. J. Phys. Chem. Lett. **11**(17), 7002–7008 (2020)
10. Yin, J., et al.: Terahertz photons inhibit cancer cells long term by suppressing nano telomerase activity. ACS Nano **18**, 4796 (2024)
11. Sun, L., et al.: Inhibition of cancer cell migration and glycolysis by terahertz wave modulation via altered chromatin accessibility. Research **2022**, 9860679 (2022)
12. Zhang, Z., et al.: Advanced terahertz refractive sensing and fingerprint recognition through metasurface-excited surface waves. Adv. Mater, 2308453 (2024)
13. Wang, P., et al.: An ultra-sensitive metasurface biosensor for instant cancer detection based on terahertz spectra. Nano Res. **16**, 7304–7311 (2023)
14. Lou, J., et al.: Calibration-free, high-precision, and robust terahertz ultrafast metasurfaces for monitoring gastric cancers. PNAS **119**(43), 2209218119 (2022)
15. Wang, P., et al.: Progress on cutting-edge infrared-terahertz biophysics. IEEE Trans. Microw. Theory Tech. **70**(11), 5117–5140 (2022)
16. Jiao, Y., et al.: Photoactive terahertz metasurfaces for ultrafast switchable sensing of colorectal cells. Mater. Horiz. **9**, 2984–2992 (2022)
17. Sun, L., et al.: A pixelated frequency-agile metasurface for broadband terahertz molecular fingerprint sensing. Nanoscale **14**, 9681–9685 (2022)
18. Wang, R., et al.: Ultrasensitive terahertz biodetection enabled by quasi-BIC-based metasensors. Small **19**, 2301165 (2023)

19. Chang, C., et al.: High-gain Thompson-scattering X-Ray free-electron laser by time-synchronic laterally tilted optical wave. Phys. Rev. Lett. **110**, 064802 (2013)
20. Gao, M., et al.: Enhanced and frequency-tunable terahertz emission from two crossed air-plasmas. Phys. Rev. Res. **5**, 023091 (2023)
21. Liu, X., et al.: Nonthermal and reversible control of neuronal signaling and behavior by midinfrared stimulation. Proc. Natl. Acad. Sci. USA **118**(10), e2015685118 (2021)
22. Zhang, J., et al.: Non-invasive, opsin-free mid-infrared modulation activates cortical neurons and accelerates associative learning. Nat Commun. **12**(1), 2730 (2021)
23. Peng, W., et al.: High-frequency terahertz waves disrupt Alzheimer's β-amyloid fibril formation. eLight **3**, 18 (2023)
24. Haydon, P.G., Nedergaard, M.: How do astrocytes participate in neural plasticity. Cold Spring Harb. Perspect. Biol. **7**, a020438 (2014)
25. Zhang, K., et al.: Fear learning induces α7-nicotinic acetylcholine receptor-mediated astrocytic responsiveness that is required for memory persistence. Nat. Neurosci. **24**(12) (2021)

Terahertz Waves Induced DSBs in Breast Cancer Cells

Junkai Yin and Zihua Song(✉)

Innovation Laboratory of Terahertz Biophysics, National Innovation Institute of Defense Technology, Beijing 100071, China
szh2016@mail.ustc.edu.cn

Abstract. Terahertz (THz) waves have shown potential in various biomedical applications due to their non-ionizing nature and unique interaction properties with biological systems. In this study, we investigate the effect of THz waves on the growth of breast cancer cells. We found that exposure to THz waves resulted in the induction of DNA double-strand breaks (DSBs) in breast cancer cells. DSBs are highly cytotoxic DNA lesions that can lead to cell death or genomic instability if not properly repaired. Using immunofluorescence assays, we observed an increase in the formation of 53BP1 foci, a marker of DSBs, in breast cancer cells treated with THz waves compared to the control group. Next, we injected the irradiated breast cancer cells subcutaneously into the mouse mammary gland and found that the tumor-forming ability of MCF-7 cells after THz irradiation was significantly decreased compared with the control group. In summary, this study provides a reference for the potential application of THz wave in cancer therapy.

Keywords: Terahertz waves · DNA double-strand breaks · Cancer therapy

1 Introduction

Terahertz (THz) waves refer to electromagnetic waves with frequencies ranging from 0.1 to 100 THz, which overlap with the long-wave and millimeter wave regions [1, 2]. THz waves can directly couple with biomolecules to induce coherent excitations and generate non-thermal effects, triggering nonlinear resonances in biomolecules and leading to conformational changes and bond breakage [3–5]. THz wave resonant absorption corresponds to unique biomolecular fingerprints, including neurotransmitters [6–8], proteins [9, 10], nucleic acids [11, 12], and ion channel [13–15], making it a suitable technique for achieving conformational changes and functional adjustments [16–19]. Recent studies have preliminarily demonstrated the reversible and non-thermal regulation of gene expression, transmission of biological information, modulation of protein function, and intervention in substance exchange at the molecular level by THz waves.

In recent years, with the development of THz sources [21, 22], THz biological effects have shown promising bioeffects [23–26], especially inhibitory effects on cancer cells [27–29]. Sun *et al.* [9] applied 83 THz waves to colon cancer cells and demonstrated long-term inhibition of cancer cells through non-pharmacological and non-invasive electromagnetic interference techniques. This approach exhibited inhibition of glycolysis in

C. Chang et al. (Eds.): YTHZ 2024, SPPHY 400, pp. 204–208, 2024.
https://doi.org/10.1007/978-981-97-3786-4_40

cancer cells, reducing cell migration by 50%, thereby confirming the inhibitory effect of THz waves on cancer cell migration in vivo. Triple-negative breast cancer cells with 60–100 THz waves was treated [29]. They found that THz waves could reshape the microtubule network into star-shaped arrangements, alter the formation of actin filaments and the localization of adhesion plaques, decrease cell migration activity and invasive capability, and induce G2/M cell cycle arrest. These findings demonstrate the potential application of high-frequency THz waves in breast cancer treatment.

In this study, we assessed the levels of 53BP1 in cells, which reflects the extent of DSBs and provides an overall evaluation of experimental results [30]. MCF-7 cells were irradiated with 33 THz waves for a duration of 21 days. Compared to the control group, the THz group showed a significant increase in DSBs. Additionally, when MCF-7 cells from the control and the irradiated groups were injected into nude mice, the tumorigenic ability of the cells exposed to long-term THz radiation was significantly reduced. Therefore, THz waves hold potential for future applications in cancer therapy.

2 Results

2.1 THz Wave Irradiation System

We used a quantum cascade laser (QCL) that emits 33 THz waves. The entire optical system consisted of the laser, a gold mirror, a focusing lens, and a displacement system (Fig. 1). The THz beam was emitted from the laser and reflected vertically to the horizontal plane. It then passed through a convex lens to focus the beam. We developed a targeted x-y two-dimensional scanning program. A hollow metal tray was fixed on the x-y scanning stage, and a culture dish was placed on the metal tray. By controlling the displacement system, we ensured that the THz beam reached the center of the culture dish. We adjusted the metal tray to position the focal plane of the THz beam at the inner bottom surface of the culture dish. The diameter of the spot was measured to be approximately 400 μm using a knife-edge method. We performed a point-by-point scanning procedure, where the scanning stage moved the culture dish while keeping the position of the spot stationary relative to the ground. We programmed the displacement control operation to set the velocity of the x-axis at 0.5 mm/s, ensuring that the cells received at least 1 s of THz stimulation. Additionally, we set the step length of the y-axis at 0.4 mm so that all cells in the culture dish received THz stimulation under this condition.

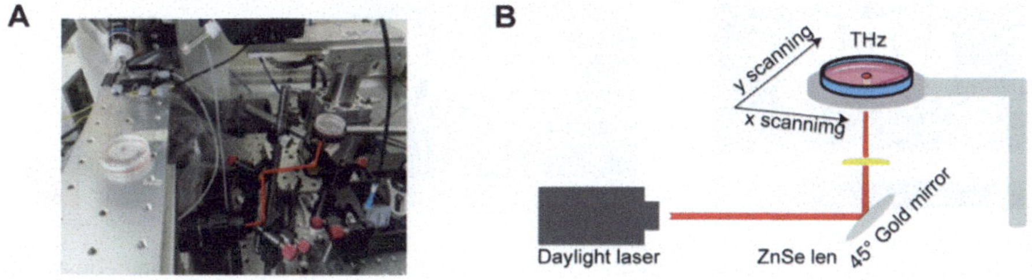

Fig. 1. A. Physical drawing of THz wave irradiation system. B. Cartoon sketch of THz wave irradiation system.

2.2 Induction of DSBs in MCF-7 Cells by 33 THz Waves.

After continuous exposure of MCF-7 cells to 33 THz waves for 21 days, we used an anti-53BP1 antibody to detect the presence of DSBs in the two types of cells. The results showed that the mean fluorescence intensity (MFI) of the 53BP1 protein in MCF-7 cells increased with the duration of exposure, indicating a significant increase in the level of DSBs. This suggests that 33 THz waves can induce DSBs in breast cancer cells (Fig. 2).

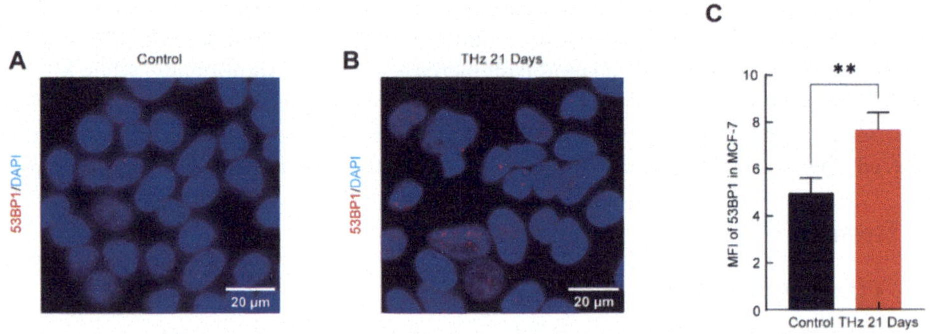

Fig. 2. A-B. Representative immunostaining images of 53BP1 protein (red fluorescence) in cells after continual THz irradiation for 21 days. Scale bar: 20 μm. C. Proportion of 53BP1-positive cells in MCF-7 cells after continual THz irradiation for 15 and 21 days. n = 15 images from 3 dishes of cells per group. Data are shown as individual values or expressed as the mean ± SEM, and significance levels are indicated as $^{**}P < 0.01$. Two-tailed unpaired Student's t test was used for C.

We exposed MCF-7 cells to continuous 33 THz waves for 21 days and then injected them into nude mice. After 28 days, we compared tumor volumes between the exposed group and the control group. We observed that MCF-7 cells exposed to 33 THz waves had a significantly reduced ability to survive in the nude mice, resulting in a significantly

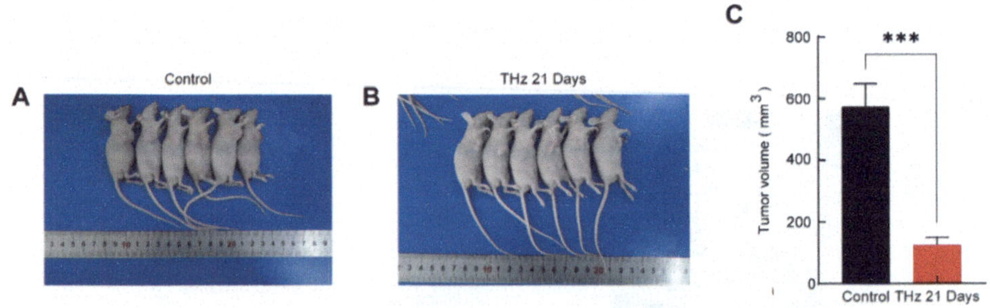

Fig. 3. A-B. MCF-7 cells were irradiated with 33 THz and transplanted subcutaneously into BALB/c nude mice after 28 days. B Tumor growth curve over 28 days *in vivo*. n = 6 mice per group. Data are means ± SEM, and significance levels are indicated as $^{***}P < 0.001$. Two-tailed unpaired Student's t test was used for C.

lower tumor volume compared to the control group. The tumor volume of the THz-treated MCF-7 cells decreased by approximately 77% compared to the control group (Fig. 3).

3 Discussion

The long-term exposure of MCF-7 cells to 33 THz waves resulted in an increased level of DNA double-strand breaks (DSBs) within the cells. This interesting and important phenomenon suggests the potential of THz waves to inhibit cell growth. However, the underlying biophysical mechanisms behind this significant biological effect are not yet clear. Further research is needed to delve deeper into understanding these mechanisms and provide theoretical support for the future application of THz waves in cancer treatment.

References

1. Liu, G., et al.: Myelin sheath as a dielectric waveguide for signal propagation in the mid-infrared to terahertz spectral range. Adv. Funct. Mater. **29**(7), 1807862 (2019)
2. Xiang, Z., et al.: A primary model of THz and far-infrared signal generation and conduction in neuron systems based on the hypothesis of the ordered phase of water molecules on the neuron surface I: signal characteristics. Sci. Bull. **65**(4), 308–317 (2020)
3. Wang, P., et al.: Progress on cutting-edge infrared-terahertz biophysics. IEEE Trans. Microwave Theory Tech. **70**(11), 5117–5140 (2022)
4. Zhang, Z., et al.: Advanced Terahertz Refractive Sensing and Fingerprint Recognition Through Metasurface-Excited Surface Waves. Adv. Mater. 2308453 (2024)
5. Lou, J., et al.: Calibration-free, high-precision, and robust terahertz ultrafast metasurfaces for monitoring gastric cancers. Proc. Natl. Acad. Sci. U.S.A. **42**(119), e2209218119 (2022)
6. Zhu, Z., et al.: Characteristic fingerprint spectrum of neurotransmitter norepinephrine with broadband terahertz time-domain spectroscopy. Analyst **144**(8), 2504–2510 (2019)
7. Tan, X., et al.: Neuromodulation of chemical synaptic transmission driven by THz photons. Research **2022**, 0010 (2022)
8. Yu, Y., et al.: Terahertz photons improve cognitive functions in posttraumatic stress disorder. Research **6**, 0278 (2023)
9. Sun, L., et al.: Inhibition of cancer cell migration and glycolysis by terahertz wave modulation via altered chromatin accessibility. Research **2022**, 9860679 (2022)
10. Peng, W., et al.: High-Frequency Terahertz Waves Disrupt Alzheimer's β-Amyloid Fibril Formation. eLight **3**(1), 18 (2023)
11. Wu, K., et al.: Terahertz wave accelerates DNA unwinding: a molecular dynamics simulation study. J. Phys. Chem. Lett. **11**(17), 7002–7008 (2020)
12. Zhang, C., et al.: Driving DNA origami assembly with a terahertz wave. Nano Lett. **22**(1), 468–475 (2022)
13. Li, Y., et al.: Terahertz wave enhances permeability of the voltage-gated calcium channel. J. Am. Chem. Soc. **143**(11), 4311–4318 (2021)
14. Zhang, J., et al.: Non-invasive, opsin-free mid-infrared modulation activates cortical neurons and accelerates associative learning. Nat. Commun. **12**(1), 2730 (2021)
15. Liu, X., et al.: Nonthermal and reversible control of neuronal signaling and behavior by midinfrared stimulation. Proc. Natl. Acad. Sci. **118**(10), e2015685118 (2021)
16. Tan, X., et al.: Minimal-invasive enhancement of auditory perception by terahertz wave modulation. Nano Res. **15**(6), 5235–5244 (2022)

17. Tan, X., et al.: A new means of energy supply driven by terahertz photons recovers related neural activity. iScience **26**(2), 105979 (2023)
18. Li, Y., et al.: Physicochemical insights on terahertz wave diminished side effects of drugs from slow dissociation. ACS Nano **16**(5), 8419–8426 (2022)
19. Yin, J., et al.: Terahertz photons inhibit cancer cells long term by suppressing nano telomerase activity. ACS Nano **18**(6), 4796–4910 (2024)
20. Xiao, T., et al.: Sensory input dependent gain modulation of the optokinetic nystagmus by mid-infrared stimulation in pigeons. Elife **12**, e78729 (2023)
21. Chang, C., et al.: High-gain thompson-scattering X-Ray free-electron laser by time-synchronic laterally tilted optical wave. Phys. Rev. Lett. **110**, 064802 (2013)
22. Gao, M., et al.: Enhanced and frequency-tunable terahertz emission from two crossed air-plasmas. Phys. Rev. Res. **5**, 023091 (2023)
23. Zhang, Q.L., Zhou, T., Chang, C., et al.: Ultrahigh-flux water nanopumps generated by asymmetric terahertz absorption. Phys. Rev. Lett. 2024 (accepted)
24. Tang, C., et al.: Integrating terahertz metamaterial and water nanodroplets for ultrasensitive detection of amyloid β aggregates in liquids. Sens. Actuator B. Chem. **329**, 129113 (2021)
25. Zhu, Z., et al.: Tunable surface wettability via terahertz electrowave controlled vicinal subnanoscale water layer. Nano Lett. **24**(10), 3243–3248 (2024)
26. Wang, R., et al.: Ultrasensitive terahertz biodetection enabled by quasi-BIC-based metasensors. Small **19**, 2301165 (2023)
27. Wang, P., et al.: An ultra-sensitive metasurface biosensor for instant cancer detection based on terahertz spectra. Nano Res. **16**, 7304–7311 (2023)
28. Jiao, Y., et al.: Photoactive terahertz metasurfaces for ultrafast switchable sensing of colorectal cells. Mater. Horiz. **9**, 2984–2992 (2022)
29. Chang, H., et al.: Quantitative proteomics reveals middle infrared radiation-interfered networks in breast cancer cells. J. Proteome Res. **14**(2), 1250–1262 (2015)
30. Fernandez, O., et al.: DNA damage-induced G2-M checkpoint activation by histone H2AX and 53BP1. Nat. Cell Biol. **4**(12), 993–997 (2002)

Mechanism of Terahertz Wave on Calcium Channels

Yuankun Sun, Shaomeng Wang, and Yubin Gong(✉)

School of Electronic Science and Engineering, University of Electronic Science and Technology of China, Chengdu 610097, Sichuan, China
ybgong@uestc.edu.cn

Abstract. L-type Ca^{2+} channels (LTCCs) play a vitalrole in excitation-contraction coupling and release of hormones from secretory cells. Abnormal Ca^{2+} fluxes arising from dysfunctional LTCCs has been associated with a list of calcium channelopathies such as hypertensive and arrhythmic. Here, we theoretically analyze the mechanism of increasing the conductance in LTCCs by a 42.5 Terahertz wave. We have proved that 42.5 THz could affect the hydrogen bonds from between Ca^{2+} inner-shell hydration waters and the side oxygens with respect to the Ca^{2+} position.

Keywords: Terahertz wave · resonate · hydrogen bond

1 Introduction

Ca^{2+} channels are key players in voltage-dependent signal transduction [1]. Activated by membrane depolarization, LTCCs permit Ca^{2+} influx into cells to maintain a stable intracellular calcium ion concentration. LTCCs are crucial for gene expression, cardiac excitability, electrical conduction [2]. Genetically-encoded channels activators along with optogenetic tools are rarely reported and are restricted by gene delivery methods and bio-safety. The Terahertz wave band (1–100 THz) overlaps with the vibrational and rotational spectra of biomolecules [3–7]. With the development of THz sources [8, 9], there have been some theoretical and experimental studies exploring the feasibility of electromagnetic-field-modulated biological functioning in the Terahertz wave band [10–25]. For example, Wu found that 44 THz wave can accelerated the unwinding of DNA hairpins [10]. Liu found that 53.6 THz wave can regulate the activity of the brain nerves [11]. Zhang have demonstrated 53.6 THz can increase the learning speed of mice by 50% [12]. Sun found that 83 THz can inhibit the migration of tumors both in vitro and in vivo [13]. In this paper, inspired by Li's research [14], we focus on studying how 42.5 THz affects the permeation of calcium channel. By means of molecular dynamics simulations, we calculated the hydrogen bonds from between Ca^{2+} inner-shell hydration waters and the side oxygens with respect to the Ca^{2+} position. The number of hydrogen bonds at different binding sites has changed.

C. Chang et al. (Eds.): YTHZ 2024, SPPHY 400, pp. 209–212, 2024.
https://doi.org/10.1007/978-981-97-3786-4_41

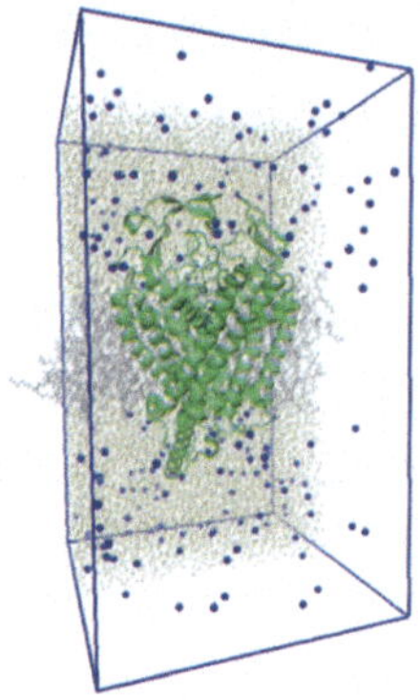

Fig. 1. The complete simulation system consisting of the simplified Cav1.2 channels.

2 Methods and Results

The simulation system (shown in Fig. 1) is constructed with the Gromacs package [26]. It consists of a $Ca_v1.2$ channel embedded between bilayer lipid membranes (with the normal along the z direction), which are solvated in a 0.15 M $CaCl_2$ solution. To reduce computation efforts, we only consider the selective filters and transmembrane sections, which line up closely along the Ca^{2+} ion pathway of the protein. Our previous work has shown that 42.5 THz can reduce the potential of mean force (PMF) for a single Ca^{2+} ion permeating the pore, especially the SF. An electric field was introduced to the system to serve as the desired electromagnetic field, with the effect of the magnetic component is negligible. We apply a pulling force (z-direction) to the calcium ions to pass through the selective filter of Cav1.2. As concluded by Li's research, 42.5 THz can reduce the LTCCs' PMF profile. Then we calculated the hydrogen bonds from between Ca^{2+} inner-shell hydration waters and the side oxygens with respect to the Ca^{2+} position. As shown in Fig. 2, there was a significant change in the number of hydrogen bonds at Site2.

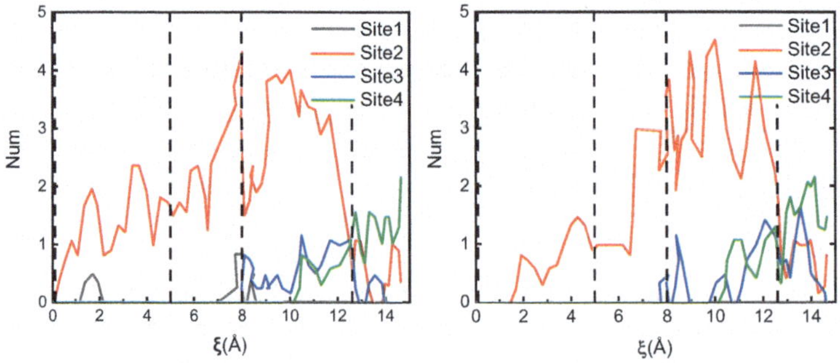

Fig. 2. Numbers of hydrogen bonds from between Ca2+ inner-shell hydration waters and the side oxygens with respect to the Ca2+ position. The left image shows the absence of THz, while the right image shows the presence of THz.

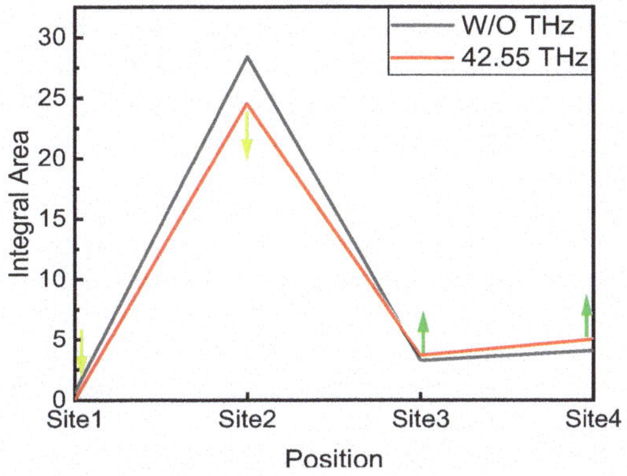

Fig. 3. Integral area of hydrogen bonds from between Ca2+ inner-shell hydration waters and the side oxygens with respect to the Ca2+ position. The left image shows the absence of THz, while the right image shows the presence of THz.

3 Discussion

In order to better understand the change in the number of hydrogen bonds, we calculated the area on the hydrogen bonds in Fig. 2, and the calculation results are shown in Fig. 3. From Fig. 3, we see clearly that the integrated areas of sites 2 and Site1 decrease under the influence of 42.5 THz, while the integrated areas of sites 3 and 4 increase. Considered that the calcium ions are mainly influenced by Site2 when passing through selective filter, the decrease in hydrogen bond integration at Site2 means that the time for calcium ions to stay at site2 under 42.5 THz will decrease, meaning that calcium ions will accelerate through the calcium channel. Our finding could enlighten inspiration for the mechanism by which terahertz affects ion channels.

References

1. Catterall, W., et al.: The chemical basis for electrical signaling. Nat. Chem. Biol. **13**, 455–463 (2017)
2. Ma, H., et al.: Excitation-transcription coupling, neuronal gene expression and synaptic plasticity. Nat. Rev. Neurosci. **24**(11), 672–692 (2023)
3. Liu, G., et al.: Myelin sheath as a dielectric waveguide for signal propagation in the mid-infrared to terahertz spectral range. Adv. Funct. Mater. **29**(7), 1807862 (2019)
4. Wang, P., et al.: Progress on cutting-edge infrared-Terahertz biophysics. IEEE Trans. Microw. Theory Tech. **70**(11), 5117–5140 (2022)
5. Xiang, Z., et al.: A primary model of THz and far-infrared signal generation and conduction in neuron systems based on the hypothesis of the ordered phase of water molecules on the neuron surface. Sci. Bull. **65**, 308 (2020)
6. Zhang, Z., et al.: Advanced terahertz refractive sensing and fingerprint recognition through metasurface-excited surface waves. Adv. Mater, 2308453 (2024)
7. Lou, J., et al.: Calibration-free, high-precision, and robust terahertz ultrafast metasurfaces for monitoring gastric cancers. PNAS **119**(43) (2022)
8. Gao, M., et al.: Enhanced and frequency-tunable terahertz emission from two crossed air-plasmas. Phys. Rev. Res. **5**, 023091 (2023)

9. Chang, C., et al.: High-gain Thompson-scattering X-ray free-electron laser by time-synchronic laterally tilted optical wave. Phys. Rev. Lett. **110**, 064802 (2013)
10. Wu, K., et al.: Terahertz wave accelerates DNA unwinding: a molecular dynamics simulation study. J. Phys. Chem. Lett. **11**(17), 7002–7008 (2020)
11. Liu, X., et al.: Nonthermal and reversible control of neuronal signaling and behavior by midinfrared stimulation. Proc. Natl. Acad. Sci. **118**(10), e2015685118 (2021)
12. Zhang, J.X., et al.: Non-invasive, opsin-free mid-infrared modulation activates cortical neurons and accelerates associative learning. Nature Comm. **12**, 2730 (2021)
13. Sun, L., et al.: Inhibition of cancer cell migration and glycolysis by terahertz wave modulation via altered chromatin accessibility. Research **2022**, 9860679 (2022)
14. Li, Y., et al.: Terahertz wave enhances permeability of the voltage-gated calcium channel. J. Am. Chem. Soc. **143**(11), 4311–4318 (2021)
15. Yu, Y., et al.: Terahertz photons improve cognitive functions in posttraumatic stress disorder. Research **6**, 0278 (2023)
16. Peng, W., et al.: High frequency terahertz waves disrupt Alzheimer's amyloid fibril formation. eLight **3,** 18 (2023)
17. Tan, X., et al.: Neuromodulation of chemical sysnaptic transmission driven by THz photons. Research **2022**, 0010 (2022)
18. Li, Y., et al.: Physicochemical insights on terahertz wave diminished side effects of drugs from slow dissociation. ACS Nano **16**(5), 8419 (2022)
19. Yin, J., et al.: Terahertz photons inhibit cancer cells long term by suppressing nano telomerase activity. ACS Nano (2024). https://doi.org/10.1021/acsnano.3c09216
20. Tan, X., et al.: Minimal-invasive enhancement of auditory perception by terhaertz wave modulation. Nano Res. **15**(6), 4127 (2022)
21. Xiao, T., et al.: Sensory input dependent gain modulation of the optokinetic nystagmus by mid-infrared stimulation in pigeons. Elife **12**, e78729 (2023)
22. Zhang, C., et al.: Driving DNA origami assembly with a Terahertz wave. Nano Lett. **22**, 468–475 (2022)
23. Yuan, Y., et al.: Midinfrared radiation accelerates DNA unwinding. ACS Photonics. https://doi.org/10.1021/acsphotonics.3c01548
24. Zhang, Q.L., Zhou, T., Chang, C., et al.: Ultrahigh-flux water nanopumps generated by asymmetric terahertz absorption. Phys. Rev. Lett. 2024 (accepted)
25. Zhu, Z., et al.: Tunable surface wettability via terahertz electrowave controlled vicinal subnanoscale water layer. Nano Lett. **24**(10), 3243–3248 (2024)
26. Abraham, M.J., et al.: GROMACS: high performance molecular simulations through multi-level parallelism from laptops to supercomputers. SoftwareX. **1–2**, 19–25 (2015)

Exploring Hub Genes of Terahertz Wave Irradiation in the Mouse Cerebral Cortex *via* Bioinformatic Analyses

Pan Liu and Zihua Song(✉)

Innovation Laboratory of Terahertz Biophysics, National Innovation Institute of Defense Technology, Beijing 100071, China
szh2016@mail.ustc.edu.cn

Abstract. In recent years, breakthroughs have been made in terahertz source and detection technologies, which have driven the research of terahertz applications within various fields. Research in terahertz biophysics has progressed rapidly, with notable advances in disease treatment, cognitive function, ion exchange, and gene regulation. The potential applications of terahertz in biomedicine are also coming to the forefront. Terahertz wave irradiation appears to regulate gene expression in biological systems at the molecular level through non-thermal effects. It is positive that terahertz wave irradiation can be specific for gene regulation, which makes it possible as a new type of non-contact therapy. In this study, the cerebral cortex of mice is irradiated at 34.5 THz based on quantum cascade laser (QCL). By RNA-Seq and based on String database, we built a protein-protein interaction(PPI) network and identified 10 hub genes, including *Cd68, Fcgr3, Itgb2, Ccl5, Aif1, Casp1, Cx3cl1, Cd33, Cd44, and Trem2.* In this study, we explored the hub genes in the mouse cerebral cortex after 34.5 terahertz wave irradiation.

Keywords: Terahertz · Terahertz Wave Irradiation · PPI Network

1 Introduction

The generalized terahertz (THz) wave frequency ranges from 0.5 to 100 THz [1]. Terahertz wave has several distinctive characteristics and has also demonstrated unique functions in the biological field. Recent studies have confirmed that the terahertz frequency range hosts the absorption peaks for many key substances involved in biological activities, such as neurotransmitters and nucleic acids [2, 3]. This makes terahertz technology suitable for detecting conformational changes and modulating functions.

THz wave has been demonstrated to exert non-thermal and reversible control over neuronal signaling and behavior [4]. THz wave has been found to significantly modulate neuronal signal transmission, enhancing the frequency, amplitude, and dynamic properties of both excitatory and inhibitory postsynaptic currents [5]. THz wave activates neurons and promotes learning in mice [6, 7]. *In vitro*, THz wave can delay the fibrillization process of amyloid-beta (Aβ) oligomers, which is promising for alleviating

C. Chang et al. (Eds.): YTHZ 2024, SPPHY 400, pp. 213–216, 2024.
https://doi.org/10.1007/978-981-97-3786-4_42

Alzheimer's disease [8]. Furthermore, the rapid development of terahertz technology in medical science is particularly notable in the detection of biomacromolecules and tissue imaging.

As THz sources [9, 10] and biological researches continue to accelerate, the role of terahertz technology as a potential therapeutic tool in the biomedical field is gradually expanding, including treatment of disease, enhancement of cognitive function, modulation of ion exchange, gene regulation [11–17] and molecules modulation [18–25]. In summary, terahertz waves are increasingly showing their potential in the field of biomedicine and are expected to become a novel form of adjunct physical therapy.

In this study, we attempted to validate the effects of terahertz wave irradiation on protein expression in the cerebral cortex through mouse experiments, providing experimental support for the beneficial effects of terahertz wave irradiation as an adjunctive physical therapy in the biomedical field.

2 Materials and Methods

2.1 Materials

This study selected C57BL/6J mice as experimental subjects. C57BL/6J male mice (SPF(Beijing) biotechnology Co., Ltd, China), aged 7–8 weeks, were used for all experimental studies. Five mice were housed in a group with ad libitum access to water and food. At a temperature of 23–25 °C in breeding room, to maintain a time rhythm, the mice were kept in a 12-h light/dark cycle.

The terahertz wave irradiation source used in the in *vivo* experiments was a 34.5 THz frequency quantum cascade laser (QCL). The optical fiber used in the experiments was a standard PIR AgCl: AgBr polycrystalline fiber with a diameter of 240 μm. Additionally, the metal cannula used in the experiments was obtained from RWD Life Science Co., Ltd.

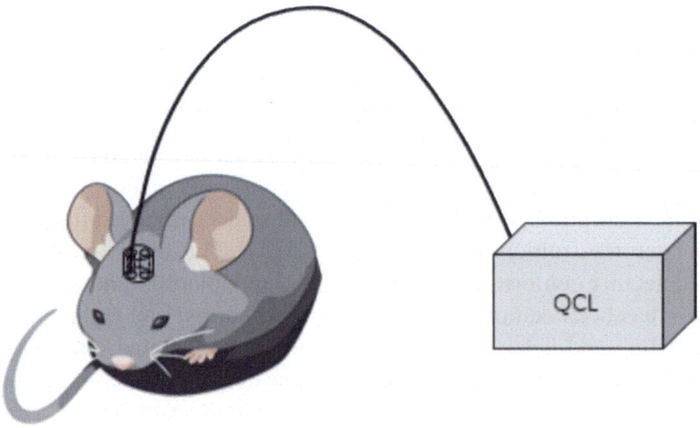

Fig. 1. Schematic of the experiment of this study

2.2 Methods

In order to conduct irradiation experiments on the target brain area, the mice were subjected to metal cannula embedding surgery. After animal was anesthetized, the metal cannula was embedded in the cerebral cortex of mice by the stereotaxic apparatus. After embedding surgery, the mice were allowed to recover for 7 days before being exposed to THz wave irradiation.

For THz wave irradiation before, mice were anesthetized with 2% isoflurane gas at a 0.5 flow rate and immobilized using a stereotaxic frame. The cap of the metal cannula was opened after the mouse was anesthetized and fixed. The optical fiber was then inserted into the metal tube to reach the designated brain area. Irradiation was performed for 10 min on each side. Mouse brain slices were sequenced by RNA-sequencing analysis.

Protein-protein interactions (PPI) are crucial for understanding how cells work and how they change in disease states. In this study, by RNA-Seq and based on String database, we built a protein-protein interaction (PPI) network. After PPI network was built, we used Cytoscape to visualize the network. Moreover, the CytoHubba plug-in was used to identify the top 10 hub genes, based on the previously constructed PPI network.

3 Results

To identify potential interactions between differentially expressed genes (DEGs), the PPI net-work was constructed for each time point and visualized using Cytoscape software. The top 10 hub genes were identified using CytoHubba. The hub genes are *Cd68, Fcgr3, Itgb2, Ccl5, Aif1, Casp1, Cx3cl1, Cd33, Cd44, and Trem2.*

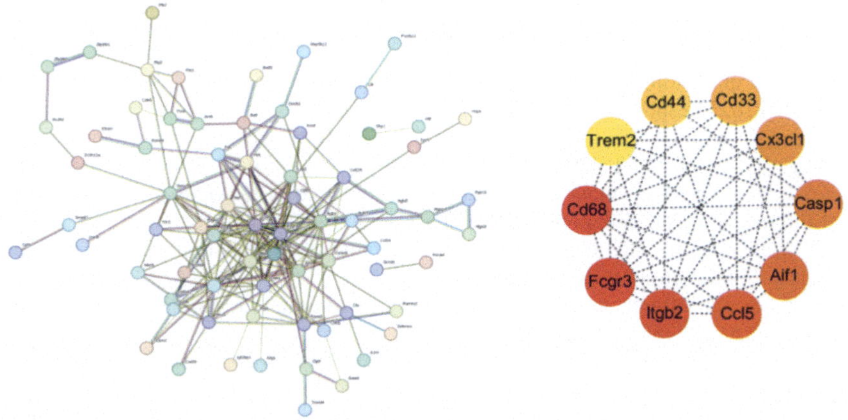

Fig. 2. PPI networks of differentially expressed genes of the control and THz-irradiated groups

References

1. Liu, G., et al.: Myelin sheath as a dielectric waveguide for signal propagation in the mid-infrared to terahertz spectral range. Adv. Funct. Mater. **29**(7), 1807862 (2019)
2. Wang, P., et al.: Progress on cutting-edge infrared-terahertz biophysics. IEEE Trans. Microw. Theory Tech. **70**(11), 5117–5140 (2022)
3. Zhang, Z., et al.: Advanced Terahertz refractive sensing and fingerprint recognition through metasurface-excited surface waves. Advanced Materials, 2308453 (2024)
4. Liu, X., et al.: Nonthermal and reversible control of neuronal signaling and behavior by mid-infrared stimulation. Proc. Natl. Acad. Sci. **118**(10), e2015685118 (2021)
5. Tan, X., et al.: Neuromodulation of chemical synaptic transmission driven by THz photons. Research **2022**, 0010 (2022)
6. Zhang, J., et al.: Non-invasive, opsin-free mid-infrared modulation activates cortical neurons and accelerates associative learning. Nat. Commun. **12**(1), 2730 (2021)
7. Yu, Y., et al.: Terahertz photons improve cognitive functions in posttraumatic stress disorder. Research **6**, 0278 (2023)
8. Peng, W., et al.: High-frequency terahertz waves disrupt Alzheimer's β-amyloid fibril formation. eLight **3,** 18 (2023)
9. Gao, M., et al.: Enhanced and frequency-tunable terahertz emission from two crossed air-plasmas. Phys. Rev. Res. **5**, 023091 (2023)
10. Chang, C., et al.: High-gain thompson-scattering X-Ray free-electron laser by time-synchronic laterally tilted optical wave. Phys. Rev. Lett. **110**, 064802 (2013)
11. Li, Y., et al.: Physicochemical insights on terahertz wave diminished side effects of drugs from slow dissociation. ACS Nano **16**(5), 8419–8426 (2022)
12. Sun, L., et al.: Inhibition of cancer cell migration and glycolysis by terahertz wave modulation via altered chromatin accessibility. Research, 9860679 (2022)
13. Li, Y., et al.: Terahertz wave enhances permeability of the voltage-gated calcium channel. J. Am. Chem. Soc. **143**(11), 4311–4318 (2021)
14. Wu, K., et al.: Terahertz wave accelerates DNA unwinding: a molecular dynamics simulation study. J. Phys. Chem. Lett. **11**(17), 7002–7008 (2020)
15. Zhang, C., et al.: Driving DNA origami assembly with a terahertz wave. Nano Lett. **22**, 468–475 (2022)
16. Yuan, Y., et al.: Midinfrared radiation accelerates DNA unwinding, ACS Photonics. https://doi.org/10.1021/acsphotonics.3c01548
17. Yin, J., et al.: Terahertz photons inhibit cancer cells long term by suppressing nano telomerase activity. ACS Nano **18**(6), 4796–4910 (2024)
18. Xiao, T., et al.: Sensory input dependent gain modulation of the optokinetic nystagmus by mid-infrared stimulation in pigeons. Elife **12**, e78729 (2023)
19. Lou, J., et al.: Optically controlled ultrafast terahertz metadevices with ultralow pump threshold. Small **17**(44), e2104275 (2021)
20. Qiao, Z., et al.: Infrared nanoimaging of nanoscale sliding dislocation of collagen fibrils. Nano Res. **15**(3), 2355–2361 (2022)
21. Lou, J., et al.: Calibration-free, high-precision, and robust terahertz ultrafast metasurfaces for monitoring gastric cancers. Proc. Natl. Acad. Sci. **119**, 43, e2209218119 (2022)
22. Zhu, Z., et al.: Characteristic fingerprint spectrum of neurotransmitter norepinephrine with broadband terahertz time-domain spectroscopy. Analyst. **144**, 2504–2510 (2019)
23. Tang, C., et al.: Integrating terahertz metamaterial and water nanodroplets for ultrasensitive detection of amyloid β aggregates in liquids. Sens. Actuators B. Chem. **329**, 129113 (2021)
24. Zhang, Q.L., Zhou, T., Chang, C., et al.: Ultrahigh-flux water nanopumps generated by asymmetric terahertz absorption. Phys. Rev. Lett. (2024). (accepted)
25. Zhu, Z., et al.: Tunable surface wettability via Terahertz electrowave controlled vicinal subnanoscale water layer. Nano Lett. **24**(10), 3243–3248 (2024)

Interaction Between Terahertz Wave and Biomolecules

Dongze Zhang(✉)

School of Physics, Peking University, Beijing 100871, China
2000011488@stu.pku.edu.cn

Abstract. Terahertz electromagnetic wave can interact with biomolecules and hence affect the characteristic and behavior of them, consequently, it may be possible to use terahertz electromagnetic wave to modulate the molecule to achieve the purpose of changing the energy states and varying the structures. By molecule dynamics simulation, the effects of terahertz wave on biomolecules can be theoretically verified, and the mechanics behind them can be understood. The idea of terahertz wave can provide a new method to control and modulate the molecules for the field of biology, medicine, etc.

Keywords: Terahertz · Biomolecule

1 Introduction

1.1 A Subsection Sample

The vibration spectrum of many biomolecules is mainly distributed in the terahertz band, so terahertz electromagnetic waves can interact with these molecules by resonance [1–3]. The interaction between terahertz wave and biomolecules will affect the structure of molecules and the functions of molecules, thus changing the characteristics and behavior of them. Research on the effect of terahertz wave stimulation on biomolecules has a promoting effect on materials [4–7], disease treatment [8–11], biotechnology [13–15] and other fields [16–23]. This paper will introduce the researches on the interaction between terahertz waves and biomolecules in recent years, and pay attention on the revelation of related mechanisms.

2 The Research

The unwinding of the double helix of the DNA molecule is vital in the replication and transcription process. Both in the research of the treatment of the DNA-related disease and the gene engineering, the unwinding process is a base approach. The conventional way to accelerate the unwinding process is to use the method of heating, and Wu *et al.* [24] proposed to use THz-EM stimulation at a specific frequency to accelerate the unwinding. The stimulation result shows that the 44.0 THz wave can shorten the time of each stage

C. Chang et al. (Eds.): YTHZ 2024, SPPHY 400, pp. 217–220, 2024.
https://doi.org/10.1007/978-981-97-3786-4_43

of the unwinding process, and the research on the hydrogen bond network also supports this conclusion. The 47.4 THz wave can also have noticeable acceleration effect, and both of the frequency are at the peak of the fingerprint peak of the vibration spectrum of purine in DNA. Hence, the mechanism is that the THz wave can break the H-bond through the resonance with the purine, and thus promote unwinding.

Ca^{2+} ions participate in many vital biological processes, so it's important to study how to promote the permeability of transmembrane calcium channels, especially the voltage-gated calcium. Li *et al.* [25] used the simulation method to study the effect of the THz wave on permeation of the calcium channel. The frequency of the EM wave is determined by the calculation of the vibrational spectra of the carboxyl and carbonyl groups, which participate in the interaction between the ions and channel. By simulating the H-bond data along the channel and the potential of mean force, it can be found that the stimulation of THz wave can affect the binding between calcium ions and functional groups. The 42.55 THz field (related to carboxyl) can promote the permeation more significantly than the 51.61 THz field (related to carbonyl), and due to forced vibrations, the functional groups can contribute to promote the permeation under non-resonant field. This research on the permeation of calcium channels can enlighten the medical treatment concerning the control of calcium ions.

Confined water plays an important role in nanostructure and material design, and the transmembrane transport of confined water is fundamental in correlation study. Zhu *et al.* [26] [27] proposed that terahertz electromagnetic stimulus can enhance the permeation of confined water nonthermally. The results of molecular dynamics simulation explicitly indicate that THz wave with specific frequency can largely increase the permeation rate of confined monolayer water and one-dimensional confined water, and each frequency is located at the peak of the vibration spectra of corresponding confined water structure as well as deviates from the bulk water peak. Furthermore, the H-bond analysis show the EM stimulation can decrease the H bond number, promote the transition of confined water to the superpermeation phase. The research provides a new route to control the transmembrane transport of confined water.

Dopamine participates in many physiologic processes as a critical neurotransmitter, and the antagonists for Dopamine D2 receptors can act as effective antipsychotic drugs. However, the slow dissociation of the drugs can cause a series of side effects. Li *et al.* [28] proposed to use THz wave interference to modulate the dissociation kinetics. The research concentrated on the D2R-resperidone (an atypical antipsychotic drug) complex. The risperidone is combined with D2R mainly through the salt bridge between the residue Asp114 and risperidone, and can be regulated by Trp100. According to the calculation, the binding free energy increases the most under the interference of 4.0 THz wave. Under the EM simulation, the conformation of risperidone will switch as a result of field resonance, making it further away from Asp114 and salt bridge become weaker. And the side chain rotation of Trp100 will also accelerate promote the dissociation of risperidone. With the Arrhenius law, the dissociation is estimated to increase 8 orders of magnitude for the 1.0 V/nm wave irritation.

3 Conclusion

In summary, we introduce the research on the interaction between terahertz waves and biomolecules in recent years. Through molecular dynamics simulation, we can clearly get the results of the influence of terahertz waves at a specific frequency on molecules, and explore the mechanism behind it. However, the above research is in the stage of theoretical calculation, and the actual effect still needs to be tested. The experimental verification of related research will provide a new and effective method to control biomolecules in the fields of medicine, biology and so on, which has positive significance for related fields.

References

1. Wang, P., et al.: Progress on cutting-edge infrared-terahertz biophysics. IEEE Trans. Microw. Theory Tech. **70**(11), 5117–5140 (2022)
2. Liu, G., et al.: Myelin sheath as a dielectric waveguide for signal propagation in the mid-infrared to terahertz spectral range. Adv. Funct. Mater. **29**(7), 1807862 (2019)
3. Xiang, Z., et al.: A primary model of THz and far-infrared signal generation and conduction in neuron systems based on the hypothesis of the ordered phase of water molecules on the neuron surface. Sci. Bull. **65**, 308 (2020)
4. Zhang, Z., et al.: Advanced terahertz refractive sensing and fingerprint recognition through metasurface-excited surface waves. Adv. Mater, 2308453 (2024)
5. Lou, J., et al.: Calibration-free, high-precision, and robust terahertz ultrafast metasurfaces for monitoring gastric cancers. PNAS **119**(43) (2022)
6. Wang, R., et al.: Ultrasensitive Terahertz biodetection enabled by Quasi-BIC-based metasensors. Small **19**, 2301165 (2023)
7. Wang, R., et al.: Electric Fano resonance-based terahertz metasensors. Nanoscale **13**, 18467–18472 (2021)
8. Yu, Y., et al.: Terahertz photons improve cognitive functions in posttraumatic stress disorder. Research **6**, 0278 (2023)
9. Yin, J., et al.: Terahertz photons inhibit cancer cells long term by suppressing nano telomerase activity. ACS Nano **18**, 4796 (2024)
10. Tan, X., et al.: Neuromodulation of chemical sysnaptic transmission driven by THz photons. Research **2022**, 0010 (2022)
11. Peng, W., et al.: High frequency terahertz waves disrupt Alzheimer's amyloid fibril formation. eLight **3**(18) (2023)
12. Sun, L., et al.: Inhibition of cancer cell migration and glycolysis by terahertz wave modulation via altered chromatin accessibility. Research **2022**, 9860679 (2022)
13. Liu, X., et al.: Nonthermal and reversible control of neuronal signaling and behavior by midinfrared stimulation. Proc. Natl. Acad. Sci. **118**(10), e2015685118 (2021)
14. Zhang, J.X., et al.: Non-invasive, opsin-free mid-infrared modulation activates cortical neurons and accelerates associative learning. Nature Comm. **12**, 2730 (2021)
15. Wang, P., et al.: An ultra-sensitive metasurface biosensor for instant cancer detection based on terahertz spectra. Nano Res. **16**, 7304–7311 (2023)
16. Chang, C., et al.: High-gain thompson-scattering X-ray free-electron laser by time-synchronic laterally tilted optical wave. Phys. Rev. Lett. **110**, 064802 (2013)
17. Gao, M., et al.: Enhanced and frequency-tunable terahertz emission from two crossed air-plasmas. Phys. Rev. Res. **5**, 023091 (2023)

18. Zhang, Q.L., et al.: Ultrahigh-flux water nanopumps generated by asymmetric terahertz absorption. Phys. Rev. Lett. 2024 (accepted)
19. Zhu, Z., et al.: Characteristic fingerprint spectrum of neurotransmitter norepinephrine with broadband terahertz time-domain spectroscopy. Analyst **144**(8), 2504–2510 (2019)
20. Tang, C., et al.: Integrating terahertz metamaterial and water nanodroplets for ultrasensitive detection of amyloid β aggregates in liquids. Sens. Actuators B. Chem. **329**, 129113 (2021)
21. Zhu, Z., et al.: Tunable surface wettability via terahertz electrowave controlled vicinal subnanoscale water layer. Nano Lett. **24**(10), 3243–3248 (2024)
22. Yuan, Y., et al.: Midinfrared radiation accelerates DNA unwinding. ACS Photonics. https://doi.org/10.1021/acsphotonics.3c01548
23. Xiao, T., et al.: Sensory input dependent gain modulation of the optokinetic nystagmus by mid-infrared stimulation in pigeons. Elife **12**, e78729 (2023)
24. Wu, K., et al.: Terahertz wave accelerates DNA unwinding: a molecular dynamics simulation study. J. Phys. Chem. Lett. **11**, 7002−7008 (2020)
25. Li, Y., et al.: Terahertz wave enhances permeability of the voltage-gated calcium channel. J. Am. Chem. Soc. **143**, 4311−4318 (2021)
26. Zhu, Z., et al.: Terahertz-light induced structural transition and superpermeation of confined monolayer water. ACS Photonics **8**, 781−786 (2021)
27. Zhu, Z., et al.: Transition to a Superpermeation Phase of Confined Water Induced by a Terahertz Electromagnetic Wave. J. Phys. Chem. Lett. 2020, 11, 256−262
28. Li, Y., et al.: Physicochemical insights on terahertz wave diminished side effects of drugs from slow dissociation. ACS Nano **16**, 8419–8426 (2022)

Possible Regulation of Terahertz Waves on TRPM8

Hongkang Qu and Zhi Zhu(✉)

University of Shanghai for Science and Technology, Shanghai 200093, China
zhuzhi@usst.edu.cn

Abstract. TRPM8 channels, referred to "cold receptors", play a crucial role in the sensory nervous system, specifically responding to cold temperatures and stimuli. Recent studies have highlighted the potential impact of terahertz waves on biomolecular structures and functions. Through the calculation of the rotation spectrum of the S6 chain within TRPM8 and subsequent analysis of its rotation mode, our research has identified a prominent absorption peak in the terahertz range associated with the collective rotation of TRPM8. Moreover, when subjected to terahertz waves at frequencies corresponding to this main absorption peak, we observed a significant increase in the kinetic energy of the S6 chain within the protein compared to conditions without terahertz radiation. This observation suggests that terahertz waves at specific frequencies can exert a biological effect on the activation of the TRPM8 protein. Our findings underscore the feasibility of regulating proteins using terahertz waves at characteristic frequencies, shedding light on the intricate mechanisms through which terahertz electromagnetic signals influence protein behavior.

Keywords: TRPM8 · Terahertz · Rotation spectrum

1 Introduction

TRPM8 is a transient receptor potential ion channel that belongs to the TRPM8 family of ion channels [1]. TRPM8 channels play an important role in the sensory nervous system, and are known as "cold receptors" because of their sensitivity to cold temperatures and cold stimuli. Structure of TRPM8 channels The TRPM8 channel consists of four subunits, each containing six transmembrane segments (S1–S6) and an extracellular macrocycle [2]. Through this tetrameric structure, the TRPM8 channel forms an ion channel pore that allows ions to pass through the cell membrane. The TRPM8 channel is primarily permeable to calcium ions and sodium ions, but is also permeable to other cations such as magnesium ions.

TRPM8 channels are activated by low-temperature stimulation, and their threshold temperature is usually between 25–28 °C. When the temperature falls below the threshold, TRPM8 channels open, allowing ions to enter the cell, generating potential changes and neural signaling. TRPM8 channels can also be activated by chemicals such as menthol and menthone, which explains why peppermint gives a cooling sensation.

C. Chang et al. (Eds.): YTHZ 2024, SPPHY 400, pp. 221–224, 2024.
https://doi.org/10.1007/978-981-97-3786-4_44

Multiple important roles in the sensory nervous system. It is a key molecule in cold perception, detecting and signaling cold stimuli. TRPM8 channels are also involved in pain transmission [3–5].

Multiple studies have demonstrated that terahertz waves can exert effects on the structure and function of proteins. Specifically, the interaction between terahertz waves and proteins has been shown to induce structural alterations [6–10]. Research indicates that the resonance frequency of terahertz waves aligns with the hydrogen bonding and vibrational modes present in proteins. Consequently, terahertz waves have the potential to disrupt the hydrogen bonding network and vibrational properties of proteins, leading to structural modifications [11–16]. These structural alterations may encompass modifications to the protein's secondary structures, such as α-helices and β-folding, alongside broader conformational changes. Furthermore, terahertz waves have been observed to impact protein function. Studies have revealed alterations in functionalities such as enzymatic activity, electrical conductivity, and photosensitivity in response to terahertz wave exposure [17–24].

In this study, we investigated the rotation spectrum of the TRPM8 protein and analyzed the kinetic energy variations following exposure to frequency-specific terahertz waves. Our analysis revealed numerous terahertz absorption peaks associated with the TRPM8 protein. Notably, these absorption peaks predominantly occur at lower frequencies, indicative of interatomic bond rotation. Additionally, the eigenfrequencies of the more pronounced absorption peaks were found to exert a discernible influence on the TRPM8 protein.

2 Methods

The initial crystal structure of the cold ion channel TRPM8 was sourced from the RCSB Protein Data Bank (PDB entry: 7wra). The model was constructed using CHARMM_GUI, wherein the protein was embedded within a hybrid membrane composed of POPC:POPE in a 1:1 ratio. To maintain overall system neutrality, sodium and chloride ions were introduced, and the salt concentration was upheld at 0.15 M. Molecular dynamics simulations were conducted employing the CHARMM36 force field and the TIP3P water model. The initial simulation system was built in CHAMGUI Membrane Builder and equilibrated for 30 ns using the six-step process provided by CHARMM_GUI. Subsequently, MD simulations were carried out to calculate the protein rotation spectra.

3 Results and Discussion

Upon analyzing the structure of the cold ion channel protein TRPM8, it was evident that the protein forms a tetramer, with the channel composed of four identical S6 chains (Fig. 1a). Notably, upon activation of the cold ion channel, we observed a distinct phenomenon where the four S6 chains undergo shifts or partial rotations induced by amino acids, resulting in an expansion of the formed pore up to a specific threshold. To elucidate this, we conducted a comprehensive analysis of the rotation spectrum of TRPM8.

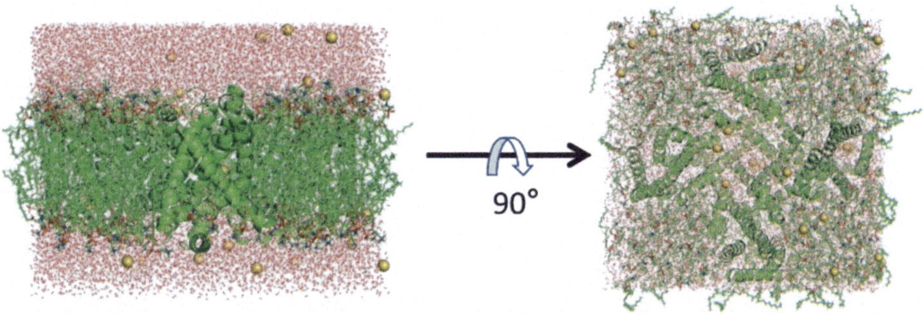

Fig. 1. Simulated model of TRPM8.

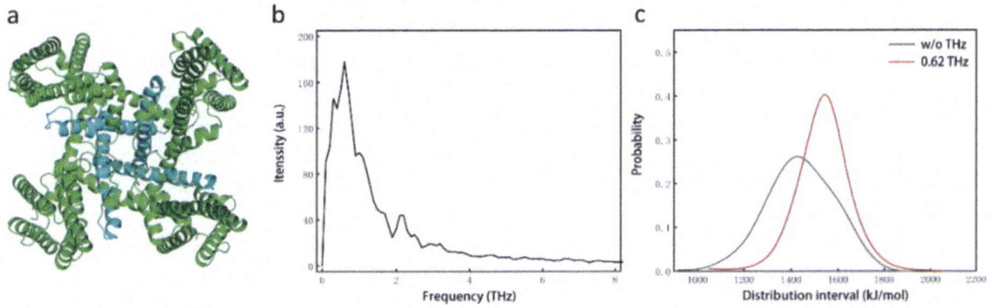

Fig. 2. (a) TRPM8 protein with S6 chain model. (b) S6 chain rotation spectra. (c) Comparison of S6 chain kinetic energy with (red) and without (black) terahertz radiation.

Our examination of the rotation spectrum revealed a wealth of terahertz information (Fig. 2b). Terahertz frequencies exhibited distinctive characteristics, with high frequencies corresponding to bond length stretching and low frequencies denotes the collective rotation of protein. Consequently, the rotation spectrum exhibited prominent peaks primarily in the low-frequency range.

Remarkably, our spectral analysis unveiled a pronounced rotation occurring at terahertz frequencies in the lower range. This led us to hypothesize that the functionality of cold ion channel proteins might be modulated upon exposure to specific rotation frequencies. Indeed, our findings demonstrated a significant increase in the kinetic energy of the S6 chain within the TRPM8 protein under terahertz irradiation compared to the control scenario (Fig. 2c).

In light of these observations, we propose that terahertz radiation of a characteristic frequency exerts regulatory influence on protein behavior. Our study underscores the intricate interplay between external stimuli, such as terahertz radiation, and protein dynamics, offering insights into potential avenues for manipulating protein function.

References

1. Bautista, D.M., Siemens, J., Glazer, J.M., et al.: The menthol receptor TRPM8 is the principal detector of environmental cold. Nature **448**(7150), 204–208 (2007)
2. Yin, Y., Wu, M., Zubcevic, L., et al.: Structure of the cold-and menthol-sensing ion channel TRPM8. Science **359**(6372), 237–241 (2018)
3. Diver, M.M., Cheng, Y., Julius, D.: Structural insights into TRPM8 inhibition and desensitization. Science **365**(6460), 1434–1440 (2019)
4. Bharate, S.S., Bharate, S.B.: Modulation of thermoreceptor TRPM8 by cooling compounds. ACS Chem. Neurosci. **3**(4), 248–267 (2012)
5. Yin, Y., Le, S.C., Hsu, A.L., et al.: Structural basis of cooling agent and lipid sensing by the cold-activated TRPM8 channel. Science **363**(6430), eaav9334 (2019)
6. Liu, G., et al.: Myelin sheath as a dielectric waveguide for signal propagation in the mid-infrared to terahertz spectral range. Adv. Funct. Mater. **29**(7), 1807862 (2019)
7. Li, Y., Chang, C., Zhu, Z., et al.: Terahertz wave enhances permeability of the voltage-gated calcium channel. J. Am. Chem. Soc. **143**(11), 4311–4318 (2021)
8. Liu, X., et al.: Nonthermal and reversible control of neuronal signaling and behavior by midinfrared stimulation. Proc. Natl. Acad. Sci. **118**(10), e2015685118 (2021)
9. Wu, K., Qi, C., Zhu, Z., et al.: Terahertz wave accelerates DNA unwinding: a molecular dynamics simulation study. J. Phys. Chem. Lett. **11**(17), 7002–7008 (2020)
10. Zhu, Z., et al.: Transition to a superpermeation phase of confined water induced by a terahertz electromagnetic wave. J. Phys. Chem. Lett. **11**(1), 256–262 (2020)
11. Li, Y., Zhu, Z., Sun, L., et al.: Physicochemical insights on terahertz wave diminished side effects of drugs from slow dissociation. ACS Nano **16**(5), 8419–8426 (2022)
12. Zhu, Z., Chen, C., Chang, C., et al.: Terahertz-light induced structural transition and superpermeation of confined monolayer water. ACS Photon. **8**(3), 781–786 (2021)
13. Peng, W., Zhu, Z., Lou, J., et al.: High-frequency terahertz waves disrupt Alzheimer's β-amyloid fibril formation. eLight **3**(1), 18 (2023)
14. Xiang, Z., et al.: A primary model of THz and far-infrared signal generation and conduction in neuron systems based on the hypothesis of the ordered phase of water molecules on the neuron surface. Sci. Bull. **65**, 308 (2020)
15. Yin, J., Wu, K., Yu, Y., et al.: Terahertz photons inhibit cancer cells long term by suppressing nano telomerase activity. ACS Nano **18**(6), 4796–4910 (2024)
16. Sun, T., Zhu, Z.: Light resonantly enhances the permeability of functionalized membranes. J. Membr. Sci. **662**, 121026 (2022)
17. Chang, C., et al.: High-gain Thompson-scattering X-ray free-electron laser by time-synchronic laterally tilted optical wave. Phys. Rev. Lett. **110**, 064802 (2013)
18. Lou, J., et al.: Calibration-free, high-precision, and robust terahertz ultrafast metasurfaces for monitoring gastric cancers. Proc. Natl. Acad. Sci. **119**(43), e2209218119 (2022)
19. Zhu, Z., Zhu, J., Chang, C., et al.: Tunable surface wettability via terahertz electrowave controlled vicinal subnanoscale water layer. Nano Lett. **24**(10), 3243–3248 (2024)
20. Zhang, Z., et al.: Advanced terahertz refractive sensing and fingerprint recognition through metasurface-excited surface waves. Adv. Mater. 2308453 (2024)
21. Zhang, Q.L., et al.: Ultrahigh-flux water nanopumps generated by asymmetric terahertz absorption. Phys. Rev. Lett. **132**(18), 184003 (2024)
22. Zhang, J., et al.: Non-invasive, opsin-free mid-infrared modulation activates cortical neurons and accelerates associative learning. Nat. Commun. **12**(1), 2730 (2021)
23. Yu, Y., et al.: Terahertz photons improve cognitive functions in posttraumatic stress disorder. Research **6**, 0278 (2023)
24. Tan, X., et al.: Neuromodulation of chemical synaptic transmission driven by THz photons. Research **6**, 1–16 (2022)

Terahertz Modulation on Biomolecules

Hongwei Wu, Pan Liu, and Zihua Song(✉)

Innovation Laboratory of Terahertz Biophysics, National Innovation Institute of Defense Technology, Beijing 10000, China
szh2016@mail.ustc.edu.cn

Abstract. Neuromodulation plays an important role in brain mechanism research and the treatment of brain functional disease. Current research methods include electrical stimulation using electric current, magnetic stimulation which effecting the entire brain, optogenetics powered by genetic engineering, ultrasonic stimulation using mechanical waves, and near-infrared stimulation through thermal effects. Here we introduce a new neural stimulation method called terahertz (THz) stimulation. Numerous biomolecules display distinctive fingerprint spectra in the THz range, attributed to intramolecular covalent bond vibrations, rotations, as well as intermolecular interactions such as hydrogen bonds and van der Waals forces. This unique characteristic allows specific THz frequency bands to engage in label-free, targeted resonant interactions with biomolecules, ultimately influencing and regulating the life activities of biological entities. Compared with traditional methods, THz stimulation has the characteristic of non-thermal effects, is non-contact, has high spatiotemporal resolution, and is minimally invasive. While the single terahertz photon energy is very low, allowing for easy control of ionizing damage to tissues, thus establishing the better means to regulate the fragile nervous system. In this paper, we discuss the possible underlying mechanisms of THz effects and their impact on cognitive improvement based on published studies.

Keywords: Neuromodulation · Terahertz · Neuron

1 Introduction

As the newly develop electromagnetic wave band, THz garnered significant attention. Over recent decades, numerous studies have focused on the effects of THz on biomolecules, including neurotransmitters [1], ion channels [2] and DNA [3, 4]. At present, we can conclude that THz effects on biomolecule is attributed to the absorption peaks of many pivotal substances involved in biological activities located in the THz frequency range, which manifest as non-thermal effects [5]. Similar to near-infrared radiation [6], THz also exhibits thermal effects due to strong water absorption, which can potentially damage cells and impact research on non-thermal effects. To eliminate thermal effects, low-power-density THz sources (sometimes pulse sources) at specific frequencies which outside the absorption peaks of hydrogen bonds between water molecules, were used [7]. Research on THz source has laid the foundation of THz stimulation to regulate biological components, from single protein to the whole

C. Chang et al. (Eds.): YTHZ 2024, SPPHY 400, pp. 225–228, 2024.
https://doi.org/10.1007/978-981-97-3786-4_45

body. The following figure illustrates several molecular mechanisms utilized in terahertz neuromodulation (Fig. 1).

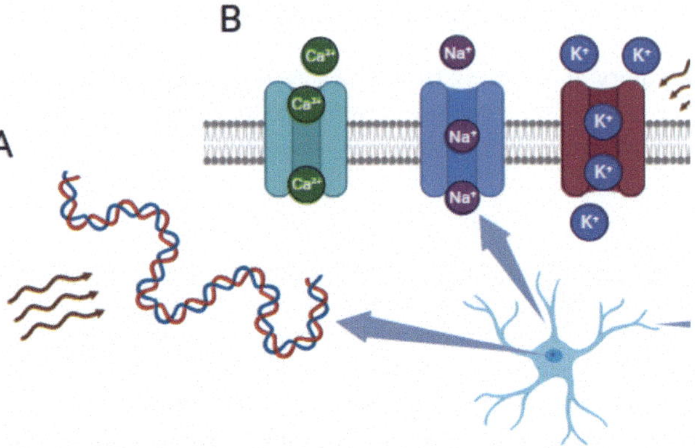

Fig. 1 Molecular mechanisms of THz bio-effects: (A) THz waves of specific frequency promotes ion channel permeability. (B) THz waves of specific frequency accelerate DNA folding (35.2 THz) and unwinding (44 THz)

2 THz Waves Effects in Cognitive Function

Existing study on cell rhythm provide the theoretical basis for understanding the THz effect on cognitive improvement, which can be applied in medical treatment and brain function enhancement.

2.1 THz-Treatment for Alzheimer's (AD)

Alzheimer's disease (AD) is a progressive neurodegenerative disease characterized by the deposition of β-amyloid (Aβ), neurofibrillary tangles, neuroinflammation, and neurodegeneration in the brain. Compare with numerous side effects of pharmacological therapy and thermally related non-pharmacological therapy, THz appears to be the most promising approach in AD treatment.

In vitro research For Aβ demonstrates that THz waves at a specific frequency of 34.88 THz could serve as a physical, efficient, non-thermal denaturation technique delaying the fibrillation process by 80%. Molecular dynamic (MD) simulations indicate that THz waves impede the kinetic progression of amyloid fibrils by transforming the structure of β-sheets into a looser configuration with more coil and bend regions [8].

In vivo research using $APP^{SWE}/PS1^{DE9}$ mice demonstrates that THz significantly reduce Aβ deposition in the AD mice both in cortex and hippocampus. Furthermore, research provides evidence that THz exposure ameliorates cognitive impairment in AD as assessed by the Y-maze test and Morris water maze [9].

2.2 THz Waves Improve Posttraumatic Stress Disorder (PTSD)

Posttraumatic stress disorder (PTSD) is a serious psychosis that can lead to cognitive impairment. The most commonly used treatment methods for PTSD include cognitive-behavioral therapy, which requires a long duration, and pharmacological therapy using antidepressants and anxiolytics, which have short-term effects. Notably, THz has emerged as a potential intervention for this disorder.

Research conducted by Yu [10] demonstrates that THz photons (34.5 THz) irradiation of the CA3 subregion (a region in hippocampus related to the initial encoding of memory) reverses the learning and memory dysfunction in PTSD rats. This study focuses on NR2B, a subunit of neurotransmitter receptor, and proves that it can increase the learning and memory abilities of mice. This is manifested through increased expression and phosphorylation of NR2B observed in both in vivo experiments and molecular dynamic simulations.

2.3 THz Waves Enhancement of Auditory Learning

Hearing serves as the foundation for cognition and behavior, intricately linked with brain function. A previous study conducted by Zhang [11] demonstrates that mid-infrared modulation (MIM) can significantly induce the firing activity of neurons in the target cortical region. Consequently, in an auditory associative learning task, mice exposed to MIM in auditory cortex exhibit a faster learning speed (approximately 50% faster) compared to control mice. Such results will probably be reproducible when utilizing THz.

Another research focus on THz auditory enhancement [12] utilizing minimal-invasive THz fibers provides evidence that THz wave can significantly increase both K^+ voltage-gated current and mechano-electrical transduce (MET) currents (approximately 50% higher) of the cochlear outer hair cells in acutely isolated mouse cochlea. Building on this foundation, an in vivo experiment with guinea pigs records auditory brainstem response signal (ABR), revealing that THz can improve hearing thresholds by an average of 10 dB.

3 Discussion

From the aforementioned research pertaining to memory, learning, and hearing, it is evident that THz stands out as the simplest nerve regulation method with the least side effects observed thus far. With the developing of THz sources [14, 15], it becomes plausible to explore the application of THz in regulating neuron transmission [16, 17], water modulation [18], and detecting biomolecules [19, 20].

While our understanding of the brain is still in its early stages, the THz neuromodulation method appears to be a promising shortcut facilitated by the intersection of information and life sciences. This approach has the potential to broaden the scope of neural research and contribute to a deeper understanding of the intricate workings of the brain.

References

1. Tan, X., et al.: Neuromodulation of chemical synaptic tTransmission driven by THz photons. Research **2022**, 10 (2022)
2. Li, Y., et al.: Terahertz wave enhances permeability of the voltage-gated calcium channel. J. Am. Che. Soc. **143**, 9 (2021)
3. Wu, K., et al.: Terahertz wave accelerates DNA unwinding: a molecular dynamics simulation study. J. Phys. Chem. Lett. **11**, 7002 (2020)
4. Zhang, C., et al.: Driving DNA origami assembly with a terahertz wave. Nano lett. **22**(1), 468 (2022)
5. Wang, P., Lou, J., Fang, G., Chang, C.: Progress on cutting-edge infrared-terahertz biophysics. IEEE Trans. Micro. Theo. Tech. **70**(11), 5117 (2022)
6. Liu, X., et al.: Nonthermal and reversible control of neuronal signaling and behavior by midinfrared stimulation. Proc. Natl. Acad. Sci. **118**(10), e2015685118 (2021)
7. Liu, M., et al.: Recent advances and research progress on microsystems and bioeffects of terahertz neuromodulation. Microsyst. Nanoeng. **9**(1), 143 (2023)
8. Peng, W., et al.: High-frequency terahertz waves disrupt Alzheimer's β-amyloid fibril formation. eLight **3**, 18 (2023).
9. Zhang, J., et al.: Terahertz irradiation improves cognitive impairments and attenuates Alzheimer's Neuropathology in the $APP^{SWE}/PS1^{DE9}$ mouse. Neurosci. Bull. (2023)
10. Yu, Y., et al.: Terahertz photons improve cognitive functions in posttraumatic stress disorder. Research **6**, 0278 (2023)
11. Zhang, J., et al.: Non-invasive, opsin-free mid-infrared modulation activates cortical neurons and accelerates associative learning. Nature Comm. **12**(1), 2730 (2021)
12. Tan, X., et al.: Minimal-invasive enhancement of auditory perception by terahertz wave modulation. Nano Res. **15**(6), 5235–5244 (2022)
13. Xiao, T., et al.: Sensory input-dependent gain modulation of the optokinetic nystagmus by mid-infrared stimulation in pigeons. eLife **12**, e78729 (2023).
14. Chang, C., et al.: High-gain Thompson-Scattering X-ray free-electron laser by time-synchronic laterally tilted optical wave. Phys. Rev. Lett. **110**, 064802 (2013)
15. Gao, M., et al.: Enhanced and frequency-tunable terahertz emission from two crossed air-plasmas. Phys. Rev. Res. **5**, 023091 (2023)
16. Liu, G., et al.: Myelin sheath as a dielectric waveguide for signal propagation in the mid-infrared to terahertz spectral range. Adv. Funct. Mater. **29**(7), 1807862 (2019)
17. Xiang, Z., et al.: A primary model of THz and far-infrared signal generation and conduction in neuron systems based on the hypothesis of the ordered phase of water molecules on the neuron surface. Sci. Bull. **65**, 308 (2020)
18. Zhang, Q.L., et al.: Ultrahigh-flux water nanopumps generated by asymmetric terahertz absorption. Phys. Rev. Lett. (2024, accepted)
19. Zhang, Z., et al.: Advanced terahertz refractive sensing and fingerprint recognition through metasurface-excited surface waves. Adv. Mater. 2308453 (2024)
20. Lou, J., et al.: Calibration-free, high-precision, and robust terahertz ultrafast metasurfaces for monitoring gastric cancers. PNAS **119**(43) (2022).

Terahertz Photons Enhance the Transmission of Calcium Ions Through NMDA Receptor

Yuan Zhong[1,2] and Rundong Jiang[2(✉)]

[1] Department of Engineering Physics, Tsinghua University, Beijing, China
zhongy20@mails.tsinghua.edu.cn

[2] Innovation Laboratory of Terahertz Biophysics, National Innovation Institute of Defense Technology, Beijing, China
rdjiang@163.com

Abstract. N-Methyl-D-aspartate (NMDA) receptors, also glutamate-gated calcium ion channels, play essential roles in synaptic transmission and plasticity and hence influence learning and memory. Biomolecules exhibit distinct spectra in the terahertz (THz) frequency range, allowing THz photons to influence their functioning and the resultant cellular activities. Previous studies have validated that THz photons in resonance with the key chemical bonds lining the voltage-gated calcium channel can enhance its permeability. In this work, we focus on the effect of specific THz electromagnetic fields on the permeation of the NMDA receptor channel. Molecular dynamics (MD) simulations demonstrate that the 35.4 and 42.4 THz fields could effectively suppress the free energy profiles of calcium ions permeating the NMDA receptor channel, with the largest decrease of 16.25 kJ/mol. The channel permeability is revealed enhanced by THz photons resonating with key amino acids especially carboxyl groups at the binding sites. The findings predict potential applications of THz radiation in treating NMDA receptor dysfunction and related diseases, offering a new physical treatment method.

Keywords: Terahertz photons · N-methyl-D-aspartic acid receptor channel · free energy · amino acids

1 Introduction

Synapses are critical structures for neuronal communication, intricately linked to the development and function of the brain. Glutamate is the primary neurotransmitter for excitatory synaptic transmission in the central nervous system [1]. NMDA receptors are glutamate-gated calcium ion channels that play essential roles in synaptic transmission and plasticity, being essential for learning and memory [2]. Calcium permeation triggered by activation of NMDA receptors channels is the critical event for initiation of neural connectivity and neuronal plasticity [2]. Dysfunction in NMDA receptors is linked to a range of disorders from ischemia to schizophrenia, which leads to cognitive deficits and abnormalities in mental behavior.

THz photons have emerged as a promising tool for biosensors [3–5], neural researches [6–9], with increasing interest in their interactions with ions in channels of

C. Chang et al. (Eds.): YTHZ 2024, SPPHY 400, pp. 229–233, 2024.
https://doi.org/10.1007/978-981-97-3786-4_46

nerve cells and triggered downstream responses by THz shooting [10–13]. Both experiments and simulations verification have showed that nerve fibers have the potential to function as dielectric waveguides for transmitting terahertz and mid-infrared waves [6]. Biomolecules have rich fingerprint spectra in the THz frequency range [7]. By resonating with their characteristic peaks, THz photons can influence the functional roles of biomolecules and the life activities of cells [14, 15]. In recent research, mice irradiated with 53.6 THz photons targeting to the auditory cortex displayed accelerated learning (approximately 50% faster) compared to control mice in an auditory-associative learning task [15]. It has also been shown that 42.55 THz photons can enhance the permeability of the voltage-gated calcium channel and 34.5 THz photons can effectively improves the cognitive function of animals with posttraumatic stress disorder through affecting NMDA receptors [16, 17].

In this work, we focus on the effect of the THz electromagnetic fields on permeation of the NMDA receptor channel. By means of MD simulations, we found that by applying specific electromagnetic fields in the THz frequency, the free energy of calcium ions through the NMDA receptor channel binding site effectively decreases. THz photons enhance the permeability of the NMDA receptor channel through resonating with twelve binding amino acids in two chains of GluN1A, especially carboxyl groups.

2 Method

The simulation system is constructed with the membrane builder of the Charmm-GUI [23]. It consists of a NMDA Receptor Channel incorporated within 1,2-Dipalmitoyl-sn-glycero-3-phosphorylcholine bilayer lipid membranes (oriented along the z-axis). The system is immersed in $CaCl_2$ solution with an ionic concentration of 0.15 mol/L.

NMDA receptors are heterotetramers comprising two GluN1 and two alternate GluN2 subunits, composed of three components: amino terminal domain (ATD), ligand-binding domain (LBD) and transmembrane domain (TMD) (shown in Fig. 1a). In recent research, the crystal structure complexed with holmium or gadolinium, show binding between the LBD-TMD linkers from the two GluN1 subunits around the center of the ion channel (shown in Fig. 1b [2])). A set of acidic residues in GluN1 located in this region is critical for the high calcium flux characteristic of NMDA receptors (shown in Fig. 1c).

The initial crystal structure of the channel protein comes from the RCSB data bank (PDB entry: 7EOS) [24], and only the amino acids 658−663 are calculated for the vibration spectrum. We employ umbrella sampling (US) methods to calculate the potential of mean force (PMF) for a single Ca^{2+} ion moving through binding residues of the channel. The MD simulations are conducted with the Gromacs code using the Charmm-36 m force fields [25].

3 Results and Discussion

With the MD simulation data of the twelve amino acids of the binding site, we can calculate the vibration/absorption spectrum as shown in Fig. 2. The spectrum (red curve) has six pronounced peaks at 29.6, 35.4, 42.4, 48.7. 51.6 and 53.3 THz.

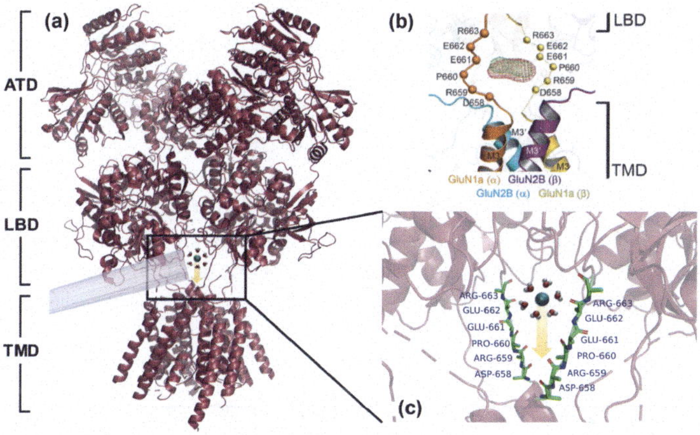

Fig. 1 Schematic of accelerated Ca^{2+} moving through the NMDA receptor channel under THz photons radiation (purple light). (a) NMDA receptor is composed of three components: ATD, LBD and TMD. (b) Holmium (green mesh) and gadolinium (red mesh) binding sites are located between the LBD-TMD linkers. (c) Close up view of the boxed region in panel a. Calcium ion (gray ball) is passing through the putative calcium binding site, which consists of amino acids 658 – 663 (green chain)

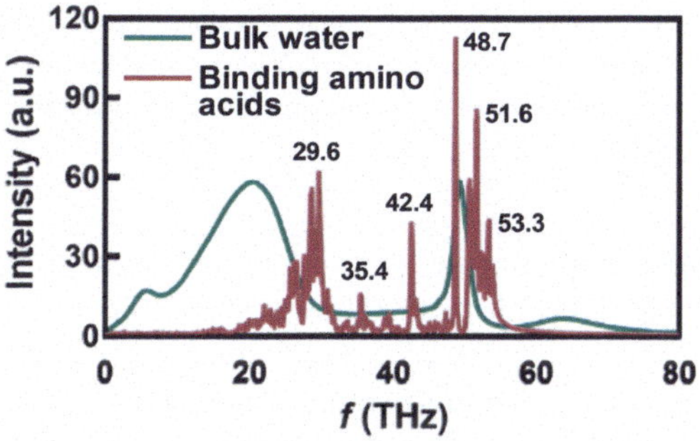

Fig. 2 Vibration/absorption spectrum of bulk water (green curve) and binding amino acids (red curve), calculated by Gromacs

We can identify the vibration modes by calculating residues in order, from amino acid 658 to 663. Specifically, the absorption peak at 29.6, 51.6 and 53.3THz arise from the vibration modes of the carbonyl ($-C=O$) groups. The absorption peak at 35.4, 42.4 and 48.7 THz is attributed to the vibration modes of the carboxyl ($-COO^-$) groups. The simulation results are very similar to those in the published article, especially the spectrum of $-$ COO $-$ groups [10].

The vibration modes of $-$ COO^- groups have a significant influence on the binding and passage of calcium ions, so we choose 35.4 and 42.4 THz as electromagnetic field frequencies to irradiate the NMDA receptor channel, considering that peak at 48.7 THz resides in strong water absorption (green curve).

We calculated the free energy for a single Ca^{2+} ion moving through the binding residues of the NMDA receptor channel with US methods. With the stimulation of the 42.4 THz field along the z-axis, the PMF profile (red curve in Fig. 3a) is suppressed significantly. The free energy profile is 16.25 kJ/mol lower than that without THz electromagnetic fields (green curve in Fig. 3a). We also applied the 35.4 THz field along the z-axis (yellow curve in Fig. 3a), but its efficacy in suppressing the free energy profile proved to be weaker compared with the 42.4 THz field. Total number of hydrogen bonds formed through the binding site increases as in Fig. 3b when it is stimulated by the THz fields. Consequently, the addition of the specific THz field significantly modifies the binding of the $-COO^-$ groups to Ca^{2+}.

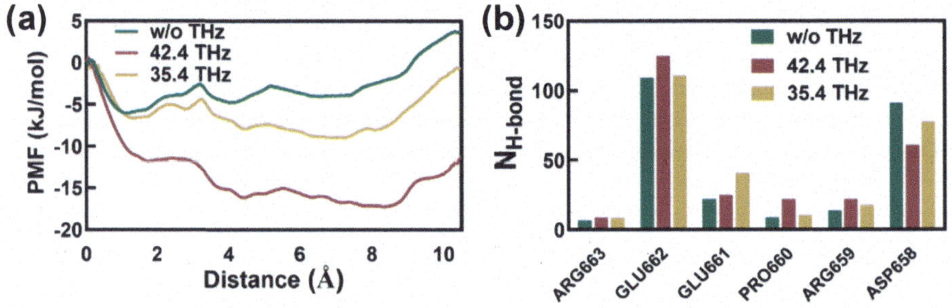

Fig. 3 (a) Free energy curves with and without external THz fields at different frequencies. The electromagnetic field intensity is the same at $E_0 = 0.6$ V/nm. (b) Total number of hydrogen bonds formed through the binding site between Ca^{2+} and residues with and without external THz fields at different frequencies

In this work, we have explored the process of accelerated Ca^{2+} moving through the NMDA receptor under THz electromagnetic fields. Our findings suggest that the permeation of Ca^{2+} is expected to increase when exposed to the 35.4 and 42.4 THz field (corresponding to $-COO^-$ vibration mode). Nowadays, THz photos are expected to inhibit cancer cell migration [18, 19], relieve neuron disease [20, 21], and modulate water molecules [22]. Our investigations predict potential applications of THz photon radiation in treating NMDA receptor dysfunction and related diseases.

References

1. Zhang, J., et al.: Distinct structure and gating mechanism in diverse NMDA receptors with GluN2C and GluN2D subunits. Nat. Struct. Mol. Biol. **30**(5), 629–639 (2023)
2. Karakas, E., et al.: Crystal structure of a heterotetrameric NMDA receptor ion channel. Science **344**(6187), 992–997 (2014)
3. Zhang, Z., et al.: Advanced terahertz refractive sensing and fingerprint recognition through metasurface-excited surface waves. Adv. Mater. 2308453 (2024)
4. Lou, J., et al.: Calibration-free, high-precision, and robust terahertz ultrafast metasurfaces for monitoring gastric cancers. PNAS **119**(43), e2209218119 (2022)

5. Wang, R., et al.: Electric Fano resonance-based terahertz metasensors. Nanoscale **13**, 18467 (2021)
6. Liu, G., et al.: Myelin sheath as a dielectric waveguide for signal propagation in the mid-infrared to terahertz spectral range. Adv. Func. Mater. **29**(7), 1807862 (2019)
7. Wang, P., et al.: Progress on Cutting-Edge Infrared-Terahertz Biophysics. IEEE Trans. Microw. Theory Tech. **70**(11), 5117–5140 (2022)
8. Xiang, Z., et al.: A primary model of THz and far-infrared signal generation and conduction in neuron systems based on the hypothesis of the ordered phase of water molecules on the neuron surface. Sci. Bull. **65**(4), 308–317 (2020)
9. Tan, X., et al.: Neuromodulation of chemical synaptic transmission driven by THz photons. Research **2022**, 0010 (2022)
10. Li, Y., et al.: Terahertz wave enhances permeability of the voltage-gated calcium channel. J. Am. Chem. Soc. **143**(11), 4311–4318 (2021)
11. Wu, K., et al.: Terahertz wave accelerates DNA unwinding: a molecular dynamics simulation study. J. Phys. Chem. Lett. **11**(17), 7002–7008 (2020)
12. Gao, M., et al.: Enhanced and frequency-tunable terahertz emission from two crossed air-plasmas. Phys. Rev. Res. **5**, 023091 (2023)
13. Chang, C., et al.: High-gain Thompson-Scattering X-ray free-electron laser by time-synchronic laterally tilted optical wave. Phys. Rev. Lett. **110**, 064802 (2013)
14. Liu, X., et al.: Nonthermal and reversible control of neuronal signaling and behavior by midinfrared stimulation. PNAS **118**(10), e2015685118 (2021)
15. Zhang, J., et al.: Non-invasive, opsin-free mid-infrared modulation activates cortical neurons and accelerates associative learning. Nat. Commun. **12**(1), 2730 (2021)
16. Li, Y., et al.: Physicochemical insights on terahertz wave diminished side effects of drugs from slow dissociation. ACS Nano **16**(5), 8419–8426 (2022)
17. Yu, Y., et al.: Terahertz photons improve cognitive functions in posttraumatic stress disorder. Research **6**, 0278 (2023)
18. Sun, L., et al.: Inhibition of cancer cell migration and glycolysis by terahertz wave modulation via altered chromatin accessibility. Research **2022**, 9860679 (2022)
19. Yin, J., et al.: Terahertz photons inhibit cancer cells long term by suppressing nano telomerase activity. ACS Nano **18**(6), 4796–4910 (2024)
20. Peng, W., et al.: High-frequency terahertz waves disrupt Alzheimer's β-amyloid fibril formation. eLight **3**(1), 18 (2023)
21. Zhu, Z., et al.: Characteristic fingerprint spectrum of neurotransmitter norepinephrine with broadband terahertz time-domain spectroscopy. Analyst **144**(8), 2504–2510 (2019)
22. Zhang, Q.L., et al.: Ultrahigh-flux water nanopumps generated by asymmetric terahertz absorption. Phys. Rev. Lett. (2024, accepted)
23. Jo, S., et al.: CHARMM-GUI: a web-based graphical user interface for CHARMM. J. Comput. Chem. **29**(11), 1859–1865 (2008)
24. Wang, H., et al.: Gating mechanism and a modulatory niche of human GluN1-GluN2A NMDA receptors. Neuron **109**(15), 2443–2456 (2021)
25. Hess, B., et al.: GROMACS 4: algorithms for highly efficient, load-balanced, and scalable molecular simulation. J. Chem. Theory Comput. **4**(3), 435–447 (2008)

Advances of Terahertz-Infrared Waves for Bio-information and Bio-effects in Neuroscience

Wenyu Peng and Yuanming Wu(✉)

Department of Biochemistry and Molecular Biology, Air Force Medical University, Xi'an 710032, China
wuym@fmmu.edu.cn

Abstract. Terahertz-infrared (THz-IR) waves are ranged between microwave and mid-infrared regions in the electromagnetic spectrum. THz-IR technology has demonstrated significant potential for biomedical applications, particularly in neuroscience. This review will primarily compile relevant research on THz-IR information transmission in the nervous system and the corresponding biological effects at specific frequencies. Finally, we will discuss how this technology can be utilized to non-invasively intervene in brain signals to benefit humanity.

Keywords: THz-IR technology · bio-information · bio-effects

1 Introduction

Guozhi Liu (*1*) proposed that the physical field of vertebrate neural signals could be the high frequency electromagnetic field from Terahertz to infrared, with the most likely frequency range being from 0.5 to 100 THz, termed as the generalized Terahertz electromagnetic wave. This conjecture has indeed been substantiated by numerous studies, encompassing the information transmission of THz-IR in nervous axons and the resulting biological effects in the nervous system.

2 THz-IR Waves for Bio-information

The myelin sheath, a lamellar structure, a lamellar structure enveloping vertebrate nerve axons (*2*), serves a multifaceted role beyond mere insulation in the nervous system (*3*). Comprised of multiple layers of nonpolar phospholipids, the myelin sheath provides insulation for low-frequency electrical signals. This unique structure facilitates a significant acceleration in signal propagation within nerves. The dissected neural axon exhibits a periodic arrangement of millimeter-long myelin segments interspersed with micrometer-long nodes of Ranvier. The retention of micrometer-sized nodes instead of continuous myelin remains a question. Chang's group has proposed a model suggesting that signals propagate through periodic relays, traversing the myelin for transmission

C. Chang et al. (Eds.): YTHZ 2024, SPPHY 400, pp. 234–237, 2024.
https://doi.org/10.1007/978-981-97-3786-4_47

and receiving amplification upon crossing the nodes of Ranvier. In 2020, Xiang et al. (*4*) delved into the electromagnetic field within the nervous system using the principles of quantum optics and electrodynamics. By assuming an ordered alignment of water molecules on neuronal surfaces and employing the Lagrangian of water molecule-field ion, they derived dynamic equations governing neural signal generation and transmission. Their findings revealed intrinsic vibration modes of electromagnetic radiation on neuronal surfaces, predominantly falling within the THz and far-infrared frequency ranges. Through atomistic molecular dynamics simulations of dimyristoylphosphatidylcholine bilayers (constituting myelin sheath membranes) in water, researchers discovered that the membrane head regions exhibited larger real and imaginary parameters compared to the tail regions across the 1–100 THz band. The spatial variation in dielectric properties suggested advantageous propagation characteristics of the phospholipid bilayer within a broad band of 55–85 THz (*5*). Gong's group (*6*) explored the electromagnetic properties of in vivo nerve fibers within the THz-IR band at a theoretical level. By analyzing the dispersion curves of THz-IR wave modes' transmission in nerve fibers, they posited that nerve fibers could function as dielectric waveguides for transmitting THz-IR waves at speeds reaching up to 108 m/s. Additionally, investigations into myelinated nerve structures using synchrotron-radiation-based Fourier-transform infrared microspectroscopy unveiled that the myelin sheath exhibited approximately a twofold higher refractive index within a specific mid-infrared to terahertz spectral range compared to the surrounding medium or the inner axon. This observation suggests that myelin may serve as an infrared dielectric waveguide (*1*).

3 THz-IR Waves' Bio-Effects in Neuroscience

In elucidating the mechanisms underlying neural signal transmission in myelinated neurons, Xiang et al. observed that the transmembrane ion current at Ranvier's node functions as an energy supplement (*7*), prompting significant research focus on the ion channel effects in THz waves. Li et al. demonstrated that by exposing a terahertz wave resonating with the stretching mode of either the –COO- or the –C = O group, the selectivity and conductance of Ca^{2+} were markedly enhanced (*8*). Through two-dimensional (2D) infrared excitation spectrum calculations, Wu et al. identified that specific THz frequency represented an optimal excitation modulation mode, resulting in DNA melting (*9*). Zhao et al. reported that the application of 48.2 THz significantly boosted the permeability of the sodium channel by a factor of 33.6. This enhancement was attributed to the transfer of absorbed terahertz photon energy by carboxylate groups in the filter region to sodium ions, facilitating their diffusion and reducing the energy barrier for crossing the channel. These studies primarily relied on molecular dynamics simulation technology. Experimental validations of the functions of THz-IR waves on the nervous system have also been conducted. Chen's group utilized in vivo single-cell electrophysiology and two-photon Ca^{2+} imaging technology to reveal that THz-IR stimulation activated cortical neurons in the brain without introducing any exogenous genes. Mice receiving THz-IR stimulation targeted at the auditory cortex during a listening task exhibited approximately a 50% faster learning speed compared to control mice (*10*). Shu's group employed cell patch-clamp technology to record mouse neocortical pyramidal cells and found that THz-IR light exerted non-thermal, long-distance modulatory effects on ion channel activity and neuronal signaling, subsequently influencing animal behavior responses to weak and strong stimuli (*11*). Specific THz waves can effectively recover cognitive functions in Posttraumatic Stress Disorder (*12*) and THz irradiation alters telomerase activity and telomere length and, thereby, exerts long-term effects on double-stranded DNA, ultimately inhibiting the proliferation of tumor cells (*13*). In our research group, we found THz wave can be effective method on tumor imaging *(14,15)*, and have also explored the bio-effects on the nervous system. We discovered that a frequency of approximately 35 THz could serve as a physical, efficient, non-thermal denaturation method, delaying the fibrotic process by 20% for Aβ oligomers in vitro, potentially offering a novel therapeutic strategy for Alzheimer's disease (*16*). Additionally, based on the resonant frequency of the Kv channel, we found that a frequency of around 36 THz could enhance voltage-gated K^+ conductance, reducing the excitability of pyramidal neurons in the anterior cingulate cortex and alleviating the neuropathic pain model in SNI mice (*17*).

4 Discussion

After reviewing all of these references, we have observed that THz-IR technology plays a significant role in bio-information transformation and the regulation of nervous system function, potentially offering valuable applications in neuroscience. However, to date, the modulation of terahertz infrared within the brain nervous system remains invasive, requiring the opening of the skull to intervene in specific brain areas. Finding a non-invasive approach is crucial for future clinical applications. Given the advancements in new THz source and metasurface technology (*18–23*), we are exploring the possibility of enhancing the field effect of the relevant brain regions to potentially develop a brain-computer interface. This avenue represents an ongoing focus of our current research efforts.

References

1. Liu, G., et al.: Myelin sheath as a dielectric waveguide for signal propagation in the mid-infrared to terahertz spectral range. Adv. Funct. Mater. **29**, 1807862 (2019)
2. Rinholm, J.E., Bergersen, L.H.: The wrap that feeds neurons. Nature **487**, 435–436 (2012)
3. Fields, R.D.: Myelin—more than insulation. Science **344**, 264–266 (2014)
4. Xiang, Z., et al.: A primary model of THz and far-infrared signal generation and con-duction in neuron systems based on the hypothesis of the ordered phase of water mole-cules on the neuron surface. Sci. Bull. **65**, 308 (2020)
5. Zhang, Z., et al.: Dielectric dispersion characteristics of the phospholipid bilayer with sub-nanometer resolution from terahertz to mid-infrared. Front. Bioeng. Biotechnol. **10**, 984880 (2022)
6. Guo, L., et al.: Electromagnetic characteristics of in vivo nerve fibers at the terahertz-far-infrared band. Front. Bioengin. Biotech. **10**, 1055232 (2022)
7. Xiang, Z., et al.: A new viewpoint and model of neural signal generation and transmission:signal transmission on unmyelinated neurons. Nano Res. **14**(3), 590–600 (2021)
8. Li, Y., et al.: Terahertz wave enhances permeability of the voltage-gated calcium channel. J. Am. Chem. Soc. **143**(11), 4311–4318 (2021)

9. Wu, K., et al.: Terahertz wave accelerates DNA unwinding: a molecular dynamics simulation study. J. Phys. Chem. Lett. **11**(17), 7002–7008 (2020)
10. Zhang, J., et al.: Non-invasive, opsin-free mid-infrared modulation activates cortical neurons and accelerates associative learning. Nature comm. **12**, 2730 (2021)
11. Liu, X., et al.: Nonthermal and reversible control of neuronal signaling and behavior by midinfrared stimulation. Proc. Natl. Acad. Sci. **118**(10), e2015685118 (2021)
12. Yu, Y., et al.: Terahertz photons improve cognitive functions in posttraumatic stress disorder. Research **6**, 0278 (2023)
13. Yin, J., et al.: Terahertz photons inhibit cancer cells long term by suppressing nano telomerase activity. ACS Nano **18**, 4796–4810 (2024)
14. Peng, W., et al.: Grade classification of human glioma using a convolutional neural network based on mid-infrared spectroscopy mapping. J. Biophotonics **15**(4), 202100313 (2022)
15. Peng, W., et al.: Identification of hepatocellular carcinoma and paracancerous tissue based on the peak area of FTIR microspectroscopy. Anal. Methods **14**(6), 3115–3124 (2022)
16. Peng, W., et al.: High-frequency terahertz waves disrupt Alzheimer's β-amyloid fibril formation. eLight **3**, 18 (2023)
17. Peng, W., et al.: High frequency terahertz stimulation alleviates neuropathic pain by inhibiting the pyramidal neuron activity in the anterior cingulate cortex of mice. bioRxiv, 2024.2003. 2006.583763 (2024)
18. Chang, C., et al.: High-gain thompson-scattering x-ray free-electron laser by time-synchronic laterally tilted optical wave. Phys. Rev. Lett. **110**(6), 064802 (2013)
19. Zhang, Q.L., et al.: Ultrahigh-flux water nanopumps generated by asymmetric terahertz absorption. Phys. Rev. Lett. (2024, accepted)
20. Lou, J., *et al.*: Calibration-free, high-precision, and robust terahertz ultrafast metasurfaces for monitoring gastric cancers. Proc. Natl. Acad. Sci. **119**(43), e2209218119 (2022)
21. Zhang, Z., et al.: Advanced terahertz refractive sensing and fingerprint recognition through metasurface-excited surface waves. Adv. Mater. 2308453 (2024)
22. Wang, P., et al.: An ultra-sensitive metasurface biosensor for instant cancer detection based on terahertz spectra. Nano Res. **16**, 7304–7311 (2023)
23. Wang, R., et al.: Ultrasensitive terahertz biodetection enabled by quasi-BIC-based metasensors. Small **19**, 2301165 (2023)

Temperature Variation with Time During the Irradiation with Terahertz Waves

Jing Ma and Zihua Song(✉)

Innovation Laboratory of Terahertz Biophysics, National Innovation Institute of Defense Technology, Beijing 100071, China
szh2016@mail.ustc.edu.cn

Abstract. The development and application of terahertz (THz) in biophysics has opened up new possibilities for studying neurons and circuits in *vivo* and in *vitro*. During THz wave transmission, heat is generated at the tip of the fiber due to the continuous supply of light sources, and it is an open question whether this affects neurons and circuits. In order to better exclude the interference generated during this process, we observed the temperature change over time at a distance of 200 μm between the tip of the optical fiber and the temperature probe. The temperature at the tip of the fiber tends to increase, with a maximum temperature of 0.59 °C, less than 1 °C higher than the initial temperature. Therefore, the THz wave has little effect on the temperature, and the effect of temperature on neurons and circuits can be ruled out, providing a basis for further related experiments.

Keywords: Terahertz · Temperature · Optical fiber

1 Introduction

The abundance of terahertz (THz) electromagnetic information in biological systems can be detected and characterized so that it can be used to diagnose the health status of organisms [1–4]. Optically tunable silicon-based THz hypersurfaces successfully identified colon cancer cells in normal, adenoma and cancerous states and were able to ensure the accuracy of the sensor performance [5]. A highly accurate platform for the detection of gastric cancer cells has been realized based on an in-house optical pumped THz probe (OPTP) spectroscopy system that completely eliminates typical measurement errors [6]. A THz supersensor based on electric and ring dipole coupling enables fast, non-destructive, label-free identification and highly sensitive quantitative detection of two types of Escherichia coli and Staphylococcus aureus [7]. Meanwhile, the interaction of THz waves with organisms may be able to cure diseases in the future. High-frequency THz waves of 34.88 THz can delay the fibrillation process of Aβ amyloid, which causes neurodegenerative diseases, by 80%, disrupting the β-sheet structure, breaking the dense conformation and enabling interventional therapies [8]. By improving the THz sources [9, 10], irradiation of THz wave at a specific wavelength was able to inhibit the of cancer cell metastasis by 50% in a metastatic xenograft mouse model, splenic injection of THz modulation (THM)-based cancer cells reduced liver metastasis of HCT-116 cancer cells by 60% [11–13].

C. Chang et al. (Eds.): YTHZ 2024, SPPHY 400, pp. 238–241, 2024.
https://doi.org/10.1007/978-981-97-3786-4_48

Based on the existing studies, THz waves irradiation have shown good results on cells, and if it is to be applied to in *vivo* animal experiments, the process requires the intervention of optical fiber. The optical fiber delivers the THz waves to the target region of the brain in a way that is superficially similar to fiber-based optogenetic techniques, but instead of introducing exogenous genes into the brain, it is only necessary to select the appropriate wavelength to achieve the relevant studies [14–26]. Then, it is still unclear whether THz waves generate heat during transmission and how the temperature changes, and thus whether it affects cells, animal tissues or organs during experiments. Therefore, we have studied the temperature of the optical fiber at a certain distance to eliminate or reduce the interference when performing in *vivo* animal experiments.

2 Materials and Methods

We used the MIRcat-QT-2400-PX-B broadband quantum cascade laser, which operates in pulsed light mode with an effective wavelength range of 5.5–11 μm and a maximum average power of 40 mW for excellent beam quality, wavelength fidelity, and fast continuous scanning.

In order to transmit the THz wave through the fiber, we connected the fiber to a quantum cascade laser and adjusted the pump current of the laser so that the output power at the tip of the fiber was 7.5 mW, and the THz wavelength was adjusted to be 8.7 μm. An optical fiber with an inner diameter of 400 μm was selected and placed at a distance of 200 μm from the temperature sensitive probe.

3 Results and Discussion

To investigate the effect of the THz wave transmitted through the fiber on the temperature over distance, and measured at the measuring point closest to the tip of the fiber (Fig. 1). At 0–2.5 min, the initial temperature was recorded several times and the average value was calculated. At 2.5 min–13 min, a specific THz wave was applied and the temperature change was recorded throughout the whole process, and after 13 min, the THz wave was removed and the temperature change was recorded. Using the temperature probe, we find that the THz wave transmitted through the optical fiber has an effect on the temperature over time (Fig. 2 and 3). At 2.5 min when the THz wave was started to be emitted, the temperature fluctuated and showed an increasing trend, increasing by 0.30 °C from the initial temperature. Then, during the continuous supply of the THz wave, the temperature increases to about 0.59 °C, and stabilizes in the period of 5.5–10 min, with small changes in temperature, fluctuating within this temperature value. During the recording process after 10 min, the temperature increased slightly, probably due to the continuous delivery of the THz wave, which led to the superposition of the temperature. However, after the THz wave was withdrawn, the temperature immediately showed a decreasing trend, indicating that the temperature of the optical fiber does not have a persistent effect on the experimental samples.

We found that the maximum value of the temperature change during the 2.5–13 min of THz wave delivery was 0.59 °C, which is a relatively small change. However, moderate heating of visible light (up to 2 °C) does not alter or inhibit neuronal activity in different regions of the brain, so temperature changes have little effect on cells and animal experimental samples [14].

In conclusion, the temperature generated by 8.7 μm THz wave during fiber optic transmission varies with the duration of irradiation. 7.5 mW of 8.7 μm THz wave irradiation produces smaller temperature variations, which have a low effect on the experimental samples and are within the tolerable range. Therefore, 7.5 mW of 8.7 μm THz wave can be one of the candidate parameters for subsequently conducting cellular and in *vivo* experiments.

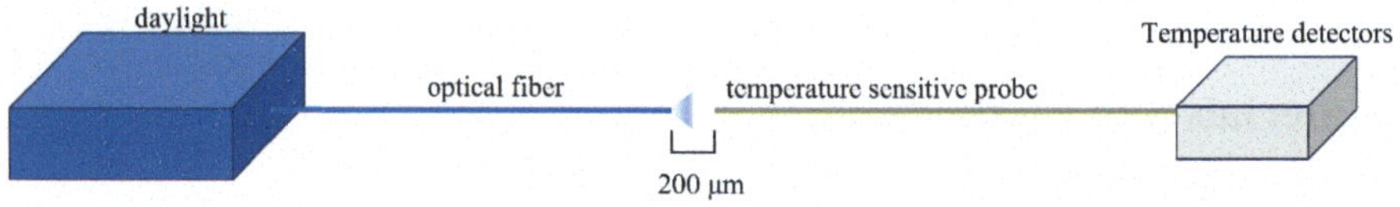

Fig. 1. Schematic diagram of the system for temperature measurement during THz wave irradiation.

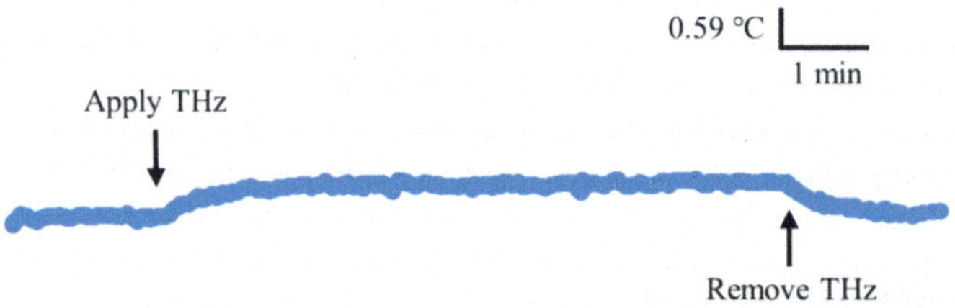

Fig. 2. Schematic diagram of temperature changes before, during and after THz wave irradiation.

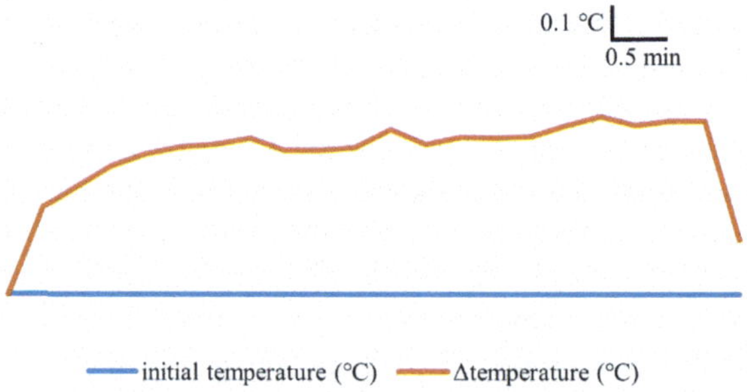

Fig. 3. Temperature changes in THz wave irradiation relative to the pre-irradiation period.

References

1. Liu, G.Z., et al:. Myelin sheath as a dielectric waveguide for signal propagation in the mid-infrared to terahertz spectral range. Adv. Funct. Mater. **29**(7) (2019)

2. Wang, P., et al.: Progress on cutting-edge infrared-terahertz biophysics. IEEE Trans. Microw. Theory Tech. **70**(11), 5117–5140 (2022)
3. Xiang, Z., et al.: A primary model of THz and far-infrared signal generation and con-duction in neuron systems based on the hypothesis of the ordered phase of water mole-cules on the neuron surface I: signal characteristics. Sci. Bull. **65**(4), 308–317 (2020)
4. Zhang, Z., et al.: Advanced terahertz refractive sensing and fingerprint recognition through metasurface-excited surface waves. Adv. Mater. e2308453 (2024)
5. Jiao, Y,, et al.: Photoactive terahertz metasurfaces for ultrafast switchable sensing of colorectal cells. Mater Horiz. **28;9**(12), 2984–2992 (2022)
6. Lou, J., et al.: Calibration-free, high-precision, and robust terahertz ultrafast metasurfaces for monitoring gastric cancers. Proc. Natl. Acad. Sci. USA **119**(43), e2209218119 (2022)
7. Ma, Z., et al.: Identification and quantitative detection of two pathogenic bacteria based on a terahertz metasensor. Nanoscale **15**(2), 515–521 (2023)
8. Chang, C., et al.: High-gain Thompson-Scattering X-ray free-electron laser by time-synchronic laterally tilted optical wave. Phys. Rev. Lett. **110**, 064802 (2013)
9. Gao, M., et al.: Enhanced and frequency-tunable terahertz emission from two crossed air-plasmas. Phys. Rev. Res. **5**, 023091 (2023)
10. Peng, W., et al.: High-frequency terahertz waves disrupt Alzheimer's β-amyloid fibril formation. eLight **3**(1) (2023)
11. Sun, L., et al.: Inhibition of cancer cell migration and glycolysis by terahertz wave modulation via altered chromatin accessibility. Research (2022)
12. Yin, J., et al.: Terahertz photons inhibit cancer cells long term by suppressing nano telomerase activity. ACS Nano **18**(4796), 2024 (2024)
13. Liu, X., et al.: Nonthermal and reversible control of neuronal signaling and behavior by midinfrared stimulation. Proc. Natl. Acad. Sci. USA **118**(10), e2015685118 (2021)
14. Zhang, J., et al.: Non-invasive, opsin-free mid-infrared modulation activates cortical neurons and accelerates associative learning. Nat. Commun. **12**(1), 2730 (2021)
15. Yu, Y., et al.: Terahertz photons improve cognitive functions in posttraumatic stress disorder. Research **6**, 0278 (2023)
16. Tan, X., et al.: Neuromodulation of chemical synaptic transmission driven by THz photons. Research, 0010 (2022)
17. Li, Y., et al.: Terahertz wave enhances permeability of the voltage-gated calcium channel. J. Am. Chem. Soc. **143**(11), 4311–4318 (2021)
18. Wu, K., et al.: Terahertz wave accelerates DNA unwinding: a molecular dynamics simulation study. J. Phys. Chem. Lett. **11**(17), 7002–7008 (2020)
19. Tan, X., et al.: Minimal-invasive enhancement of auditory perception by terahertz wave modulation. Nano Res. **15**(6), 5235–5244 (2022)
20. Li, Y., et al.: Physicochemical insights on terahertz wave diminished side effects of drugs from slow dissociation. ACS Nano **16**(5), 8419–8426 (2022)
21. Xiao, T., et al.: Sensory input dependent gain modulation of the optokinetic nystagmus by mid-infrared stimulation in pigeons. Elife **12**, e78729 (2023)
22. Yuan, Y., et al.: Midinfrared radiation accelerates DNA unwinding. ACS photonics. https://doi.org/10.1021/acsphotonics.3c01548
23. Zhang, Q.L., et al.: Ultrahigh-flux water nanopumps generated by asymmetric terahertz absorption. Phys. Rev. Lett. (2024, accepted)
24. Zhu, Z., et al.: Tunable surface wettability via terahertz electrowave controlled vicinal subnanoscale water layer. Nano Lett. **24**(10), 3243–3248 (2024)
25. Zhu, Z., et al.: Characteristic fingerprint spectrum of neurotransmitter norepinephrine with broadband terahertz time-domain spectroscopy. Analyst **144**(8), 2504–2510 (2019)
26. Tang, C., et al.: Integrating terahertz metamaterial and water nanodroplets for ultrasensitive detection of amyloid β aggregates in liquids. Sens. Actuators B. Chem. **329**, 129113 (2021)

Possible Terahertz Signals in Visual Processes

Shiyu Gu[1] and Yangmei Li[2](✉)

[1] College of Optical-Electrical and Computer Engineering, University of Shanghai for Science and Technology, Shanghai 200093, China
[2] Innovation Laboratory of Terahertz Biophysics, National Innovation Institute of Defense Technology, Beijing 100071, China
meiyang@mail.ustc.edu.cn

Abstract. Rhodopsin consisting of an opsin and its ligand (i.e., retinal) is critical to vision formation. When the rhodopsin is stimulated by light, the retinal undergoes a structural isomerization that activates the process of visual signaling. This photoisomerization involves the interaction between the opsin and retinal. Studies have shown that terahertz (THz) spectroscopy is a potential non-invasive means to identify the structure and rotation-vibration of biomolecules. More specifically, the rotational spectra of opsin and retinal in the low frequency THz band might indicate their dynamic conformations. Herein, we calculated the rotational inertia and spectra of the opsin and retinal. We found that both the opsin and retinal exhibit main absorption peaks in the low frequency THz band, which correspond to the collective rotational modes of the molecules. Thus, we suggest that the interactions and structural changes between opsin and retinal can be characterized by low-frequency THz spectra. The study of the rotational spectra of opsin and retinal is important for understanding the function of the visual system, disease mechanisms, and the development of drug therapy. It sheds light on the structure and interactions of these molecules and advances the field of vision science and medicine.

Keywords: Moment of inertia · Rotational spectrum · Terahertz

1 Introduction

Vision is one of the crucial ways through which humans perceive the world, and the functioning of the visual system relies on the interaction between two key molecules: opsin and retinal. [1, 2] As the primary components of photoreceptors, opsin play a critical role in visual signal transduction through their binding with retinal. [3–5] To gain a deeper understanding of this biological process, researchers have been devoted to exploring the structure, dynamics, and interaction mechanisms between opsin and retinal.

Rotational spectroscopy and rotational inertia are essential tools and concepts in the study of molecular rotational behaviors. Rotational spectroscopy reveals the characteristics and behaviors of internal molecular rotations by measuring the frequency distribution of radiation absorbed or emitted during rotational motion. On the other hand, rotational

C. Chang et al. (Eds.): YTHZ 2024, SPPHY 400, pp. 242–246, 2024.
https://doi.org/10.1007/978-981-97-3786-4_49

inertia describes the intrinsic rotational properties of molecules and is closely related to their structure and mass distribution. By studying the rotational spectroscopy and rotational inertia of opsin and retinal, we can gain insights into the rotational characteristics of these two molecules and their mutual interaction modes.

Simultaneously, THz waves, as a unique frequency range in the electromagnetic spectrum between infrared and microwave, possess distinctive physical and chemical properties. THz waves are characterized by high penetrating power, non-ionization, harmlessness and specific sensitivity to many substances, and thus have a wide potential for applications in the fields of material structure, chemical reactions, biomolecular dynamics and nanomaterials. [6–12] The study of THz waves is important for understanding the fundamental properties and interactions of matter. [13–20] It can reveal information about molecular and crystal structures, the electronic and vibrational dynamics of solid materials, and the structural and dynamical behavior of biomolecules. In recent years, researchers have discovered a certain correlation between the rotational behavior of opsin and retinal and THz waves. This has sparked interest in the biological significance and potential applications of the interaction between opsin, retinal, and THz waves.

In this work, we calculate the rotational spectra and rotational inertia between opsin and retinal, and investigate and interpret their relationship. Through a deep understanding of these relationships, we will uncover the essential mechanisms of opsin and retinal in the visual system, providing robust support for further biological research and applications.

2 Methods

The initial crystal structure of rhodopsin protein comes from the RCSB data bank (PDB entry: 1U19) [21]. We used Charmm-GUI's [22] Membrane Builder to construct rhodopsin between bilayer lipid membranes (normal along the Z-axis) immersed in a KCl solution with an ionic concentration of 0.15 mol/liter to simulate the real environment (Fig. 1). Through molecular dynamics simulations, the rotational spectra of the retinal and opsin were calculated based on an analytical method of angular momentum.

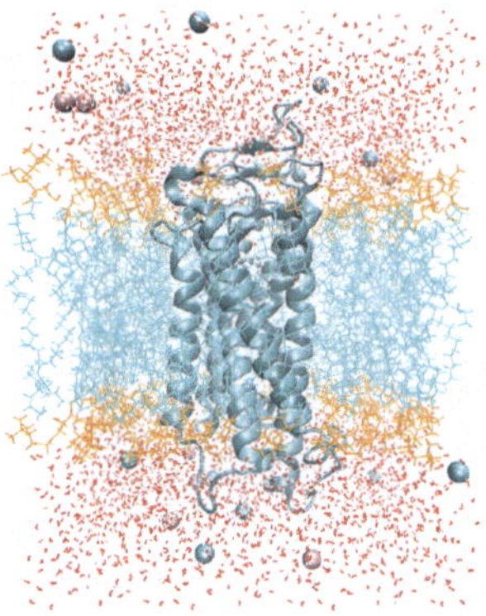

Fig. 1. Model of the rhodopsin system.

3 Results and Discussion

Rhodopsin is due to the composition of opsin and retinal (Fig. 2a). We calculated the rotational inertia of opsin and retinal separately and found that the rotational inertia of opsin is much larger than that of retinal (Fig. 2b). From the rotation spectra of opsin and retinal, we observed abundant THz information in both structures (Fig. 2c and d). The characteristic frequency of opsin is at 0.5 THz. However, the characteristic frequency of retinal is at 5.8 THz. The difference in the spectra suggests that the frequency of the light absorbed or emitted by the molecules is related to the rotational energy level spacing. For larger molecules, the rotational energy level spacing is smaller and the frequency of light absorbed or emitted is lower. For smaller molecules, the rotational intervals are larger and the frequency of light absorbed or emitted is higher.

The rotational energy level spacing is the difference in the energy of the molecules as they rotate. Larger molecules have smaller rotational energy level intervals in their rotation spectra because they have a larger rotational moment of inertia, while smaller molecules have larger rotational energy level intervals because they have a smaller rotational moment of inertia. Therefore, the difference between the rotation spectra of large and small molecules is mainly reflected in the magnitude of the rotational moment of inertia and the rotational energy level spacing, resulting in different frequencies of absorbed or emitted light.

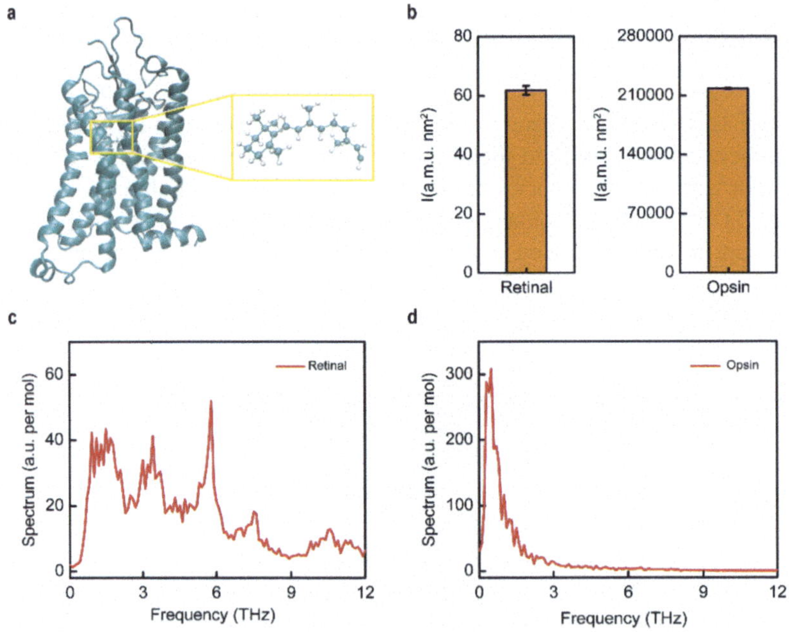

Fig. 2. Structures of retinal and opsin (a). Moment of inertia of retinal and opsin (b) Spectra of rotation of retinal (c) and opsin (d).

By studying the rotation spectra of opsin and retinal, we can obtain the characteristic frequency distributions of their absorbed or emitted light during rotation. Measurement and analysis of these frequencies can reveal the rotational behavior within the molecule, including the speed of rotation, the orientation of the rotation axis, and the symmetry of the rotation pattern. In addition, rotation spectroscopy can provide information about the structure and conformation of a molecule, since the structure of a molecule directly affects its rotational moment of inertia and the size of its rotational energy level spacing.

References

1. Hubbard, R., Kropf, A.: The action of light on rhodopsin. Proc. Natl. Acad. Sci. U. S. A. **44**(2), 130–139 (1958)
2. Palczewski, K.: G Protein–coupled receptor rhodopsin. Annu. Rev. Biochem. **75**, 743–767 (2006)
3. Smith, S.O.: Mechanism of activation of the visual receptor rhodopsin. Annu. Rev. Biophys. **52**, 301–317 (2023)
4. Athanasiou, D., et al.: The molecular and cellular basis of rhodopsin retinitis pigmentosa reveals potential strategies for therapy. Prog. Retinal Eye Res. **62**, 1–23 (2018)
5. Jones, B.W., Kondo, M., Terasaki, H., et al.: Retinal remodeling. Jpn. J. Ophthalmol. **2012**(56), 289–306 (2012)
6. Liu, G., et al.: Myelin sheath as a dielectric waveguide for signal propagation in the mid-infrared to terahertz spectral range. Adv. Funct. Mater. **29**(7), 1807862 (2019)
7. Li, Y., Chang, C., Zhu, Z., et al.: Terahertz wave enhances permeability of the voltage-gated calcium channel. J. Am. Chem. Soc. **143**(11), 4311–4318 (2021)
8. Liu, X., et al.: Nonthermal and reversible control of neuronal signaling and behavior by midinfrared stimulation. Proc. Natl. Acad. Sci. U.S.A. **118**, e2015685118 (2021)
9. Wu, K., Qi, C., Zhu, Z., et al.: Terahertz wave accelerates DNA unwinding: a molecular dynamics simulation study. J. Phys. Chem. Lett. **11**(17), 7002–7008 (2020)
10. Zhu, Z., et al.: Transition to a superpermeation phase of confined water induced by a terahertz electromagnetic wave. J. Phys. Chem. Lett. **11**(1), 256–262 (2019)
11. Li, Y., Zhu, Z., Sun, L., et al.: Physicochemical insights on terahertz wave diminished side effects of drugs from slow dissociation. ACS Nano **16**(5), 8419–8426 (2022)
12. Zhu, Z., Chen, C., Chang, C., Song, B.: Terahertz-light induced structural transition and superpermeation of confined monolayer water. ACS Photonics **8**(3), 781–786 (2020)
13. Peng, W., Zhu, Z., Lou, J., Chen, K., Yuanming, W., Chang, C.: High-frequency terahertz waves disrupt Alzheimer's β-amyloid fibril formation. eLight **3**(1) (2023)
14. Xiang, Z., et al.: A primary model of THz and far-infrared signal generation and conduction in neuron systems based on the hypothesis of the ordered phase of water molecules on the neuron surface. Sci. Bull. **65**, 308 (2020)
15. Yin, J., Wu, K., Yu, Y., et al.: Terahertz photons inhibit cancer cells long term by suppressing Nano telomerase activity. ACS Nano **18**(6), 4796–4910 (2024)
16. Sun, T., Zhu, Z.: Light resonantly enhances the permeability of functionalized membranes. J. Membr. Sci. **662**, 121026 (2022)
17. Lou, J., et al.: Calibration-free, high-precision, and robust terahertz ultrafast metasurfaces for monitoring gastric cancers. Proc. Natl. Acad. Sci. **119**(43), e2209218119 (2022)
18. Zhu, Z., Zhu, J., Chang, C., et al.: Tunable surface wettability via Terahertz electrowave controlled vicinal subnanoscale water layer. Nano Lett. **24**(10), 3243–3248 (2024)
19. Zhang, J., et al.: Non-invasive, opsin-free mid-infrared modulation activates cortical neurons and accelerates associative learning. Nature Comm. **12**(1), 2730 (2021)

20. Zhang, Q.L., et al.: Ultrahigh-flux water nanopumps generated by asymmetric terahertz absorption. Phys. Rev. Lett. **132**(18), 184003 (2024)
21. Okada, T., Sugihara, M., Bondar, A.-N., Elstner, M., Entel, P., Buss, V.: The retinal conformation and its environment in rhodopsin in light of a new 2.2 Å crystal structure. J. Molecular Biol. **342**(2), 571–583 (2004)
22. Lee, J., et al.: CHARMM-GUI membrane builder for complex biological membrane simulations with glycolipids and lipoglycans. J. Chem. Theory Comput.Comput. **15**, 775 (2018)

Effects of Terahertz Wave Irradiation on Transcriptomics of the Mouse Cerebral Cortex

Pan Liu and Zihua Song(✉)

Innovation Laboratory of Terahertz Biophysics, National Innovation Institute of Defense Technology, Beijing 100071, China
szh2016@mail.ustc.edu.cn

Abstract. As terahertz (THz) biological research continues to accelerate, the role of THz technology as a potential therapeutic tool in the biomedical field is gradually expanding. Recent studies have confirmed that the absorption peaks of many key substances in biological activities, such as neurotransmitters and nucleic acids, are located in the THz frequency range, making it a suitable technology to achieve conformational changes and functional adjustments. In this study, the effects of THz wave irradiation on the gene level of mice were investigated by RNA-Seq, irradiating the cerebral cortex of mice at 34.5 THz based on quantum cascade laser (QCL). The differential gene enrichment was obtained after GO and KEGG analysis, and a total of 189 differential genes were obtained by screening conditions. The results showed that after 34.5 THz irradiation, the mouse cerebral cortex showed differential gene expression, 170 genes were up-regulated and 19 genes were down-regulated, involving multiple signaling pathways such as TNF, which has potential applications in the treatment of brain function and brain diseases.

Keywords: Terahertz Wave Irradiation · Gene · RNA-Seq

1 Introduction

Terahertz technology is emerging in areas such as biomedicine [1, 2], and high-sensitivity biosensing [3–8]. THz wave photon will not destroy the biomolecules due to low energy without ionization, and thus the beam intensity is low when it can be safely acted on the human body [2].

In recent years, the researches on THz sources [9, 10] and biophysics [11] have progressed rapidly, and significant progress has been made in the treatment of diseases [12–15], gene regulation [16–18], ion channels [19, 20], and protein spectrum characterization [21]. THz waves irradiation has been shown to produce non-thermal and reversible control of neuronal signaling and behavior [22, 23]. THz waves irradiation activates mouse cortical neurons and accelerates associative learning in mice [24, 25].

In this study, we attempt to validate the effects of THz wave irradiation on gene transcriptomics in the cerebral cortex through mouse experiments, providing experimental support for the beneficial role of THz wave irradiation as an adjunctive physical therapy in biomedical fields.

C. Chang et al. (Eds.): YTHZ 2024, SPPHY 400, pp. 247–250, 2024.
https://doi.org/10.1007/978-981-97-3786-4_50

2 Materials and Methods

C57BL/6J male mice (SPF(Beijing) biotechnology Co., Ltd, China), aged 7–8 weeks, were used for all experimental studies. Mice were group-housed five per cage; all mice had ad libitum access to food and water. They were housed at a stable temperature (23–25°C) with a 12-h light/dark cycle (lights on from 7:00 a.m. to 7:00 p.m.)

The laser used in the in *vivo* experiments was a 34.5 THz frequency quantum cascade laser (QCL). The optical fiber was a standard PIR AgCl:AgBr polycrystalline fiber. The metal cannula was purchased from RWD Life Science Co., Ltd (RWD, 62049). After the cannula was fixed, the optical fiber was then inserted into the metal tube to reach the designated brain area. Irradiation was performed for 10 min on each side.

In this study, RNA-Seq was used to screen control and irradiated groups for differential genes. Mouse brain slices were sequenced by RNA-Seq to screen for differentially expressed genes, and the differentially expressed genes were biologically annotated by Gene Ontology (GO) database and Kyoto Encyclopedia of Genes and Genomess (KEGG) database to explore the effects of THz irradiation on gene transcription and related signaling pathways in the cerebral cortex.

3 Results

The heat map and volcano map of differentially expressed genes showed the distribution of differentially expressed genes in the cerebral cortex of control and THz-irradiated mice. 189 differentially expressed genes were found in control and irradiated mice, among which the number of up-regulated genes was 170 (90%), and the number of down-regulated genes was 19 (10%). Compared with control mice, the expression of *Hba-a2, Hbb-bs, Actb, Hba-a1, and Ftl1* were significantly upregulated, and the expression of *Nlrc5, Cd22, Clec7a, Gpr65, and Runx* were significantly downregulated in the THz-irradiated group (Fig. 1).

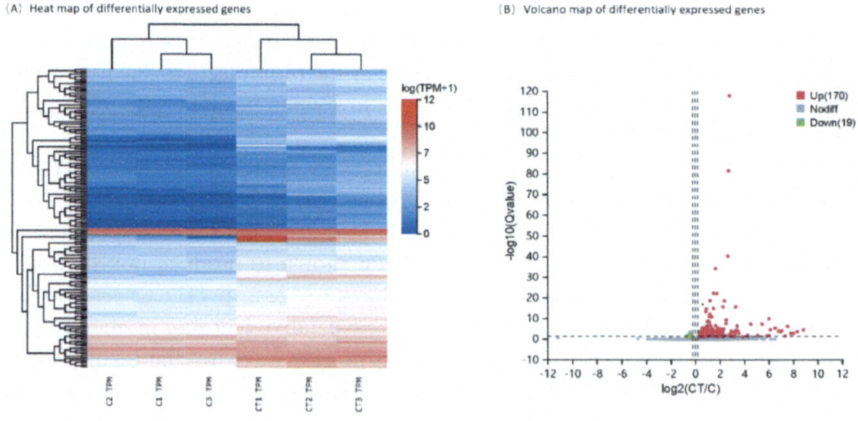

Fig. 1. Differentially expressed genes of the control and THz-irradiated groups

GO analysis showed that the differentially expressed genes were involved in 1994 biological processes, 296 cellular components and 429 molecular functions. Among them, BP includes inflammatory response, immune response, immune system process, etc.; CC includes extramal side of plasma membrane, cell surface, extracellular space, etc.; MF includes protein binding, protein-containing complex binding, TAP complex binding and other functions. CC includes extramal side of plasma membrane, cell surface, extracellular space, etc.; MF includes protein binding, protein-containing complex binding, TAP complex binding and other functions.

KEGG results showed a total of 221 signaling pathways were activated in THz-irradiated mice compared to the control group, including: Malaria, TNF signaling pathway, Leishmaniasis, Rheumatoid arthritis, Human cytomegalovirus infection, Fuild shear stress and atherosclerosis, Osteoclast differentiation, Phagosome, Tuberculosis, IL-17 signaling pathway and other pathways (Fig. 2).

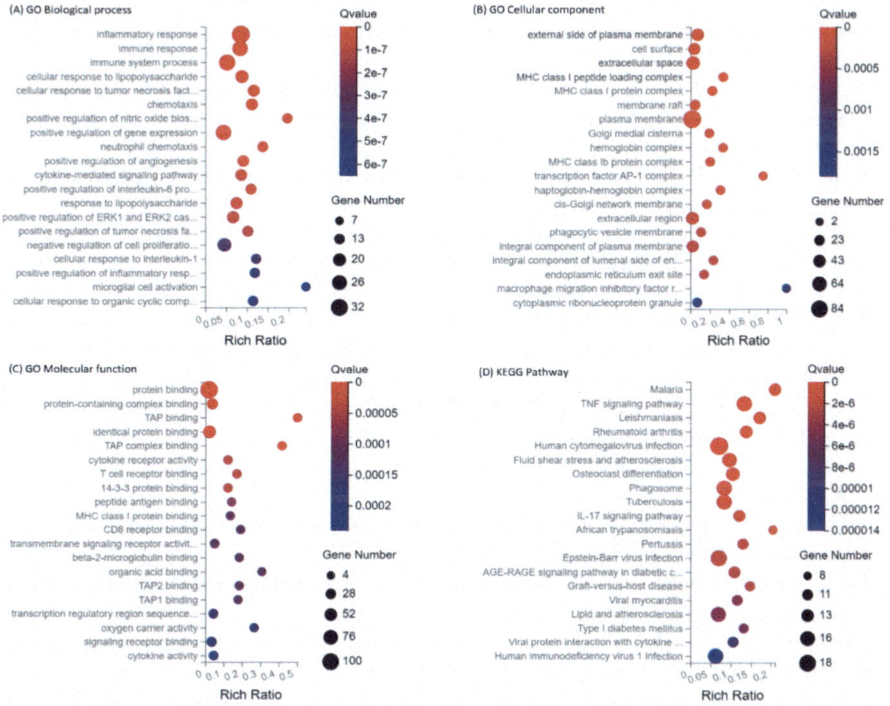

Fig. 2. GO and KEGG enrichment bubble map of differentially expressed genes in control and THz-irradiated groups

References

1. Conti Nibali, V., Havenith, M.: New insights into the role of water in biological function: Studying solvated biomolecules using terahertz absorption spectroscopy in conjunction with molecular dynamics simulations. JACS **136**(37), 12800–12807 (2014)

2. Wilmink, G.J., Grundt, J.E.: Invited review article: current state of research on biological effects of terahertz radiation. JIMTW **32**(10), 1074–1122 (2011)
3. Wang, P., et al.: Progress on cutting-edge infrared-terahertz biophysics. IEEE Trans. Microw. Theory Tech. **70**(11), 5117–5140 (2022)
4. Liu, G., et al.: Myelin sheath as a dielectric waveguide for signal propagation in the mid-infrared to terahertz spectral range. Adv. Funct. Mater. **29**(7), 1807862 (2019)
5. Zhang, Z., et al.: Advanced Terahertz refractive sensing and fingerprint recognition through metasurface-excited surface waves. Adv. Mater., 2308453 (2024)
6. Lou, J., et al.: Calibration-free, high-precision, and robust terahertz ultrafast metasurfaces for monitoring gastric cancers. Proc. Natl. Acad. Sci. **119**(43), e2209218119 (2022)
7. Lou, J., et al.: Optically Controlled Ultrafast Terahertz Metadevices with Ultralow Pump Threshold. Small **17**(44), e2104275 (2021)
8. Wang, R., et al.: Ultrasensitive terahertz biodetection enabled by quasi-BIC-based metasensors. Small, 2301165 (2023)
9. Chang, C., et al.: High-gain Thompson-scattering X-Ray free-electron laser by time-synchronic laterally tilted optical wave. Phys. Rev. Lett. **110**, 064802 (2013)
10. Gao, M., et al.: Enhanced and frequency-tunable terahertz emission from two crossed air-plasmas. Phys. Rev. Res. **5**, 023091 (2023)
11. Xiang, Z., et al.: A primary model of THz and far-infrared signal generation and conduction in neuron systems based on the hypothesis of the ordered phase of water molecules on the neuron surface. Sci. Bull. **65**, 308 (2020)
12. Peng, W., et al.: High-frequency terahertz waves disrupt Alzheimer's β-amyloid fibril formation. eLight 3,18 (2023)
13. Sun, L., et al.: Inhibition of cancer cell migration and glycolysis by terahertz wave modulation via altered chromatin accessibility. Research, 9860679 (2022)
14. Yin, J., et al.: Terahertz photons inhibit cancer cells long term by suppressing nano telomerase. ACS Nano **18**(6), 4796–4910 (2024)
15. Li, Y., et al.: Physicochemical insights on terahertz wave diminished side effects of drugs from slow dissociation. ACS Nano **16**(5), 8419–8426 (2022)
16. Wu, K., et al.: Terahertz wave accelerates DNA unwinding: a molecular dynamics simulation study. J. Phys. Chem. Lett. **11**(17), 7002–7008 (2020)
17. Zhang, C., et al.: Driving DNA origami assembly with a terahertz wave. Nano Lett. **22**, 468–475 (2022)
18. Yuan, Y., et al.: Midinfrared radiation accelerates DNA unwinding. ACS Photonics (2024). https://doi.org/10.1021/acsphotonics.3c01548
19. Li, Y., et al.: Terahertz wave enhances permeability of the voltage-gated calcium channel. J. Am. Chem. Soc. **143**(11), 4311–4318 (2021)
20. Zhang, Q.L., et al.: Ultrahigh-flux water nanopumps generated by asymmetric terahertz absorption. Phys. Rev. Lett. (2024)
21. Zhu, Z., et al.: Characteristic fingerprint spectrum of neurotransmitter norepinephrine with broadband terahertz time-domain spectroscopy. Analyst **144**(8), 2504–2510 (2019)
22. Liu, X., et al.: Nonthermal and reversible control of neuronal signaling and behavior by mid-infrared stimulation. Proc. Natl. Acad. Sci. **118**(10), e2015685118 (2021)
23. Tan, X., Zhong, Y., Li, R.: Neuromodulation of chemical synaptic transmission driven by THz photons. Research **2022**, 0010 (2022)
24. Zhang, et al.: Non-invasive, opsin-free mid-infrared modulation activates cortical neurons and accelerates associative learning. Nat. Commun. **12**(1), 2730 (2021)
25. Yu, Y., et al.: Terahertz photons improve cognitive functions in posttraumatic stress disorder. Research **6**, 0278 (2023)

Effects of Terahertz Wave Irradiation on ATP Synthase

Tingrong Zhang[1,2], Mengyao Lei[1], Zhuoyang Lu[1(✉)], and Jiangang Long[1(✉)]

[1] School of Life Science, Xi'an Jiaotong University, Xi'an, China
{luzhuoyang,jglong}@xjtu.edu.cn

[2] Innovation Laboratory of Terahertz Biophysics, National Innovation Institute of Defense Technology, Beijing, China

Abstract. Energy metabolism is a necessary process to maintain life processes. Observing the cellular energy metabolism process reveals the pivotal roles of ATP in cellular life. Compared to plants, animals primarily rely on ATP synthase to generate the required ATP without the ability of plants to utilise light energy. The absorption characteristics of biological macromolecules are mainly located in the terahertz frequency. When terahertz waves act on ion channels, effective regulation is achieved. As the site of ATP generation, ATP synthase produces ATP through proton translocation coupling, and it may also be regulated by terahertz waves. With the help of molecular dynamics, the characteristic absorption spectrum of the proton translocation site of yeast ATP synthase was roughly calculated. Combined with the existing theoretical results, the 34.8 THz photons may affect the ATP production by ATP synthase. Then, the ATP content with and without 34.9 THz was examined. The increase in the intracellular ATP content existed after the effect of 34.9 THz. Therefore, this special light energy can be utilized by animals to support the ongoing processes of energy metabolism in possible.

Keywords: Terahertz · ATP synthase · molecular dynamics

1 Introduction

Adenosine triphosphate (ATP) is the central molecule in cellular energy metabolism. ATP is primarily synthesized through the electron transfer chain on the ATP synthase in inner mitochondrial membrane. It is a type of adenosine triphosphatase (ATPase), specifically F-Type F_OF_1-ATP synthase. F_1 consists of three catalytic sites surrounding a central rotor that rotates to synthesis ATP. F_O is a transmembrane proton turbine with a core consisting of the subunit a and the c_{10} ring, which is a ring of 10 subunit c (see Fig. 1A) [1]. The other subunits make composition of the rotor domain. The process of proton transport through the F_O region which is reversible. This process of transport is mainly through the residue glutamate on c_{10} ring and subunit a. The proton binds to the -COO^- group of the glutamate at residue 223 of subunit a and enters the channel of mitochondrial intermembrane space. Then proton transfers to the residue of glutamate in the c_{10} ring, rotating through about eight glutamates, reaching and relocating to the -COO^- group of the glutamate at residue 162 of subunit a. Finally, proton reaches channel of the mitochondrial matrix [2]. The process completes.

C. Chang et al. (Eds.): YTHZ 2024, SPPHY 400, pp. 251–255, 2024.
https://doi.org/10.1007/978-981-97-3786-4_51

The physical field of nerve information should be in the high frequency electromagnetic field from terahertz (THz) to infrared, named as generalized THz wave. Biological macromolecules usually exhibit collective vibrations in the electro-magnetic field in the infrared to THz spectral range [3–8]. It was found that myelin sheath of neuro may be a dielectric waveguide for signal propagation in THz range [9]. With the development of THz sources [10, 11], in neurodegenerative diseases treatment, 34.88 THz could delay the amyloid-β (Aβ) oligomers fibrotic process by 80% as an efficient technique in *vitro* without side effect [12]. For drug delivery, THz wave modulation promoted the dissociation between the selective D2R antagonists and one risperidone ligand to diminish the side effects [13]. In the field of cancer treatment, 83.6 THz could inhibit cancer cell migration by 50% in *vivo* and *vitro* with reducing glycolysis by 60% [14]. According to the calculation of Wu, 44 THz can unwind the double helix of the DNA molecule at low temperature by breaking the hydrogen bongs connection through MD [15]. It could also be found that 34.5 THz photons could improve the cognitive function of animal with posttraumatic stress disorder through affecting the glutamate receptors [16]. With the technology of THz wavelength modulation, mice exhibit a faster learning speed than control mice during the auditory associative learning task [17]. THz photons can also long-termly inhibit cancer cells through telomerase activity suppression as a non-drug and non-invasive strategy [18]. Various researches have suggested that THz may affect the ion transport. The action potential waveforms were shortened by accelerating the repolarization due to the increase in voltage-gated K^+ channel without changing of temperature [19]. Li had founded that Ca^{2+} permeated more rapidly under 42.55 THz field and the -COO- and -C = O group of the selectivity filter made contribution to the permeation [20]. The similar results can be found in other ion channels.

2 Methods

2.1 MD Simulation

The model was based on the open-state crystal structure of ATP synthase of yeast from RCSB data bank (PDB ID: 6CP6). Some deletions were applied to this structure to low the computational effort. The remaining portion which was sufficient to allow proton transport contained the c_{10} ring and subunit a in the F_O region (see Fig. 1B). Meanwhile, to make sure the coherence of each subunit c, only the residues 1–74 were preserved. The simulation system was constructed with the membrane builder of the CHARMM-GUI web server [21]. The system contained the rest region of ATP synthase embedded between bilayer palmitoyl-oleoyl-phosphatidylcholine (POPC) membranes dissolved in KCl solution with the ionic concentration of 0.15 mol/L. MD simulations were performed using GROMACS (version 2019.4 & 2022.5) software with a CHARMM36m force field [21, 22]. The simulation process was based on the running package provided by CHARMM-GUI with modifications. After the MD production, the results regarding the absorption spectrum of the -COOH and -COO^- groups in residue of subunit c were calculated by *gmx velacc* command.

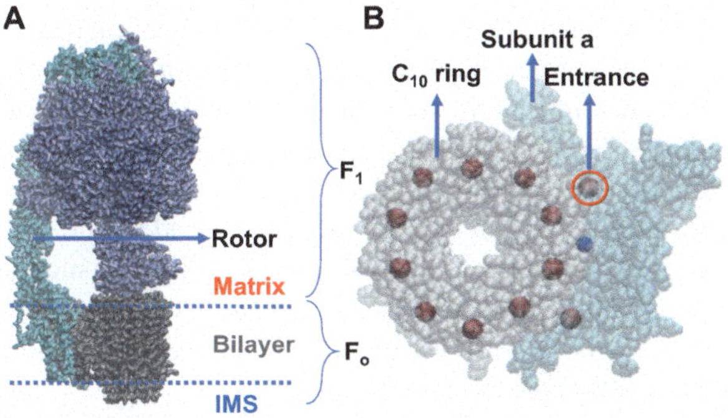

Fig. 1. A. The structure of ATP synthase in yeast. The color cyan, gray and iceblue represented the domain of rotor, F_O and F_1, relatively. B. Top view of F_O region under modification. The glutamate is in red and the arginine is in dark blue.

2.2 Detection of ATP Content

ATP content was measured by fluorescence analysis using ATP-Red 1 dye. Dye concentration and staining time referred to Wang [23]. Co-staining with ATP-Red 1 and Hoechst 33342 was performed when HT22 cells were in healthy growth condition. Irradiation was performed from the bottom of the petri dish after staining was completed, with half of the bottom covered with blackout paper as a control. After irradiation, fluorescence microscopy was carried out immediately, and the red fluorescence intensity was photographed and counted in a single cell. The fluorescence intensity was obtained from Image J and counted through graphpad prism software.

3 Results and Discussion

Researches had suggested that the -COO^- and -C = O group could be affected by the THz wavelength. During the process of proton transport by ATP synthase, the entrance residue glutamate which owns one -COO- group on its side chain on the subunit a is crucial for proton translocation. Therefore, the infrared absorption spectra of the -COO^- group of entrance at different protonation states were calculated. Based on the research of Song, the energy released during the hydrolysis of phosphoanhydride-bond of dNTPs is in form of THz photons. While the residue is at -COO^- state, an absorption peak appeared at 34.9 THz which is closely to reported results. However, the absorption peak at 34.9 THz disappeared after changing to the protonated state -COOH (see Fig. 2A). This suggests that the 34.9 THz frequency is absorbed by the side-chain -COO^- group of glutamate at the proton transport site instead of -COOH group, which may affect proton binding and consequently influence the proton transport process.

Based on the calculation results, ATP content was detected in the HT22 cell line with and without exposure to THz waves. With the ATP-Red Live fluorescent probe and Hoechst 33342 live cell staining for co-staining, the changes in ATP content in cells were observed after 34.9 THz irradiation. The ATP-Red Live probe specifically stained ATP, appearing as red. The higher fluorescence intensity indicates higher ATP content at the position. Hoechst 33342 stained the cell nuclei for cell localization (see Fig. 2B, C). After statistically analyzing the average fluorescence intensity of single cells (see Fig. 2D), it was found that the red fluorescence intensity of single cells significantly increased after 34.9 THz irradiation at, implying an increase in cellular ATP content. This result suggested that cellular ATP content changed after exposure to 34.9 THz. However, it had not been yet certain whether this change was directly caused by the impact of 34.9 THz on ATP synthase. So that, further detection of related indicators such as the enzymatic activity of ATP synthase must be necessary. Meanwhile, the structure of ATP synthase used in MD needs to be replaced with the mammals' along with the potential difference model to mimic the real structure in *vivo*.

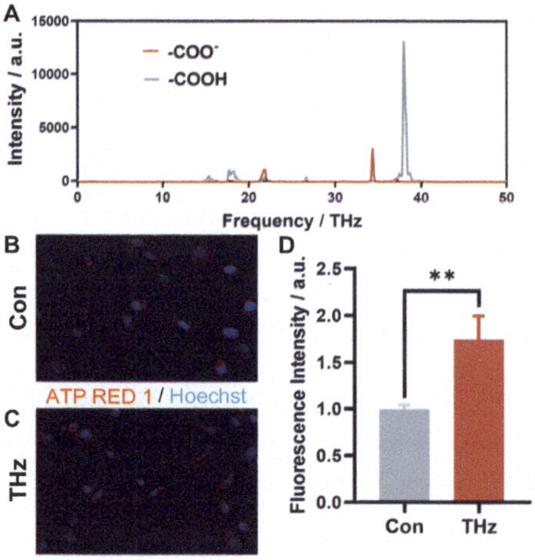

Fig. 2. A. The infrared absorption spectra of the -COO$^-$/-COOH group of the entrance residue glutamate. B, C. The fluorescence images of HT22 cells with and without THz waves. D. Quantification of the fluorescence intensity of the ATP content in HT22 cells. The results are presented as the mean ± SE.

References

1. Srivastava, A.P., et al.: High-resolution cryo EM analysis of the yeast ATP synthesis in a live memory. Science **360**(6389) (2018)
2. Guo, H., et al.: Atomic model for the dimeric Fo region of mitochondrial ATP synthase. Science **358**(6365), 936–940 (2017)
3. Wang, P., et al.: Progress on cutting-edge infrared-terahertz biophysics. IEEE Trans. Microw. Theory Tech. **70**(11), 5117–5140 (2022)
4. Xiang, Z., et al.: A primary model of THz and far-infrared signal generation and conduction in neuron systems based on the hypothesis of the ordered phase of water molecules on the neuron surface. Sci. Bull. **65**(4), 308–317 (2020)
5. Zhang, Z., et al.: Advanced terahertz refractive sensing and fingerprint recognition through metasurface-excited surface waves. Adv. Mater. 2308453 (2024)

6. Lou, J., et al.: Calibration-free, high-precision, and robust terahertz ultrafast metasurfaces for monitoring gastric cancers. Proc. Natl. Acad. Sci. U.S.A. **119**(43) (2022)
7. Wang, R., et al.: Electric Fano resonance-based terahertz metasensors. Nanoscale **13**(44), 18467–18472 (2021)
8. Lou, J., et al.: Optically controlled ultrafast terahertz metadevices with ultralow pump threshold. Small **17**(44), 2104275 (2021)
9. Liu, G., et al.: Myelin sheath as a dielectric waveguide for signal propagation in the mid-infrared to terahertz spectral range. Adv. Funct. Mater. **29**(7), 1807862 (2019)
10. Gao, M., et al.: Enhanced and frequency-tunable terahertz emission from two crossed air-plasmas. Phys. Rev. Res. **5**(2), 023091 (2023)
11. Chang, C., et al.: High-gain thompson-scattering x-ray free-electron laser by time-synchronic laterally tilted optical wave. Phys. Rev. Lett. **110**(6), 064802 (2013)
12. Peng, W., et al.: High-frequency terahertz waves disrupt Alzheimer's β-amyloid fibril formation. eLight 3, 18 (2023)
13. Li, Y., et al.: Physicochemical insights on terahertz wave diminished side effects of drugs from slow dissociation. ACS Nano **16**(5), 8419–8426 (2022)
14. Sun, L., et al.: Inhibition of cancer cell migration and glycolysis by terahertz wave modulation via altered chromatin accessibility. Research **2022**, 9860679 (2022)
15. Wu, K., et al.: Terahertz wave accelerates DNA unwinding: a molecular dynamics simulation study. J. Phys. Chem. Lett. **11**(17), 7002–7008 (2020)
16. Yu, Y., et al.: Terahertz photons improve cognitive functions in posttraumatic stress disorder. Research **6**, 0278 (2023)
17. Zhang, J., et al.: Non-invasive, opsin-free mid-infrared modulation activates cortical neurons and accelerates associative learning. Nat. Commun. **12**(1), 2730 (2021)
18. Yin, J., et al.: Terahertz photons inhibit cancer cells long term by suppressing nano telomerase activity. ACS Nano **18**(6), 4796–4910 (2024)
19. Liu, X., et al.: Nonthermal and reversible control of neuronal signaling and behavior by mid-infrared stimulation. Proc. Natl. Acad. Sci. U.S.A. **118**(10), e2015685118 (2021)
20. Li, Y., et al.: Terahertz wave enhances permeability of the voltage-gated calcium channel. J. Am. Chem. Soc. **143**(11), 4311–4318 (2021)
21. Jo, S., et al.: CHARMM-GUI: A web-based graphical user interface for CHARMM. J. Comput. Chem. **29**(11), 1859–1865 (2008)
22. Hess, B., et al.: GROMACS 4: algorithms for highly efficient, load-balanced, and scalable molecular simulation. J. Chem. Theory Comput. **4**(3), 435–447 (2008)
23. Wang, L., et al.: A multisite-binding switchable fluorescent probe for monitoring mitochondrial ATP level fluctuation in live cells. Angew. Chem. Int. Ed. Engl.. Chem. Int. Ed. Engl. **55**(5), 1773–1776 (2016)

The Molecular Mechanism of Neural Signal Transduction Based on Terahertz Irradiation

Qiwei Wang(✉)

ShanghaiTech University, Shanghai 201210, China
wangqw2022@shanghaitech.edu.cn

Abstract. With the development of terahertz sources and detection devices, an increasing number of terahertz irradiation biological effects have been discovered. Terahertz irradiation possesses non-ionizing and non-invasive characteristics, and due to the abundance of terahertz information in biological organisms and the known characteristic frequencies of many biomolecules and chemical bonds, it can serve as an effective means of modulating neural activity. This paper, focusing on resonance energy injection as the core framework, mainly introduces the most representative research findings on the biological effects and mechanisms of terahertz irradiation on calcium ion channels, potassium ion channels, and neurotransmitter-specific receptor binding, and provides an outlook on the future development of terahertz neurobiology.

Keywords: Terahertz resonance energy injection · Ion channel irradiation effect · Neurotransmitter and receptor irradiation effect

1 Introduction

As research into the electromagnetic spectrum and its energy transitions in interactions with matter deepens, exploration has progressed from the excitation of electrons in atoms and molecules by ultraviolet and visible light radiation, to vibrational transitions excited by near-infrared, and further to those excited by mid to far-infrared. Accompanied by the input of terahertz wave resonance energy at characteristic frequencies, unique terahertz irradiation biological effects are produced [1–5]. This paper briefly introduces significant advancements in the modulation of neural biological activity functions by terahertz electromagnetic irradiation.

The transmembrane transport of ions through ion channels plays a crucial role in biological neural activity and function. Calcium ions flow into cells through ion channels, and the change in concentration triggers the release of neurotransmitters at chemical synapses for communication (see Fig. 1). Moreover, the transmembrane transport of potassium and sodium ions through ion channels can generate and transmit action potentials across the cell membrane, facilitating the communication of electrical signals within neurons [6–8]. Therefore, studying how to artificially regulate the opening and closing of ion channels, the specific binding of neurotransmitters to post-synaptic receptors, and other processes is of great importance.

C. Chang et al. (Eds.): YTHZ 2024, SPPHY 400, pp. 256–260, 2024.
https://doi.org/10.1007/978-981-97-3786-4_52

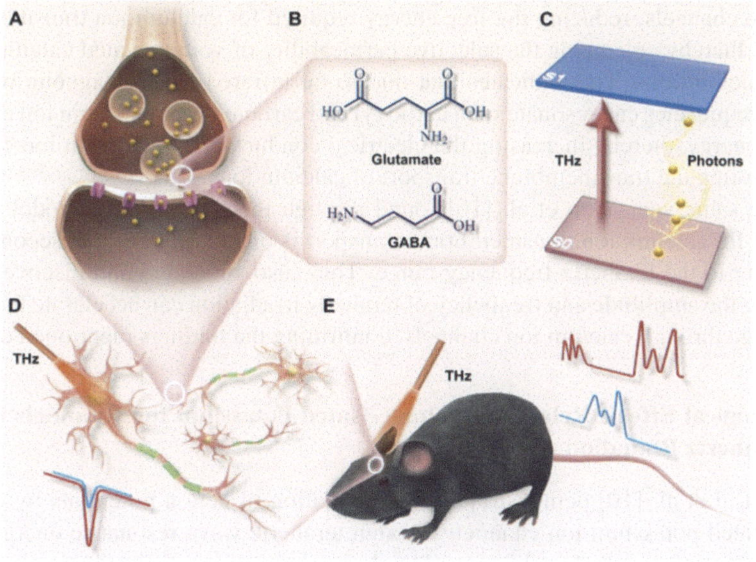

Fig. 1. (A) Synaptic connection diagram. (B) Molecular structure of glutamate and GABA. (C) THz can stimulate theneurotransmitter to release photons. (D) THz photons enhance postsynaptic currents in vitro. (E) THz photons alter neuronal excitability in vivo (in the actual experiment, the brain of the mice was cut open). The blue and redtraces in (D) and (E) represent synaptic and Ca^{2+} signals from recorded neurons [8] (Open Access Journal).

2 Biological Effects Related to Ion Channels Under Terahertz Radiation

The intrinsic vibration modes of the surfaces of nerve cells fall within the terahertz to far-infrared frequency range [2]. Terahertz irradiation can activate or inhibit ion channel proteins in cell membranes [9]. More profoundly, it can regulate neural electrical activity through the transmembrane transport of ions. Additionally, terahertz electromagnetic irradiation can generate hydrophilic pores that can close on phospholipid membranes [10], reducing the free energy required for the transmembrane transport of drug molecules and ions, thus producing a regulatory effect on biological neural activity similar to that described above.

2.1 Biological Effects Related to Voltage-Gated Calcium Ion Channels Under Terahertz Radiation

In 2021, Li et al. [11] for the first time studied the resonance energy input into calcium ion channel proteins at the atomic level through molecular dynamics simulation. The study constructed a molecular dynamics physical model of voltage-gated calcium channels and simulated the irradiation of terahertz waves at the resonance frequencies of 42.55 THz for carboxyl groups and 51.61 THz for carbonyl groups within the channels. The results showed that irradiation could change the free energy distribution of calcium ions within the channels, reducing the free energy required for calcium ion transmembrane transport, thereby enhancing the selective permeability of voltage-gated calcium channels to calcium ions. This indicates that mid to far-infrared electromagnetic waves of specific frequencies can resonate with carboxyl and carbonyl groups within ion channels to inject energy, thereby increasing the electrical conductance of calcium ion channels and promoting the transmembrane transport of calcium ions.

In the same year, Guo et al. [16] found through their constructed model that the timescale for calcium ion transmembrane transport is on the order of picoseconds, corresponding to the terahertz frequency range. They also simulated and discovered that increasing the amplitude and frequency of terahertz irradiation can accelerate the rate of ion passage through calcium ion channels, confirming the findings mentioned earlier.

2.2 Biological Effects Related to Voltage-Gated Potassium Ion Channels Under Terahertz Radiation

In 2021, Liu et al. [10] demonstrated the modulation of action potentials by opening voltage-gated potassium ion channels through terahertz wave resonance energy injection, marking an important application of terahertz electromagnetic irradiation in the field of brain science. The experiment utilized a quantum cascade laser combined with patch-clamp technique to record action potential waveforms in pyramidal cells from mouse prefrontal cortex slices. The study employed electromagnetic waves at the carbonyl resonance frequency of 53.53 THz (5.6 μm mid-infrared wave) for irradiation, observing an increase in potassium ion flow through voltage-gated potassium channels under non-thermal conditions, leading to narrower action potential waveforms in pyramidal cells, thus modulating neural signals. The regulatory effect disappeared when the 53.53 THz electromagnetic wave irradiation was removed and reappeared upon re-irradiation, indicating its reversibility. For the startle response to UV light-induced C-bend in zebrafish, irradiation at a wavelength of 5.6 μm suppressed the startle response to weak UV light stimulation while enhancing the response to strong UV light stimulation. These phenomena indicate that terahertz electromagnetic irradiation in this frequency range can modulate neural signals in a non-thermal, non-contact, reversible, and non-destructive manner.

In the same year, Zhang et al. [11] by performing craniotomy on mice and stimulating intact thin cranial bones, irradiated mouse brain neurons with 53.53 THz electromagnetic waves (5.6 μm mid-infrared wave) of the same frequency mentioned above. They found that the 53.53 THz electromagnetic radiation induced discharge in cortical neurons controlling auditory functions, significantly improving the mice's auditory-related learning ability by 50%, with the effect disappearing after the stimulus was removed, indicating its reversibility. Compared to the popular optogenetic techniques, this method does not require the introduction of exogenous genes, such as specific photosensitive protein genes, to activate neurons and is non-invasive. Therefore, terahertz genetics might serve as an effective means to enhance human cognitive and learning abilities or to treat neurological diseases.

3 Biological Effects Related to Neurotransmitters Under Terahertz Radiation

For the communication of neurotransmitter release at chemical synapses, in 2019, Zhi et al. [16] discovered that terahertz irradiation at 0.16 THz (50 mW, 6 min) / 0.17 THz (10 mW, 60 min) can reduce neurotransmitter levels and alter protein expression levels in primary neurons and neuron-like cells in rat brain regions. In 2023, Yu et al. [12] used electromagnetic waves at a frequency of 34.5 THz for deep brain stimulation in a non-thermal and non-damaging manner, targeting the CA3 subregion of the hippocampus in mice. They studied the effects of terahertz irradiation on neurotransmitters and their specific receptor molecules. Through various behavioral experiments, including novel object recognition, they found that 34.5 THz radiation could improve learning and memory abilities in PTSD animals. The mechanism behind this effect was a reduction of about 50% in the glutamate neurotransmitter content in the hippocampal CA3 subregion after terahertz irradiation, and an increase of about 50% in the phosphorylation level of the NMDA receptor subunit NR2B, with the total NR2B level also increasing by about 40%. Molecular dynamics simulations revealed that the binding free energy between glutamate receptors and glutamate decreased by about 31.98 kJ/mol under irradiation at this frequency. These results indicate that terahertz irradiation promotes the binding between glutamate and NR2B receptors, thereby increasing the phosphorylation level of NR2B. This experiment provides a new method to enhance the binding ability between glutamate and NR2B subunits through terahertz irradiation, offering new insights into artificially regulating the binding of neurotransmitters to specific receptors, which may also solve the problem of Alzheimer disease [13].

4 Summary and Prospect

Given the known oscillation frequencies of a vast number of biological molecules and chemical bonds [14, 15], the biological effects of terahertz waves hold broad research prospects. Further investigation into the mechanisms is needed, such as establishing a standard database for biological effects induced by terahertz irradiation at different characteristic frequencies, developing tunable terahertz sources, advancing research on quantum effects within organisms under terahertz irradiation, and exploring the microscopic physical mechanisms behind brain cognition and information transmission at the atomic and molecular levels.

References

1. Malmivuo, J., Plonsey, R.: Bioelectromagnetism: Principles and Applications of Bioelectric and Biomagnetic Fields. Oxford University Press, Oxford (1995)
2. Zapara, T.A., Treskova, S.P., Ratushniak, A.S.: Effect of antioxidants on the interaction of terahertz (submillimeter) laser radiation and neuronal membrane. J. Surf. Invest. X-ray, Synchrotron Neutron Tech. **9**(5), 869–871 (2015). https://doi.org/10.1134/S1027451015050195

3. Wang, P., et al.: Progress on cutting-edge infrared-terahertz biophysics. IEEE Trans. Microw. Theory Tech. **70**(11), 5117–5140 (2022)
4. Wu, K., et al.: Terahertz wave accelerates DNA unwinding: a molecular dynamics simulation study. J. Phys. Chem. Lett. **11**(17), 7002–7008 (2020)
5. Yin, J., et al.: Terahertz photons inhibit cancer cells long term by suppressing nano telomerase activity. ACS Nano **18**, 4796 (2024)
6. Xiang, Z., et al.: A primary model of THz and far-infrared signal generation and conduction in neuron systems based on the hypothesis of the ordered phase of water molecules on the neuron surface. Sci. Bull. **65**, 308 (2020)
7. Liu, G., et al.: Myelin sheath as a dielectric waveguide for signal propagation in the mid-infrared to terahertz spectral range. Adv. Funct. Mater. **29**(7), 1807862 (2019)
8. Tan, X., et al.: Neuromodulation of chemical sysnaptic transmission driven by THz photons. Research **6**, 1–16 (2022)
9. Li, Y., et al.: Terahertz wave enhances permeability of the voltage-gated calcium channel. J. Am. Chem. Soc. **143**, 4311 (2021)
10. Liu, X., et al.: Nonthermal and reversible control of neuronal signaling and behavior by midinfrared stimulation. Proc. Natl. Acad. Sci. U. S. A. **118**, e2015685118 (2021)
11. Zhang, J., et al.: Non-invasive, opsin-free mid-infrared modulation activates cortical neurons and accelerates associative learning. Nat. Commun. **12**, 2730 (2021)
12. Yu, Y., et al.: Terahertz photons improve cognitive functions in posttraumatic stress disorder. Research **6**, 0278 (2023)
13. Peng, W., et al.: High frequency terahertz waves disrupt Alzheimer's amyloid fibril formation. eLight **3**, 18 (2023)
14. Zhang, Z., et al.: Advanced terahertz refractive sensing and fingerprint recognition through metasurface-excited surface waves. Adv. Mater., 2308453 (2024)
15. Lou, J., et al.: Calibration-free, high-precision, and robust terahertz ultrafast metasurfaces for monitoring gastric cancers. Proc. Natl. Acad. Sci. 119, 43, e2209218119 (2022)
16. Guo, L., et al.: Numerical simulation of calcium ion channel regulation by unipolar terahertz field. In: 2021 46th IRMMW-THz, Chengdu, China, pp. 1-2 (2021)
17. Zhi, T.S., et al.: Exposure effects of terahertz waves on primary neurons and neuron-like cells under nonthermal conditions. Biomed. Environ. Sci. **32**(10), 739–754 (2019)

Terahertz Photons Up-Regulated Acetylcholine by Increasing the Expression of Choline Acetyltransferase

Yun Yu[1,2](✉)

[1] School of Life Science and Technology, Xi'an Jiaotong University, Xi'an 710049, China
18811591255@163.com

[2] Innovation Laboratory of Terahertz Biophysics, National Innovation Institute of Defense Technology, Beijing 100071, China

Abstract. Acetylcholine (ACH) is a crucial excitatory neurotransmitter in the nervous system and plays an important role in cognitive function. Terahertz (THz) photons have been shown to enhance neuronal excitability and improve cognitive function in animals. In this study, we examined the impact of terahertz photons on the expression of the excitatory neurotransmitter acetylcholine (ACH). Our findings revealed that exposure to terahertz photons led to increased ACH levels in the hippocampus of animals with PTSD, while showing no effect on the medial prefrontal cortex (mPFC). Subsequent investigations demonstrated that THz photon irradiation elevated the expression of choline acetyltransferase (ChAT), a pivotal protein involved in acetylcholine synthesis, without affecting acetylcholinesterase (AChE), which is responsible for ACH degradation. These results suggest that the heightened ACH levels may be attributed to the increased expression of ChAT induced by terahertz photon irradiation. This study offers a potential mechanism for the impact of terahertz photons on the nervous system.

Keyword: Acetylcholine · THz photons · choline acetyltransferase

1 Introduction

The frequency domain for the response of living organisms consistently lies in the spectrum ranging from 0.1 to 100 terahertz (THz). Within this frequency range, THz photons exhibit numerous biological applications [1–3]. In tumor therapy, specific frequencies of THz photons can alter chromatin accessibility and telomerase activity in tumor cells, thereby inhibiting tumor cell migration and proliferation [4, 5]. Additionally, due to the characteristic frequency of the tumor cells, the recognition of different tumor cells can be achieved [6, 7]. The resonant absorption of biological macromolecules to THz photons can induce changes in protein structure, thereby affecting neurotransmitters [8–12] and ion channels, and further influencing the activity and function of the nervous system [13–15]. Other studies have shown that THz photons can also affect nucleic acid molecules, promoting DNA molecular unwinding and driving DNA origami [16–18]. All these studies have demonstrated that THz photons have reversible regulatory effects on organisms at both the protein and molecule levels [19–33].

C. Chang et al. (Eds.): YTHZ 2024, SPPHY 400, pp. 261–265, 2024.
https://doi.org/10.1007/978-981-97-3786-4_53

Recent studies have demonstrated that, with the development of novel THz sources and the emergence of innovative methods of THz stimulation [25, 26], THz photons have the potential to influence deep brain regions in animals. For instance, 34.5 THz photons have been found to enhance glutamate binding to NR2B, thereby improving cognitive function in animals with PTSD. This was accompanied by an increase in the relative expression of the NR2B receptor [27]. Furthermore, cortical irradiation with THz photons has been shown to enhance the learning ability of animals. Studies indicate that 53.6 THz photons can specifically induce neural discharge activity in the irradiated region, resulting in a 50% improvement in the animal's learning speed [28]. These findings suggest that THz photons can heighten the excitability of the nervous system by acting on neurotransmitters, subsequently influencing the cognitive function of animals.

As an excitatory neurotransmitter, acetylcholine (ACH) plays a crucial role in the central nervous system, regulating the encoding, consolidation, and retrieval of attention and memory [34]. It is catalyzed by choline acetyltransferase (ChAT), acts on acetylcholine receptors, and is hydrolyzed by acetylcholinesterase (AChE) [35]. Restoring the function of cholinergic neurons in the nervous system is of great significance for improving learning and memory abilities. In this study, 34.5 THz photons were used to irradiate the CA3 subarea in PTSD animals, and it was found that these photons increased the expression level of hippocampal acetylcholine. The expression of ChAT was increased without affecting the expression of AChE. Therefore, this study provides a novel insight into the effects of THz photons on the nervous system and enhances the excitability of the nervous system.

2 Results

2.1 THz Photons Incerased the Acetylcholine Levels in the Hippocampus of PTSD Animals

The animal model of PTSD was established following the method described by Yun *et al.* and was exposed to 34.5 THz photons. The power density of the THz wave was 44 W/cm^2, and the irradiation was targeted at the hippocampal CA3 subregion. The bilateral subregions were irradiated for 10 min over three days. Following irradiation, hippocampal CA3 tissue was isolated, homogenized, and the acetylcholine levels in the brain tissue were measured using ELISA kit. The results indicated a significant decrease in acetylcholine levels in the brain tissue of the model group, while exposure to THz waves led to a significant increase in acetylcholine levels in the brain tissue of the model group ($p < 0.05$). In the medial prefrontal cortex (mPFC), there were no significant changes in acetylcholine levels in the model and THz groups (Fig. 1).

2.2 THz Photons Irradiation Increased the Expression of ChAT

Acetylcholine levels are influenced by the enzymes acetylcholinesterase and acetyltransferase. Measuring the expression levels of these two enzymes can reveal the cause of changes in acetylcholine levels. Acetylcholinesterase and acetylcholinesterase, which

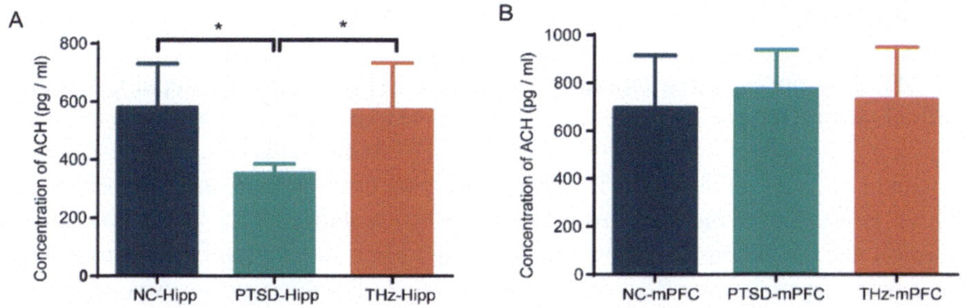

Fig. 1. *In vivo* experiments showed that 34.5 THz photons could elevate ACH levels in irradiated brain regions. A-B. The relative expression of ACH in different brain regions was measured by ELISA kit. NC represents the negative control group, PTSD represents the model group, THz represents the group treated with THz photons, Hipp represents the hippocampus, and mPFC represents the medial prefrontal cortex. Data are presented as individual values or expressed as the mean ± SD, with significance levels indicated as * $P < 0.05$.

synthesize acetylcholine by acetylation and degrade it by esterification, work together to control acetylcholine levels in the brain. We utilized the Western blot method to further detect the expression of these two enzymes. The results indicated that after irradiation, the expression of AChE did not change significantly, but the expression of ChAT increased significantly ($p < 0.05$). This result suggests that THz photons radiation increased the synthesis of acetylcholine without affecting its degradation (Fig. 2).

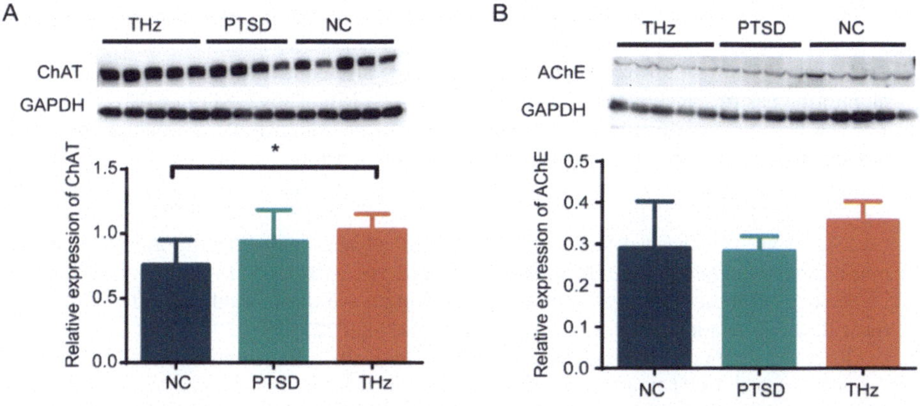

Fig. 2. *In vivo* experiments showed that THz photons could elevate the relative expression of ChAT in irradiated brain regions. A-B. The relative expression of ChAT and AChE in hippocampus. NC denotes the negative control group, PTSD represents the model group, THz indicates the THz photons treated group. Data are presented as individual values or expressed as the mean ± SD, with significance levels indicated as * $P < 0.05$.

3 Discussion

After 34.5 THz photons irradiation, the level of ACH in the hippocampal CA3 region of animals with PTSD was restored, and the relative expression of ChAT was increased, with no observed effect on the relative expression of AChE. This result suggests that the mechanism of ACH recovery by THz waves may be through increasing the expression of ChAT. However, the mechanisms underlying the regulation of ChAT by THz photons irradiation need to be further investigated. As an important excitatory neurotransmitter in the nervous system, the effect of THz waves on cholinergic neurons was explored in this study, providing novel insights and a possible mechanism for the impact of THz photons on the nervous system.

References

1. Liu, G., et al.: Myelin sheath as a dielectric waveguide for signal propagation in the mid-infrared to terahertz spectral range. Adv. Funct. Mater. **29**(7), 1807862 (2019)
2. Xiang, Z., et al.: A primary model of THz and far-infrared signal generation and conduction in neuron systems based on the hypothesis of the ordered phase of water molecules on the neuron surface I: signal characteristics. Sci. Bull. **65**(4), 308–317 (2020)
3. Wang, P., et al.: Progress on cutting-edge infrared-terahertz biophysics. IEEE Trans. Microwave Theory Tech. **70**(11), 5117–5140 (2022)
4. Sun, L., et al.: Inhibition of cancer cell migration and glycolysis by terahertz wave modulation via altered chromatin accessibility. Research **2022**, 9860679 (2022)
5. Yin, J., et al.: Terahertz photons inhibit cancer cells long term by suppressing nano telomerase activity. ACS Nano **18**(6), 4796–4910 (2024)
6. Zhang, Z., et al.: Advanced terahertz refractive sensing and fingerprint recognition through metasurface-excited surface waves. Adv. Mater., 2308453 (2024)
7. Lou, J., et al.: Calibration-free, high-precision, and robust terahertz ultrafast metasurfaces for monitoring gastric cancers. Proc. Natl. Acad. Sci. **42**(119), e2209218119 (2022)
8. Zhu, Z., et al.: Characteristic fingerprint spectrum of neurotransmitter norepinephrine with broadband terahertz time-domain spectroscopy. Analyst **144**(8), 2504 (2019)
9. Tan, X., et al.: Neuromodulation of chemical synaptic transmission driven by THz photons. Research **2022**, 0010 (2022)
10. Liu, X., et al.: Nonthermal and reversible control of neuronal signaling and behavior by midinfrared stimulation. Proc. Natl. Acad. Sci. **118**(10), e2015685118 (2021)
11. Li, Y., et al.: Physicochemical insights on terahertz wave diminished side effects of drugs from slow dissociation. ACS Nano **16**(5), 8419–8426 (2022)
12. Li, Y., et al.: Terahertz wave enhances permeability of the voltage-gated calcium channel. J. Am. Chem. Soc. **143**(11), 4311–4318 (2021)
13. Tan, X., et al.: Minimal-invasive enhancement of auditory perception by terahertz wave modulation. Nano Res. **15**(6), 5235–5244 (2022)
14. Tan, X., Gao, M., Chang, C.: A new means of energy supply driven by terahertz photons recovers related neural activity. iScience **26**(2), 105979 (2023). https://doi.org/10.1016/j.isci.2023.105979
15. Peng, W., Zhu, Z., Lou, J., Chen, K., Yuanming, W., Chang, C.: High-frequency terahertz waves disrupt Alzheimer's β-amyloid fibril formation. eLight **3**(1),(2023). https://doi.org/10.1186/s43593-023-00048-0
16. Wu, K., et al.: Terahertz wave accelerates DNA unwinding: a molecular dynamics simulation study. J. Phys. Chem. Lett. **11**(17), 7002–7008 (2020)

17. Zhang, C., et al.: Driving DNA origami assembly with a terahertz wave. Nano Lett. **22**(1), 468–475 (2022)
18. Yuan, Y., et al.: Midinfrared radiation accelerates DNA unwinding. ACS Photonics **11**(4), 1473–1479 (2024). https://doi.org/10.1021/acsphotonics.3c01548
19. Wang, P., et al.: An ultra-sensitive metasurface biosensor for instant cancer detection based on terahertz spectra. Nano Res. **16**, 7304–7311 (2023)
20. Tang, C., et al.: Integrating terahertz metamaterial and water nanodroplets for ultrasensitive detection of amyloid β aggregates in liquids. Sens. Actuator B. Chem. **329**, 129113 (2021)
21. Zhu, Z., et al.: Broadband Terahertz signatures and vibrations of dopamine. Analyst **145**(18), 6006–6013 (2020)
22. Xiao, T., et al.: Sensory input dependent gain modulation of the optokinetic nystagmus by mid-infrared stimulation in pigeons. Elife **12**, e78729 (2023)
23. Zhang, Q.L., et al.: Ultrahigh-flux water nanopumps generated by asymmetric terahertz absorption. Phys. Rev. Lett. (2024)
24. Zhu, Z., et al.: Tunable surface wettability via terahertz electrowave controlled vicinal subnanoscale water layer. Nano Lett. **24**(10), 3243–3248 (2024)
25. Chang, C., et al.: High-gain thompson-scattering X-Ray free-electron laser by time-synchronic laterally tilted optical wave. Phys. Rev. Lett. **110**, 064802 (2013)
26. Gao, M., et al.: Enhanced and frequency-tunable terahertz emission from two crossed air-plasmas. Phys. Rev. Res. **5**, 023091 (2023)
27. Yu, Y., et al.: Terahertz photons improve cognitive functions in posttraumatic stress disorder. Research **6**, 0278 (2023)
28. Zhang, J., et al.: Non-invasive, opsin-free mid-infrared modulation activates cortical neurons and accelerates associative learning. Nat. Commun. **12**(1), 2730 (2021)
29. Wang, R., et al.: Ultrasensitive terahertz biodetection enabled by Quasi-BIC-based metasensors. Small **19**, 2301165 (2023)
30. Jiao, Y., et al.: Photoactive terahertz metasurfaces for ultrafast switchable sensing of colorectal cells. Mater. Horiz. **9**, 2984–2992 (2022)
31. Sun, L., et al.: A pixelated frequency-agile metasurface for broadband terahertz molecular fingerprint sensing. Nanoscale **14**, 9681–9685 (2022)
32. Wang, R., et al.: Electric Fano resonance-based terahertz metasensors. Nanoscale **13**, 18467–18472 (2021)
33. Lou, J., et al.: Optically controlled ultrafast terahertz metadevices with ultralow pump threshold. Small **17**, 2104275 (2021)
34. Yann, S., et al.: Hippocampal acetylcholine modulates stress-related behaviors independent of specific cholinergic inputs. Mol. Psychiatry **27**(3), 1829–1838 (2022)
35. Miao, J., et al.: An optimized acetylcholine sensor for monitoring in vivo cholinergic activity. Nat. Methods **17**(11), 1139–1146 (2020)

Terahertz Wave Tunneling Through Subwavelength Rectangular Channel

Haomin Zeng(✉)

National Innovation Institute of Defense Technology, Beijing 100071, China
zenghm@whu.edu.cn

Abstract. Since the vibration and rotation spectrums of many biomolecules are in the terahertz band, the close relation between terahertz waves and many biological effects has been widely demonstrated. Whereas highly localized terahertz stimulation for subcellular study is challenging because the wavelength of terahertz waves is commonly much larger than the size of cells and molecules. Therefore we focus on the subwavelength wave tunneling effect in the terahertz band. The transmission-line model is investigated and then a specific wave tunneling waveguide aimed at 34THz is proposed. Finite element simulations are carried out for verification. The results show that the air-filled rectangular metallic tunneling waveguide can achieve a transmittance close to unity at the design wavelength with a bandwidth of about 0.06 THz. The transmittance is independent of the length of the tunneling channel, showing the ability to transmit confined terahertz waves over long distances.

Keywords: Terahertz biophysics · Subwavelength electromagnetic tunneling · Transmission-line model

1 Introduction

Terahertz waves have been shown functional in many disciplines such as biophysics [1], biosensors [2–5], and bioeffects [6–10]. They are believed to be the information carriers of the nervous system and are engaged in the whole process of signal generation, transmission, amplification, and storage [11]. Preliminary theoretical studies have shown that terahertz waves can be generated and transmitted in neurons, and the nodes of Ranvier can serve as a relay amplifier for information [12–17]. Some other experimental studies revealed the correlation between terahertz waves and neurotransmitters, ion channels, and neural activity [18–24]. Despite the inspiring advances that have been made, it is still very challenging to directly couple terahertz waves into neurons due to the wavelengths of terahertz waves are typically much larger than neuron diameters. Therefore subwavelength terahertz waveguide is necessary for further studies of the roles of terahertz waves in the nervous system.

The subwavelength electromagnetic wave tunneling was first proposed in the parallel plate waveguide filled with ε-near-zero (ENZ) materials [25] and soon demonstrated experimentally by two different groups [26, 27]. Further studies clarified the effect of

C. Chang et al. (Eds.): YTHZ 2024, SPPHY 400, pp. 266–270, 2024.
https://doi.org/10.1007/978-981-97-3786-4_54

finite conductivity of the metal walls and generalized the theoretical model to rectangular waveguides [28, 29]. The interests of these previous works are restricted to microwave bands and mainly studied wave tunneling between two regular waveguides. In this study, we investigate the possibility of squeezing terahertz waves through a subwavelength channel without another successive waveguide.

2 Transmission-Line Model and Simulation Results

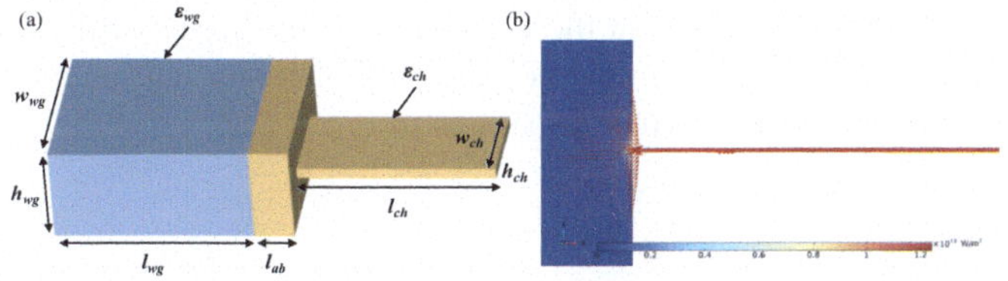

Fig. 1. (a) Structure of the whole tunneling waveguide. (b) Cross-sectional distribution of the Poynting vector.

Considering a rectangular waveguide followed by an abruption section and then a narrow wave tunneling channel, as depicted in Fig. 1. The metallic walls of the whole structure are assumed to be perfectly electrical conducting (PEC). The reflection coefficient at the entrance port of the channel can be expressed as [29]:

$$R = \frac{\eta_{ch} - \eta_{wg} + iB\eta_{ch}\eta_{wg}}{\eta_{ch} + \eta_{wg} - iB\eta_{ch}\eta_{wg}}, \tag{1}$$

where η is the normalized characteristic line impedance and B is the load susceptance of the abruption section. Their value can be calculated by the following formula:

$$\eta = h/w \cdot \sqrt{\mu/\varepsilon_{eff}}, \tag{2}$$

$$B = \frac{k_0 h_{ab}\sqrt{\varepsilon_{ch}}}{\pi}[2\ln\frac{h_{wg}^2 - h_{ch}^2}{4h_{wg}h_{ch}} + (\frac{h_{wg}}{h_{ch}} + \frac{h_{ch}}{h_{wg}})\ln\frac{h_{wg} + h_{ch}}{h_{wg} - h_{ch}}], \tag{3}$$

respectively. The effective permittivity in Eq. (2) is defined as:

$$\varepsilon_{eff} = \varepsilon_0[\varepsilon_r - \frac{c^2}{4f^2w^2}]. \tag{4}$$

The minimum reflection $R_{min} = -B\eta_{wg}/(2i + B\eta_{wg})$ is achieved when the impedance matching condition $\eta_{ch} = \eta_{wg}$ is satisfied. Based on the theoretical model, we designed a terahertz wave tunneling waveguide with $l_{wg} = l_{ch} = 8.8\mu m$, $l_{ab} = 34nm$; $w_{wg} = 6.2\mu m$, $w_{ch} = 4.4\mu m$; $h_{wg} = 2.2\mu m$, $h_{ch} = 34nm$. The materials filled in different sections are assumed to be the same, and their relative permittivity $\varepsilon_{r,\,wg} = \varepsilon_{r,\,ch} = 1$. Finite element simulations were carried out to verify the tunneling effect of the waveguide.

As shown in the Poynting vector plot in Fig. 1 (b), despite the large height contrast in the interface, the electromagnetic field can be squeezed into the narrow tunneling channel and propagate along the channel. The transmittance of the wave tunneling channel is frequency dependent and approaches 1 around 34THz, as illustrated in Fig. 2 (a). The peak frequency and transmittance of the theoretical model and the simulation results are in good agreement. The only difference is that the bandwidth of the simulation results is narrower. The FWHM value is 0.14THz and 0.06THz for theoretical model and simulation results, respectively.

The location and value of the transmittance peak are determined by the width and permittivity of the waveguide and are irrelevant to the length of the tunneling channel. The energy losses mainly occurred at the entrance port of the tunneling channel, once the field has been squeezed into the channel, the field intensity will maintain to the end regardless of the length of the channel, as shown in Fig. 2 (b).

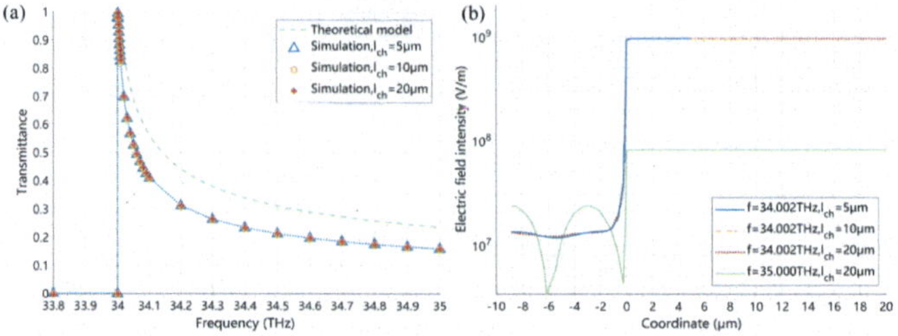

Fig. 2. (a) Transmittance of the terahertz wave tunneling waveguide given by the theoretical model and finite element simulations. (b) Variation of the electric field intensity along the wave propagation direction.

3 Conclusion

The electromagnetic wave tunneling effect in the terahertz band is studied with the utilization of the transmission-line model. A specific waveguide for wave tunneling in 34THz was proposed and verified by finite element simulations, showing high transmittance irrelevant to the channel length. By modifying the permittivity of materials inside and the dimensions of the waveguide, the wave tunneling can realized in a wide range of terahertz bands.

References

1. Wang, P., et al.: Progress on cutting-edge infrared-terahertz biophysics. IEEE Trans. Microw. Theory Tech. **70**, 5117–5140 (2022)
2. Zhang, Z., et al.: Advanced terahertz refractive sensing and fingerprint recognition through metasurface-excited surface waves. Adv. Mater. 2308453 (2024)
3. Lou, J., et al.: Calibration-free, high-precision, and robust terahertz ultrafast metasurfaces for monitoring gastric cancers. Proc. Natl. Acad. Sci. **119**, e2209218119 (2022)

4. Lou, J., et al.: Optically controlled ultrafast terahertz metadevices with ultralow pump threshold. Small **17**, 2104275 (2021)
5. Wang, R., et al.: Ultrasensitive terahertz biodetection enabled by Quasi-BIC-based metasensors. Small **19**, 2301165 (2023)
6. Yu, Y., et al.: Terahertz photons improve cognitive functions in posttraumatic stress disorder. Research **6**, 0278 (2023)
7. Peng, W., et al.: High-frequency terahertz waves disrupt Alzheimer's β-amyloid fibril formation. eLight 3, 18 (2023)
8. Wu, K., et al.: Terahertz wave accelerates DNA unwinding: a molecular dynamics simulation study. J. Phys. Chem. Lett. **11**, 7002–7008 (2020)
9. Yin, J., et al.: Terahertz photons inhibit cancer cells long term by suppressing nano telomerase activity. ACS Nano. 3c09216 (2024)
10. Chang, C., et al.: High-gain thompson-scattering X-Ray free-electron laser by time-synchronic laterally tilted optical wave. Phys. Rev. Lett. **110**, 064802 (2013)
11. Liu, G.: The conjectures on physical mechanism of vertebrate nervous system. Chin. Sci. Bull. **63**, 3864–3865 (2018)
12. Xiang, Z., et al.: A primary model of THz and far-infrared signal generation and conduction in neuron systems based on the hypothesis of the ordered phase of water molecules on the neuron surface I: signal characteristics. Sci. Bull. **65**, 308–317 (2020)
13. Liu, Y., et al.: Amplification of terahertz/infrared field at the nodes of Ranvier for myelinated nerve. Sci. China Phys. Mech. Astron. **63**, 274211 (2020)
14. Liu, G., et al.: Myelin sheath as a dielectric waveguide for signal propagation in the mid-infrared to terahertz spectral range. Adv. Funct. Mater. **29**, 1807862 (2019)
15. Xiang, Z., et al.: A new viewpoint and model of neural signal generation and transmission: signal transmission on myelinated neuron. Chin. Phys. B **29**, 108701 (2020)
16. Xiang, Z., et al.: A new viewpoint and model of neural signal generation and transmission: signal transmission on unmyelinated neurons. Nano Res. **14**, 590–600 (2021)
17. Zhu, Z., et al.: Terahertz-light induced structural transition and superpermeation of confined monolayer water. ACS Photonics **8**, 781–786 (2021)
18. Zhang, J., et al.: Non-invasive, opsin-free mid-infrared modulation activates cortical neurons and accelerates associative learning. Nat. Commun. **12**, 2730 (2021)
19. Liu, X., et al.: Nonthermal and reversible control of neuronal signaling and behavior by midinfrared stimulation. Proc. Natl. Acad. Sci. **118**, e2015685118 (2021)
20. Li, Y., et al.: Terahertz wave enhances permeability of the voltage-gated calcium channel. J. Am. Chem. Soc. **143**, 4311–4318 (2021)
21. Zhu, Z., et al.: Broadband terahertz signatures and vibrations of dopamine. Analyst **145**, 6006–6013 (2020)
22. Zhu, Z., et al.: Characteristic fingerprint spectrum of neurotransmitter norepinephrine with broadband terahertz time-domain spectroscopy. Analyst **144**, 2504–2510 (2019)
23. Cheng, C., et al.: Broadband terahertz recognizing conformational characteristics of a significant neurotransmitter γ-aminobutyric acid. RSC Adv. **9**, 20240–20247 (2019)
24. Li, Y., et al.: Physicochemical insights on terahertz wave diminished side effects of drugs from slow dissociation. ACS Nano **16**, 8419–8426 (2022)
25. Silveirinha, M., et al.: Tunneling of electromagnetic energy through subwavelength channels and bends using ε-near-zero materials. Phys. Rev. Lett. **97**, 157403 (2006)
26. Liu, R., et al.: Experimental demonstration of electromagnetic tunneling through an epsilon-near-zero metamaterial at microwave frequencies. Phys. Rev. Lett. **100**, 023903 (2008)
27. Edwards, B., et al.: Experimental verification of epsilon-near-zero metamaterial coupling and energy squeezing using a microwave waveguide. Phys. Rev. Lett. **100**, 033903 (2008)

28. Silveirinha, M.G., et al.: Theory of supercoupling, squeezing wave energy, and field confinement in narrow channels and tight bends using ε near-zero metamaterials. Phys. Rev. B **76**, 245109 (2007)
29. Alù, A., et al.: Transmission-line analysis of ε-near-zero–filled narrow channels. Phys. Rev. E **78**, 016604 (2008)

The Molecular Mechanism of Targeted Therapy Based on Terahertz Irradiation

Qiwei Wang(✉)

ShanghaiTech University, Shanghai 201210, China
wangqw2022@shanghaitech.edu.cn

Abstract. Long-wave infrared irradiation and terahertz irradiation, due to their low-energy photon characteristics, have broad prospects for safe treatment in the field of biomedicine. This article introduces the reparative effects of terahertz irradiation on burned skin and nerve tissues, as well as its applications in enhancing sperm vitality and treating cardiovascular diseases. Subsequently, it analyzes in detail the intervention of characteristic frequency terahertz waves on cancer cells and discusses their mechanisms. Furthermore, a detailed analysis is provided on the use of terahertz irradiation to modulate drugs without altering their chemical structure, along with its effects and mechanisms. Finally, prospects for the application of terahertz irradiation in biomedicine are outlined.

Keywords: Terahertz irradiation effect · Biological tissue repair · Tumor treatment · Antipsychotic drugs

1 Introduction

As the global population ages, humanity faces an increasing number of challenges related to disease treatment. Research in life sciences and medicine is becoming one of today's most critical frontier areas [1–4]. Terahertz (THz) photon energy is significantly lower than the energy of various chemical bonds within biological organisms. Consequently, terahertz electromagnetic waves have non-ionizing properties when irradiating living organisms, biological tissues, cells, and biomolecules (such as DNA) [5–8]. This makes them an effective technological means for modulating the functions of living organisms, producing effects similar to life sciences and medicine, modulating the nerves or brain functions (see Fig. 1) [9–18]. This text will briefly introduce the representative achievements and research progress of the biological effects of terahertz irradiation in the field of life sciences and medicine [15–28].

2 Terahertz Irradiation Biomedical Effects

2.1 Biological Tissue Regeneration

In terms of biological tissue rehabilitation, terahertz electromagnetic irradiation has been proven to promote the regeneration and repair of damaged tissues. For instance, Ostrovskiy et al. used a 0.15 THz electromagnetic field to irradiate patients with burns,

C. Chang et al. (Eds.): YTHZ 2024, SPPHY 400, pp. 271–275, 2024.
https://doi.org/10.1007/978-981-97-3786-4_55

finding that the irradiation promoted the recovery of the local burned skin tissue [13]. Similarly, Chen and colleagues irradiated the sciatic nerve tissue of rats after surgery with a 0.3–100 THz electromagnetic field, finding that irradiation accelerated the recovery of the damaged sciatic nerve tissue [14]. These results suggest that terahertz electromagnetic irradiation has potential application value in the rehabilitation of biological tissues.

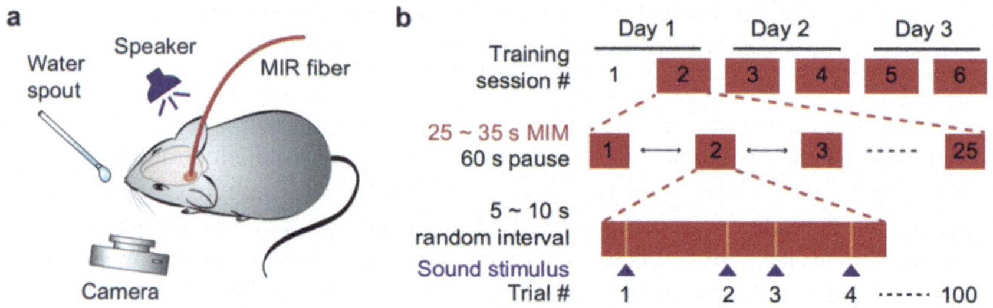

Fig. 1. MIM in the auditory cortex accelerates learning during a sound-licking associative training task. (a) Schematic illustrating the general scheme for the experiments using MIM in a sound-water associative training task. (b) Detail of the experimental protocol for MIM application throughout the training Sessions [10] (Open Access Journal).

2.2 Disease Treatments

In disease treatment, terahertz electromagnetic irradiation has also been found to have various functional effects. For example, Wei et al. used a 0.1–3 THz electromagnetic field to irradiate the sperm cells of patients with insufficient sperm vitality, finding that the irradiation significantly enhanced sperm vitality [3]. This indicates that terahertz electromagnetic irradiation may have a positive effect on improving human fertility. Moreover, Kirichuk et al. used a 0.24 THz frequency, 1 mW/cm^2 power terahertz electromagnetic field to irradiate the whole blood of patients with unstable angina, finding that the viscosity of the blood decreased and the deformability of red blood cells increased after irradiation [4]. This suggests that terahertz electromagnetic irradiation may aid in the treatment of unstable angina.

2.3 Cancer Cell Intervention

In the intervention of cancer cells, Titova et al. in 2013 [26] found that broadband terahertz irradiation ranging from 0.1 to 2 THz (peak frequency 0.5 THz, pulse energy 0.1 J) could induce an increase in the expression levels of tumor-suppressing proteins and damage-repair proteins in artificial skin tissue. That same year, through whole-genome analysis, they discovered that intense pulsed terahertz irradiation could induce differential expression of genes associated with non-melanoma skin cancer in artificial skin tissue, providing a potential means of intervention for cancer cells [27, 28].

Sun et al. in 2022 [19] used a 3.6 μm (frequency of 83 THz) terahertz wave, finding that terahertz irradiation had a non-thermal inhibitory effect on cancer cell migration and glycolysis, and explained the molecular mechanism of this effect by changing chromatin accessibility. The study first used molecular dynamics simulations to find that this wavelength of terahertz electromagnetic radiation could significantly reduce the free energy of binding between histones and DNA, increasing their affinity, leading to a more compact chromatin condensation state and reduced chromatin accessibility. Subsequently, cell migration experiments and metabolic analysis found that irradiation at 3.6 μm (0.3 mW) inhibited cancer cell migration and glycolysis by 50% and 60%, respectively. In a liver metastasis xenograft mouse model established by splenic injection of HCT-116 colon cancer cells, it was found that terahertz irradiation inhibited the metastasis of colon cancer cells by 60%. By combining multi-omics studies, it was found that terahertz electromagnetic irradiation enhanced the affinity between chromatin and DNA, changing the accessibility of the genome chromatin and adjusting genes related to migration in tumor cells, such as reducing the expression of the HIF-1α gene, thereby inhibiting tumor cell migration and glycolysis. Notably, terahertz irradiation did not affect cell proliferation in the experiments, meaning it could be a non-destructive method to directly inhibit cancer cell metastasis and metabolism without causing chemical toxicity damage to the body, thus providing a new method for cancer treatment.

In 2024, Yin et al. [20] used a 33 THz electromagnetic field to irradiate breast cancer cells, finding that three weeks of long-term irradiation significantly inhibited the activity of nanoscale telomerase in breast cancer cells, accelerating the consumption of telomeres in tumor cells, leading to telomere shortening, thereby triggering a telomere crisis, causing DNA damage and apoptosis, and thus inhibiting the growth of tumor cells in the body (see Fig. 1). At the molecular level, molecular dynamics simulations found that the terahertz waves resonated with the zipper amino acids in the telomerase structure, thereby reducing DNA synthesis, while the hydrogen bond connections between telomerase and DNA were reduced, further lowering the free energy of binding between DNA and RNA, explaining the molecular mechanism by which terahertz waves inhibit cancer cell growth by reducing the rate of DNA synthesis. The study also found that long-term exposure to 33 THz radiation of human normal breast epithelial cells and human osteosarcoma cells (ALT-positive cells) did not show a significant difference in apoptosis rate and telomere length compared to the control group, indicating that long-term exposure to 33 THz radiation has a minimal impact on the survival ability of normal cells and ALT-positive cells, and can be considered safe. In the future, terahertz optical fibers can be used to deliver radiation deep into the body to reduce the absorption of terahertz waves by water molecules, for the treatment of cancers with high telomerase activity.

2.4 Pharmacokinetics

Li et al. in 2022 [21] used 4.0 THz electromagnetic irradiation to modulate the dissociation kinetics of high-affinity antipsychotic drugs from their receptors, to mitigate the drugs' side effects. Through molecular dynamics simulations and energy calculations, it was found that 4.0 THz electromagnetic irradiation could resonantly excite the salt bridge connections between the dopamine D2 receptor and the drug risperidone, significantly weakening their binding, thereby reducing the binding affinity between the drug and the receptor, facilitating the faster dissociation of the drug from the receptor. On the other hand, terahertz irradiation caused the tryptophan side chain located at the exit of the drug-receptor binding pocket to rotate, thereby expanding the dissociation pathway and accelerating the drug's dissociation, making terahertz irradiation act similarly to enzyme catalysis to increase the drug's dissociation rate by eight orders of magnitude. This method provides new insights for drug design and optimization without altering the chemical structure of the drugs. Additionally, this terahertz irradiation method can also be applied to the treatment of other metabolic diseases related to GPCR dysfunction, such as diabetes and obesity.

3 Summary and Prospect

The future applications of terahertz radiation in the biomedicine field are filled with possibilities. As a non-invasive diagnostic tool, terahertz technology has the potential to provide safer imaging methods than X-rays, capable of distinguishing subtle tissue differences. In tumor treatment, terahertz radiation could reduce damage to healthy tissues through its precise targeting ability. Moreover, the unique response of biological macromolecules to the terahertz band could promote the development of new therapies for diseases that currently lack effective treatments. In drug development, terahertz spectroscopy could be used to improve the quality control of pharmaceuticals, ensuring their safety and efficacy. With technological advancements, terahertz radiation is expected to play an increasingly important role in the biomedical field, opening up new research and treatment avenues. However, it is important to note that most of these applications are still in the research stage and may require further refinement before they can be widely applied in clinical settings.

References

1. Ostrovskiy, N.V., et al.: Joint 30th International Conference on Infrared and Millimeter Waves & 13th International Conference on Terahertz Electronics Williamsburg, USA 19–23, 301 (2005)
2. Chen, T.Y., et al.: Evid. Based Complement. Alternat. Med. **2015**, 207245 (2015)
3. Wei, C., et al.: Biomed. Opt. Express **9**, 3998 (2018)
4. Kirichuk, V.F., et al.: Bull. Exp. Biol. Med. **2015**, 207245 (2008)
5. Wang, P., et al.: Progress on cutting-edge infrared-terahertz biophysics. IEEE Trans. Microw. Theory Tech. **70**(11), 5117–5140 (2022)
6. Liu, G., et al.: Myelin sheath as a dielectric waveguide for signal propagation in the mid-infrared to terahertz spectral range. Adv. Funct. Mater. **29**(7), 1807862 (2019)
7. Xiang, Z., et al.: A primary model of THz and far-infrared signal generation and conduction in neuron systems based on the hypothesis of the ordered phase of water molecules on the neuron surface. Sci. Bull. **65**, 308 (2020)
8. Wu, K., et al.: Terahertz wave accelerates DNA unwinding: a molecular dynamics simulation study. J. Phys. Chem. Lett. **11**(17), 7002–7008 (2020)
9. Liu, X., et al.: Nonthermal and reversible control of neuronal signaling and behavior by midinfrared stimulation. Proc. Natl. Acad. Sci. 118(10), e2015685118 (2021)
10. Zhang, J.X., et al.: Non-invasive, opsin-free mid-infrared modulation activates cortical neurons and accelerates associative learning. Nature Comm. **12**, 2730 (2021)

11. Yu, Y., et al.: Terahertz photons improve cognitive functions in posttraumatic stress disorder. Research **6**, 0278 (2023)
12. Tan, X., et al.: Neuromodulation of chemical sysnaptic transmission driven by THz photons. Research **2022**, 0010 (2022)
13. Zhang, Q.L., et al.: Ultrahigh-flux water nanopumps generated by asymmetric terahertz absorption. Phys. Rev. Lett. (2024)
14. Zhu, Z., et al.: Characteristic fingerprint spectrum of neurotransmitter norepinephrine with broadband terahertz time-domain spectroscopy. Analyst **144**(8), 2504–2510 (2019)
15. Tang, C., et al.: Integrating terahertz metamaterial and water nanodroplets for ultrasensitive detection of amyloid β aggregates in liquids. Sens. Actuators B. Chem. **329**, 129113 (2021)
16. Zhu, Z., et al.: Tunable surface wettability via terahertz electrowave controlled vicinal subnanoscale water layer. Nano Lett. **24**(10), 3243–3248 (2024)
17. Yuan, Y., et al.: Midinfrared radiation accelerates DNA unwinding. ACS Photonics (2024). https://doi.org/10.1021/acsphotonics.3c01548
18. Xiao, T., et al.: Sensory input dependent gain modulation of the optokinetic nystagmus by mid-infrared stimulation in pigeons. Elife **12**, e78729 (2023)
19. Sun, L., et al.: Inhibition of cancer cell migration and glycolysis by terahertz wave modulation via altered chromatin accessibility. Research (2022)
20. Yin, J., et al.: Terahertz photons inhibit cancer cells long term by suppressing nano telomerase activity. ACS Nano, ACS Nano **18**(6), 4796–4810 (2024)
21. Li, Y., et al.: Physicochemical insights on terahertz wave diminished side effects of drugs from slow dissociation. ACS Nano, 8419–8426 (2022)
22. Wang, P., et al.: An ultra-sensitive metasurface biosensor for instant cancer detection based on terahertz spectra. Nano Res. **16**, 7304–7311 (2023)
23. Lou, J., et al.: Calibration-free, high-precision, and robust terahertz ultrafast metasurfaces for monitoring gastric cancers. Proc. Natl. Acad. Sci. 119(43), e2209218119 (2022)
24. Chang, C., et al.: High-gain thompson-scattering X-Ray free-electron laser by time-synchronic laterally tilted optical wave. Phys. Rev. Lett. **110**, 064802 (2013)
25. Zhang, Z., et al.: Advanced terahertz refractive sensing and fingerprint recognition through metasurface-excited surface waves. Adv. Mater., 2308453 (2024)
26. Titova, L.V., et al.: Intense THz pulses cause H2AX phosphorylation and activate DNA damage response in human skin tissue. Biomed. Opt. Express **4**(4), 559–568 (2013)
27. Titova, L.V., et al.: Intense picosecond THz pulses alter gene expression in human skin tissue in vivo. Bellingham: Inter Soc. Opt. Photonics **8585**, 85850Q (2013)
28. Titova, L.V., et al.: Intense THz pulses down-regulate genes associated with skin cancer and psoriasis: a new therapeutic avenue? Sci. Rep. **3**(1), 2363 (2013)

Bioinformatics Analysis of Terahertz Photons on Breast Cancer Cells

Jingzhou Liu(✉) and Junkai Yin

Innovation Laboratory of Terahertz Biophysics, National Innovation Institute of Defense Technology, Beijing 100071, China
Liujingzhou06@163.com

Abstract. Although it has been shown that terahertz (THz) waves can lead to epigenetic changes in cancer cells, little is known about the therapeutic effects of THz waves on cancer. Here, the gene expression changes of MCF-7 cells after THz exposure are showed using differentially expressed genes analysis. GO enrich analysis showed that endocytosis and exocytosis are related to 33THz stimulation. Kyoto Encyclopedia of Genes and Genomes (KEGG) analysis showed that 33THz photons irradiation may have significant effect on estrogen signaling pathway. Estrogen receptors play an important role in endocytosis, growth and migration of breast cancer cells. Therefore, this study speculates that 33THz wave may affects tumor growth and migration by manipulating estrogen receptors and estrogen signaling pathway. In summary, this article provides a possible application of 33THz wave on breast cancer treatment.

Keywords: Terahertz Photons · Breast Cancer Cells · Differentially Expressed Genes Analysis

1 Introduction

1.1 A Subsection Sample

Terahertz (THz) wave modulation (THM) have been widely studied for cancer diagnostics and imaging [1]. However, THz-based cancer treatment has been less studied. Yin et.al. Found a resonant peak at approximately 33 THz to inhibit cancer cells by suppressing telomerase [2]. Sun et.al. Explored that 3.6μm THM significantly increased binding affinity between histone and DNA, thus altered genomic chromatin accessibility and further resulted in the changes of gene expression associating with the migration and glycolysis [3]. The authors injected the THM-pretreated cancer cells that decreased a successful metastasis *in vivo*. The results confirmed that the 3.6μm THM could suppress the migration and glycolysis in cancer cells and displayed THM-induced epigenetic change, such as altered chromatin accessibility. Besides, THz can effectively interact with purine bonds of DNA and alter the structures [4, 5]. These results indicate a potential medical field using THM-induced gene changes in the cancer treatment.

To explore more possibility of THM for cancer treatment, we exposed MCF-7 cells to 33 THz and performed RNA sequencing along with unexposed controls. Important gene expression changes after THM are analyzed in this article.

C. Chang et al. (Eds.): YTHZ 2024, SPPHY 400, pp. 276–280, 2024.
https://doi.org/10.1007/978-981-97-3786-4_56

2 Methods

The THz radiation system is commercially available (Daylight) [6]. We selected 33THz to irradiate MCF-7 cells by the previously described scanning method [7]. The THz power was 5 ± 0.5 mW, and the power density was 4 W/cm^2. The spot diameter used was 400 μm. The duration of the laser pulse was 300 ns and the repetition rate was 100 kHz.

3 Results and Discussion

The DEGs are showed in Fig. 1. After THz exposure, 711 DEGs were identified, more of which were downregulated (445) than upregulated (266), indicating that 33THz obviously affected the gene expression.

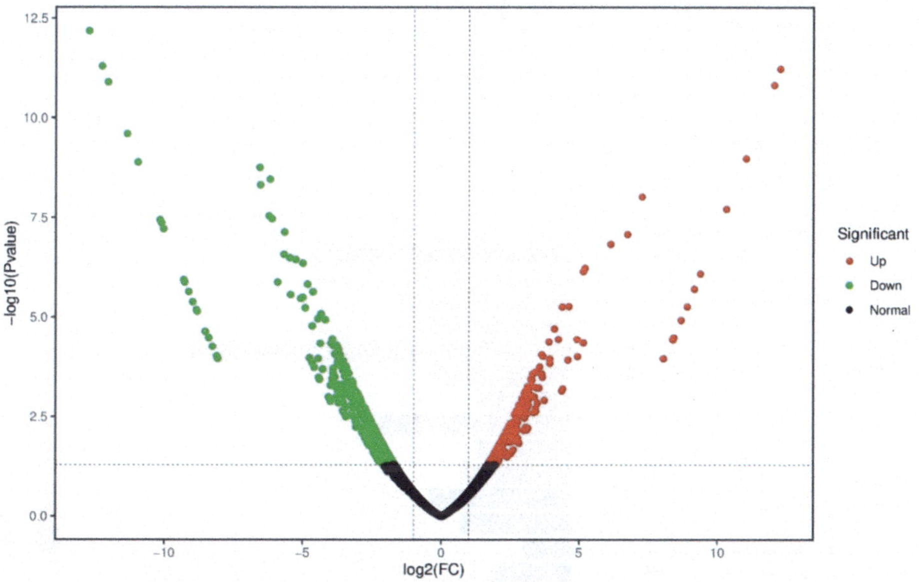

Fig. 1. Volcano map of DEGs (Control vs THz). Red dots: remarkedly up-regulated genes; green dots: remarkedly down-regulated genes; black dots: no differentially expressed genes.

From GO enrich analysis, we found that most of the statistically differentially expressed genes were enriched in endocytosis and regulated exocytosis, which could be caused by the change of plasma membrane and structural molecule activity (Fig. 2).

Furthermore, Kyoto Encyclopedia of Genes and Genomes (KEGG) analysis showed that 33THz exposure may affect estrogen signaling pathway (Fig. 3).

The vibrations of biomacromolecules are usually within the 0.5-100THz wave [8–10], thus the THz wave have be used to enhance bond vibrations of biomolecules [11–29]. For example, THM could affect the permeability of calcium channel [30] and increase

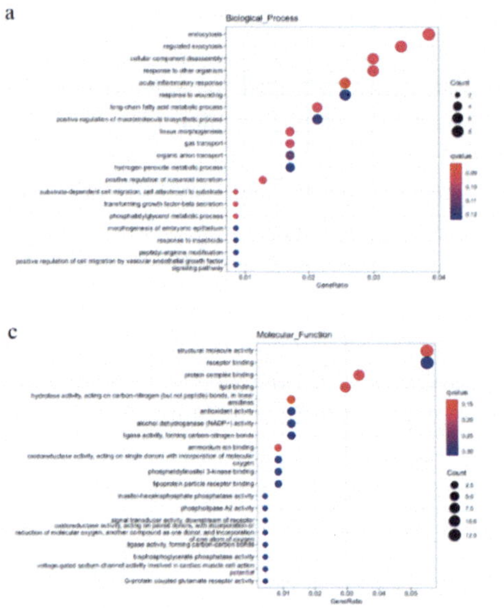

Fig. 2. GO analysis for DEGs. The dots represent GO terms; the size indicates the count of genes enriched; the color indicates the qvalue.

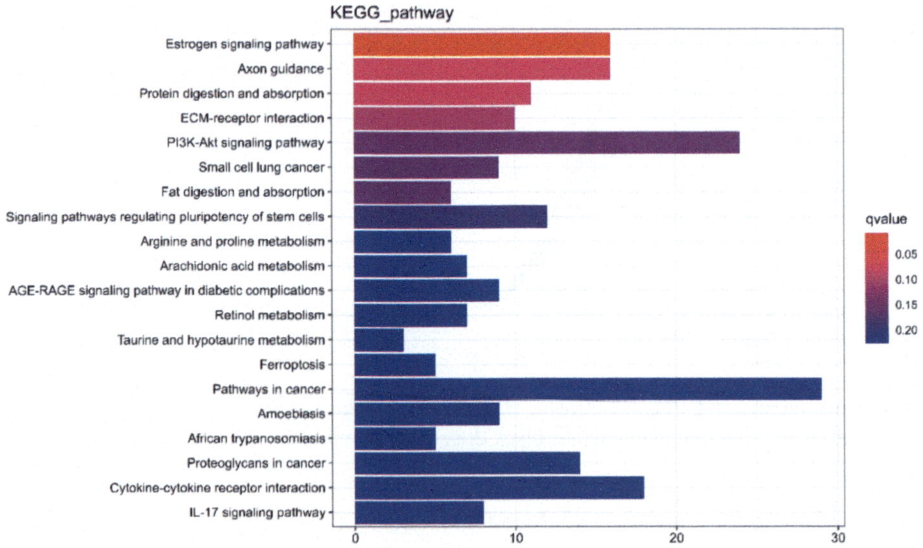

Fig. 3. KEGG analysis of DEGs. The bars represent KEGG terms; the size indicates the count of genes enriched; the color indicates the qvalue.

the binding between glutamate and glutamate receptors [31]. Estrogen receptors (ERs) is expressed on breast cancer cells and play an important role in tumor growth and migration [32]. Membrane ERs are susceptible to multiple stimuli and interact with caveolin-1, a scaffold protein which is closely related to endocytosis [33]. Therefore, we speculate that 33THz may manipulate membrane ERs and further affect the endocytosis. In summary, this article provides a possible target for cancer treatment, which requires further experimental confirmation.

References

1. Wang, P., et al.: Progress on cutting-edge infrared-terahertz biophysics. IEEE Trans. Microw. Theory Tech. **70**(11), 5117–5140 (2022)
2. Yin, J., et al.: Terahertz photons inhibit cancer cells long term by suppressing nano telomerase. ACS Nano **18**(6), 4796–4910 (2024)
3. Sun, L., et al.: Inhibition of cancer cell migration and glycolysis by terahertz wave modulation via altered chromatin accessibility. Research **2022**, 9860679 (2022)
4. Wu, K., et al.: Terahertz wave accelerates DNA unwinding: a molecular dynamics simulation study. J. Phys. Chem. Lett. **11**(17), 7002–7008 (2020)
5. Zhang, C., et al.: Driving DNA origami assembly with a terahertz wave. Nano Lett. **22**, 468–475 (2022)
6. Zhang, J., et al.: Non-invasive, opsin-free mid-infrared modulation activates cortical neurons and accelerates associative learning. Nat. Commun. **12**(1), 2730 (2021)
7. Liu, X., et al.: Nonthermal and reversible control of neuronal signaling and behavior by midinfrared stimulation. PNAS **118**(10) (2021)
8. Liu, G., et al.: Myelin sheath as a dielectric waveguide for signal propagation in the mid-infrared to terahertz spectral range. Adv. Funct. Mater. **29**(7), 1807862 (2019)
9. Xiang, Z., et al.: A primary model of THz and far-infrared signal generation and conduction in neuron systems based on the hypothesis of the ordered phase of water molecules on the neuron surface I: signal characteristics. Sci. Bull. **65**(4), 308–317 (2020)
10. Chang, C., et al.: High-gain thompson-scattering X-Ray free-electron laser by time-synchronic laterally tilted optical wave. Phys. Rev. Lett. **110**, 064802 (2013)
11. Zhang, Z.; et al.: Advanced terahertz refractive sensing and fingerprint recognition through metasurface-excited surface waves. Adv. Mater., 2308453 (2024)
12. Lou, J., et al.: Calibration-free, high-precision, and robust terahertz ultrafast metasurfaces for monitoring gastric cancers. PNAS **119**(43) (2022)
13. Tan, X., et al.: Minimal-invasive enhancement of auditory perception by terahertz wave modulation. Nano Res. **15**(6), 5235–5244 (2022)
14. Li, Y., et al.: Physicochemical insights on terahertz wave diminished side effects of drugs from slow dissociation. ACS Nano **16**(5), 8419–8426 (2022). https://doi.org/10.1021/acsnano.2c02952
15. Tan, X., et al.: Neuromodulation of chemical synaptic transmission driven by THz photons. Research **2022**, 0010 (2022)
16. Peng, W., et al.: High-frequency terahertz waves disrupt Alzheimer's β-amyloid fibril formation. eLight **3**, 18 (2023)
17. Lou, J., et al.: Optically controlled ultrafast terahertz metadevices with ultralow pump threshold. Small **17**(44), e2104275 (2021)
18. Qiao, Z., Xue, M., Zhao, Y., et al.: Infrared nanoimaging of nanoscale sliding dislocation of collagen fibrils. Nano Res. **15**(3), 2355–2361 (2022)
19. Wang, P., et al.: An ultra-sensitive metasurface biosensor for instant cancer detection based on terahertz spectra. Nano Res. **16**, 7304–7311 (2023)
20. Zhang, Q.L., et al.: Ultrahigh-flux water nanopumps generated by asymmetric terahertz absorption, Phys. Rev. Lett. (2024)

21. Zhu, Z., et al.: Characteristic fingerprint spectrum of neurotransmitter norepinephrine with broadband terahertz time-domain spectroscopy. Analyst **144**(8), 2504–2510 (2019)
22. Tang, C., et al.: Integrating terahertz metamaterial and water nanodroplets for ultrasensitive detection of amyloid β aggregates in liquids. Sens. Actuators B. Chem. **329**, 129113 (2021)
23. Zhu, Z., et al.: Tunable surface wettability via terahertz electrowave controlled vicinal subnanoscale water layer. Nano Lett. **24**(10), 3243–3248 (2024)
24. Yuan, Y., et al.: Midinfrared radiation accelerates DNA unwinding. ACS Photonics **11**(4), 1473–1479 (2024). https://doi.org/10.1021/acsphotonics.3c01548
25. Xiao, T., et al.: Sensory input dependent gain modulation of the optokinetic nystagmus by mid-infrared stimulation in pigeons. Elife **12**, e78729 (2023)
26. Wang, R., et al.: Ultrasensitive terahertz biodetection enabled by quasi-BIC-based metasensors. Small **19**, 2301165 (2023)
27. Jiao, Y., et al.: Photoactive terahertz metasurfaces for ultrafast switchable sensing of colorectal cells. Mater. Horiz. **9**, 2984–2992 (2022)
28. Sun, L., et al.: A pixelated frequency-agile metasurface for broadband terahertz molecular fingerprint sensing. Nanoscale **14**, 9681–9685 (2022)
29. Wang, R., et al.: Electric Fano resonance-based terahertz metasensors. Nanoscale **13**, 18467–18472 (2021)
30. Li, Y., et al.: Terahertz wave enhances permeability of the voltage-gated calcium channel. J. Am. Chem. Soc. **143**(11), 4311–4318 (2021)
31. Yu, Y., et al.: Terahertz photons improve cognitive functions in posttraumatic stress disorder. Research **6**, 0278 (2023)
32. Sampayo, R.G., et al.: Fibronectin rescues estrogen receptor α from lysosomal degradation in breast cancer cells. J. Cell Biol. **217**(8), 2777–2798 (2018)
33. Marczell, I., et al.: Membrane-bound estrogen receptor alpha initiated signaling is dynamin dependent in breast cancer cells. Eur. J. Med. Res. **23**(1), 31 (2018)

Research on Mechanism of Neural Signal Generation and Transmission

Songyuan Yang(✉)

Department of Engineering Physics, Tsinghua University, Beijing, China
2540913445@qq.com

Abstract. This paper mainly introduces a study about mechanisms underlying neural signal generation and transmission, placing a novel focus on the integration of neural electromagnetic fields within the terahertz to infrared frequency range. Through theoretical analysis and simulation, we explore the implications of terahertz frequencies in modulating these processes, encompassing aspects of neural regulation and energy provision. These conclusion sheds light on the broader impact of terahertz frequencies on biological functions.

Keywords: Neuron · THz · Neuron signal generation and conduction · Neuromodulation · Energy supply

1 Introduction

In the realm of neuroscience, unraveling the intricacies of neural signal generation and transmission stands as a cornerstone of research. While classical models like Hodgkin-Huxley [1] have provided fundamental insights, they fall short when confronted with the nuanced complexities of neuronal communication. This paper introduces a groundbreaking perspective, postulating that the electromagnetic field associated with neural signals spans the terahertz to infrared frequencies. We delve into the implications of this hypothesis by exploring signal dynamics in both unmyelinated and myelinated neurons, thereby uncovering the pivotal roles of terahertz and far-infrared signals [2–6]. These insights include their interaction with neurotransmitters and potential as novel mechanisms in neural signal generation and conduction, offering a fresh lens to view the interplay of electromagnetic fields and neurological functions.

2 A New Perspective and Model on the Generation and Transmission of Neural Signals

In this study, we investigate the neural signal generation and transmission mechanisms from a physical perspective. The key insights include [2]: (1) The neural signals in vertebrates are hypothesized to be high-frequency electromagnetic fields, ranging from terahertz to infrared frequencies, primarily between 0.5 and 100 terahertz. (2) The generation, transmission, and integration of these signals in biological neurons occur at micro

C. Chang et al. (Eds.): YTHZ 2024, SPPHY 400, pp. 281–284, 2024.
https://doi.org/10.1007/978-981-97-3786-4_57

and nanoscales, possibly extending to individual molecular interactions, in line with electromagnetic field theories and quantum mechanics. (3) The quantum energy states of ions and molecules within the nervous system are integral to the process of signal generation, amplification, and coupling, as well as in the storage of biological information. (4) Ranvier nodes on neuronal axons are identified as critical 'relay amplifiers' for nerve signals, marked by strong auto-regulation and substantial energy consumption within the nervous system. Neurons constitute the essential units for information storage, processing, and exchange. (5) Ambient temperature is shown to have a substantial effect on the generation and transmission of terahertz signals within the nervous system, implying a temperature-dependent rate of signal transmission. In extreme temperature conditions, the efficiency of terahertz signal transmission is compromised. Furthermore, the abundant terahertz electromagnetic information in living organisms can be harnessed for diagnosing various biological states, including health conditions. This could lead to the development of innovative diagnostic methods, encompassing both active and passive modalities. Notably, passive diagnostics, being non-intrusive, offer significant benefits. Additionally, the interaction of terahertz electromagnetic signals with living organisms under certain conditions suggests their potential as a focal point for future therapeutic interventions and as a crucial factor in disease prevention.

In this study, the generation and transmission of neural signals are analyzed through kinetic equations derived from the interaction of water molecules, fields, and ions. The properties of high-frequency signals are elucidated using perturbation theory and numerical methods. This approach facilitates the calculation of dispersion relations, and the time and frequency domain waveforms of the magnetic field. Our findings are multifold [3]. The formation of a long-range ordered structure by water molecules, characterized by a substantial coupling coefficient ($g > 1$), results in vibrational eigenmodes within the terahertz to infrared frequency range. Disturbances in the system can excite these eigenmodes, with significant external perturbations leading to the activation of several nonlinear modes. Furthermore, the eigenmodes of the system are influenced by physical parameters, evidenced by variations in the jump frequency between different energy levels of water molecules, the water molecule-field coupling coefficient, and the ion oscillation frequency. Changes in these parameters lead to distinct shifts in the dispersion curves of the eigenmodes. Analytical and numerical solutions converge in scenarios of weak perturbations, where nonlinear terms are inconsequential. However, in the presence of strong perturbations, nonlinear effects become crucial, manifesting in the excitation of various nonlinear modes within the system.

This research examines signaling in unmyelinated nerves over various time scales. On longer time scales, the model concurs with the circuit model's findings. However, on shorter time scales, the nervous system exhibits electromagnetic oscillations in the terahertz and infrared spectrum. Interestingly, while the fast-oscillating signals exhibit a rapidly changing energy envelope curve, the overall energy variation is more gradual [4]. In contrast, myelinated nerves display distinct characteristics due to the presence of myelin. The energy replenishment during signal transmission is confined to the Ranvier nodes, significantly reducing energy consumption and enhancing transmission efficiency. Yet, the myelin length is constrained by a threshold value, l_T. Generally, signal attenuation occurs along the myelin, with amplification at the Ranvier nodes. If the myelin length surpasses l_T, signal decay persists instead of recovery at the Ranvier nodes. Under normal physiological conditions, l_T typically measures in millimeters [5]. Additionally, both high and low ambient temperatures are found to be detrimental to neural signaling [4, 5].

3 Terahertz Photon Driven Neural Regulation and Energy Supply

Chemical synapses employ a range of neurotransmitters, including acetylcholine, dopamine, glutamate, and γ-aminobutyric acid, to regulate postsynaptic currents, with several being utilized in treating neurological disorders. A novel neuromodulation approach arises from the discovery that 30–45-THz photons resonate with these neurotransmitter molecules, inducing energy absorption and higher energy state transitions. Experimental evidence supports that terahertz irradiation significantly modifies neuronal signal firing, enhancing the properties of excitatory and inhibitory postsynaptic currents [7].

ATP, an essential energy storage molecule, is pivotal in various biological functions. The release of terahertz photons during ATP phosphate bond disruption has been shown to stimulate life activities. This process, avoiding foreign substances or electrode implantation, effectively restores diminished action potential firing frequencies due to inadequate ATP levels [8]. This suggests terahertz photons' direct role in regulating brain information processing and energy metabolism.

4 Conclusion

In this paper, we delve into the electromagnetic field's role, particularly terahertz frequencies, in the neural signal generation and transmission pathway. Terahertz is located at the character spectra of bio-molecules [9–15], thus, THz photons influences the neuronal activities [16–21] and neuronal transmission [22].

In addition, we further explore terahertz photons' impact on neuromodulation and energy dynamics. The capacity of terahertz irradiation to stimulate synaptic molecules into higher energy states, thereby altering the postsynaptic current profile, is examined. This observation suggests that terahertz photons of specific wavelengths could offer innovative means for directly modulating neuronal excitability and synaptic transmission, both in laboratory settings and within living organisms. Moreover, terahertz radiation might fulfill roles akin to ATP in energizing neural activities. This leads to the hypothesis that neural communication or physiological signal transduction could fundamentally involve the transmission of endogenous terahertz waves and other high-frequency electromagnetic signals, with the infusion of external terahertz photons potentially modifying or regulating a range of biological processes.

References

1. Hodgkin, A.L., Huxley, A.F.: A quantitative description of membrane current and its application to conduction and excitation in nerve. J. Physiol. **117**, 500–544 (1952)
2. Liu, G., et al.: Myelin sheath as a dielectric waveguide for signal propagation in the mid-infrared to terahertz spectral range. Adv. Funct. Mater. **29**(7), 1807862 (2019)
3. Xiang, Z., et al.: A primary model of THz and far-infrared signal generation and conduction in neuron systems based on the hypothesis of the ordered phase of water molecules on the neuron surface I: signal characteristics. Sci. Bull. **65**(4), 308–317 (2020)
4. Xiang, Z., et al.: A new viewpoint and model of neural signal generation and transmission: signal transmission on unmyelinated neurons. Nano Res. **14**, 590–600 (2021)
5. Xiang, Z., et al.: A new viewpoint and model of neural signal generation and transmission: signal transmission on myelinated neuron. Chin. Phys. B **29**(10) (2020)
6. Liu, Y., et al.: Amplification of terahertz/infrared field at the nodes of Ranvier for myelinated nerve. Sci. China Phys. Mech. Astron. **63**(7), 274211 (2020)
7. Tan, X., et al.: Neuromodulation of chemical synaptic transmission driven by THz photons. Research **2022**, 0010 (2022)
8. Tan, X., et al.: A new means of energy supply driven by terahertz photons recovers related neural activity. iScience **26**(2), 105979 (2023)
9. Wang, R., et al.: Electric Fano resonance-based terahertz metasensors. Nanoscale **13**, 18467–18472 (2021)
10. Lou, J., et al.: Optically controlled ultrafast terahertz metadevices with ultralow pump threshold. Small **17**, 2104275 (2021)
11. Zhang, Z., et al.: Advanced terahertz refractive sensing and fingerprint recognition through metasurface-excited surface waves. Adv. Mater, 2308453 (2024)
12. Chang, C., et al.: High-gain thompson-scattering X-Ray free-electron laser by time-synchronic laterally tilted optical wave. Phys. Rev. Lett. **110**, 064802 (2013)
13. Lou, J., et al.: Calibration-free, high-precision, and robust terahertz ultrafast metasurfaces for monitoring gastric cancers. PNAS **119**(43) (2022)
14. Jiao, Y., et al.: Photoactive terahertz metasurfaces for ultrafast switchable sensing of colorectal cells. Mater. Horiz. **9**, 2984–2992 (2022)
15. Wang, P., et al.: Progress on cutting-edge infrared-terahertz biophysics. IEEE Trans. Microw. Theory Tech. **70**(11), 5117–5140 (2022)
16. Li, Y., et al.: Terahertz wave enhances permeaility of the voltage-gated calcium channel. J. Am. Chem. Soc. **143**, 9 (2021)
17. Zhang, J.X., et al.: Non-invasive, opsin-free mid-infrared modulation activates cortical neurons and accelerates associative learning. Nature Comm. **12**, 2730 (2021)
18. Yu, Y., et al.: Terahertz photons improve cognitive functions in posttraumatic stress disorder. Research **6**, 0278 (2023)
19. Peng, W., et al.: High frequency terahertz waves disrupt Alzheimer's amyloid fibril formation. eLight **3**,18 (2023)
20. Wu, K., et al.: Terahertz wave accelerates DNA unwinding: a molecular dynamics simulation study. J. Phys. Chem. Lett. **11**(17), 7002–7008 (2020)
21. Yin, J., et al.: Terahertz photons inhibit cancer cells long term by suppressing nano telomerase. ACS Nano (2024). https://doi.org/10.1021/acsnano.3c09216
22. Liu, X., et al.: Nonthermal and reversible control of neuronal signaling and behavior by midinfrared stimulation. Proc. Natl. Acad. Sci. **118**(10), e2015685118 (2021)

Possible Interaction of Terahertz Signals in Neurons with Tryptophan Multimers

Bin Zhou(✉)

Tsinghua University, Beijing 100084, China
zhou-b21@mails.tsinghua.edu.cn

Abstract. Aromatic amino acids (AAAs) as a major constituent of microtubules in cells play an important role in multiple neural signaling processes. The microtubule has been shown to possibly transmit biophotons of cell mitochondria through its ordered structure. Recent studies show that terahertz waves presumably take part in transmission, modulation and enhancement of neural signals, the processes of which are shown inseparable to the amino acid multimers. Here we calculated the infrared spectra of tryptophan, an AAA which is prevalent in microtubule and building its scaffold. By calculating the vibrational modes of both the monomer and dimer of the tryptophan, we discovered two new and high-intensity peaks in the THz range interpreting the inter-molecular vibration, indicating energy convergence towards intermolecular interaction. The rest of the peaks presents a phase shift similar to the exciplex structure. Therefore, we propose that the multimer structure of tryptophan in tubulins may promote terahertz interaction in microtubules. This work presents a possibility of terahertz signal in neurons interacting with microtubules, demonstrating how terahertz electromagnetic signals may participate in the complex process of neural signal transmission.

Keywords: Aromatic Amino Acid · Tryptophan · Terahertz Transmission

1 Introduction

Aromatic amino acids (AAA), or amino acids with aroma groups, mainly including tyrosine (Tyr), phenylalanine (Phe) and tryptophan (Trp), are important constituents of various compounds and structures inside human body. AAAs are the precursor of multiple neurotransmitters and other chemical compounds involved in the nervous system, including 5-hydroxytryptamine(5-HT), dopamine, histamine, and norepinephrine [1].

The cytoskeleton of eukaryotic cells functions as both the backbone and the transmission channel of the cell. The cytoskeleton mainly consists of three polymers, microtubules, actin-filaments and intermediate filaments. Microtubule is mainly involved in transferring proteins and organelles across the cell, and has a high density of AAAs. Recent studies have shown that the ordered structure of microtubules and the accumulation of AAAs make it possible for microtubules to transfer biophotons across the cell [2]. Since aromatic compounds are known to form exciplex and transfer photons, some proposed that the AAAs in microtubules, especially tryptophan molecules can also transfer biophotons through similar mechanisms.

C. Chang et al. (Eds.): YTHZ 2024, SPPHY 400, pp. 285–288, 2024.
https://doi.org/10.1007/978-981-97-3786-4_58

Terahertz waves have been conjectured to underlie neural signaling mechanisms [3], including the generation, transmission, modulation, and storage of neural signals. With the development of novel probing and modulating technique [4–9], recent studies have shown the potential of terahertz waves in transmitting through neurons [10–12], modulating neurotransmitters, ion channels and other bio-molecules [13–20], as well as affecting the learning behavior of animals [21–24]. Vibrational energy levels of biomolecules lie in the terahertz range, including tryptophan, the AAA which builds up the microtubule network of cells. Therefore, it's possible that terahertz waves could interact with microtubules through the resonance of tryptophan multimers.

In this work, we calculated the infrared spectrum of tryptophan monomer and tryptophan dimer. We discovered rich terahertz absorption peaks in both structures. We observed most peaks with similar frequency while two main peaks are exclusively found in dimers, corresponding to the monomer resonance and molecule vibration in tryptophan dimer. Such findings support the idea that the multimer structure of tryptophan could promote terahertz interaction through intermolecular resonance.

2 Methods

The initial crystal structure of the tubulin comes from the RCSB data bank (PDB entry: 6O2S) [25, 26]. We emphasize the tryptophan residues in each tubulin dimer (see Fig. 1). The tryptophan monomer and dimer structure are constructed and methylamino group and acetyl group are added in order to simulate realistic tubulin environment. The structures are optimized and IR spectra are calculated using B3LYP/6-311G (d, p) basis group in density functional theory [27]. Diffusion function is added for the dimer structure.

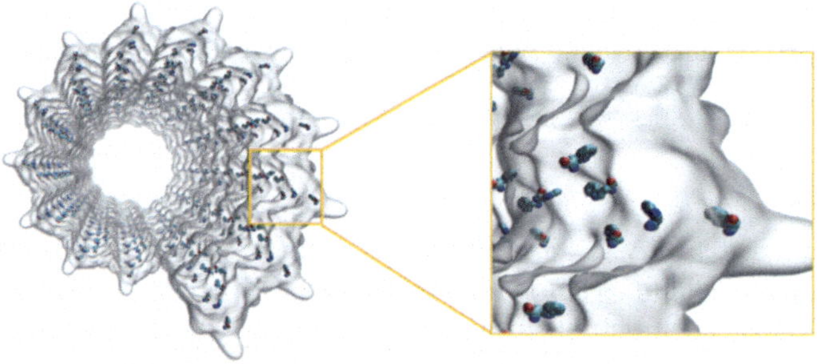

Fig. 1. The structure of tubulin ensembles and tryptophan within.

3 Results and Discussion

From the structure of tubulin, we discovered an ordered structure of tryptophan multimers in the tubulin acting as the main backbone of the structure. Therefore, inspired by the exciplex model in chemistry, we assume tryptophan multimers may exhibit similar features allowing the tubulin to interact with THz waves. We calculate the structure and IR spectra of tryptophan monomer and dimer. The optimized structures exhibit a similar ordered fashion as within the tubulin (see Fig. 2a, 2b).

From the IR spectra of monomer and dimer, we observe rich THz information in both structure (see Fig. 2c). Most peaks in monomer are conserved in the dimer structure, with tiny phase shift similar to the exciplex structure. However, the two strongest peaks are exclusive to the dimer structure. Both vibration modes correspond to resonance between two monomers.

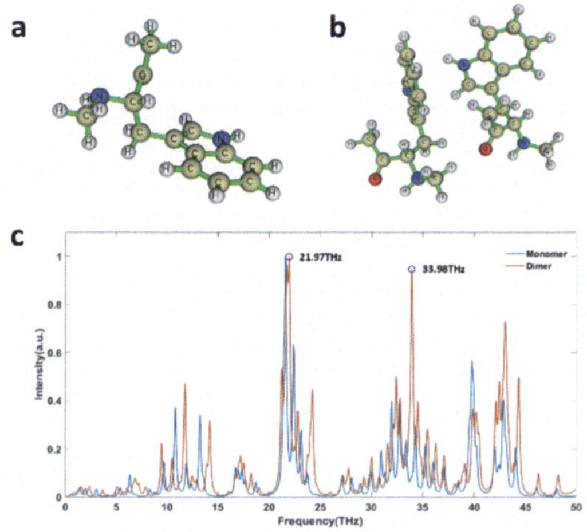

Fig. 2. The optimized structure of tryptophan monomer (a) and dimer (b). The IR spectra (c) are of similar structure, with strongest peaks of dimers marked in circle.

The difference in spectra shows that strong resonant behavior in THz range is present when multimer structure is considered. Therefore, we propose that the dimer structure is capable of interacting with more THz energy than monomer structure. The results imply that the ordered multimer structure in the tubulin may promote THz interaction inside neurons, with main interaction in the group resonance range. Our work presents a new possibility of THz interaction inside the neurons, demonstrating how THz waves may contribute to the neural signaling mechanism.

References

1. Li, B.X., et al.: Site-selective tyrosine bioconjugation via photoredox catalysis for native-to-bioorthogonal protein transformation. Nat. Chem. **13**(9), 902–908 (2021)
2. Celardo, G.L., et al.: On the existence of superradiant excitonic states in microtubules. New J. Phys. **21**(2), 023005 (2019). https://doi.org/10.1088/1367-2630/aaf839
3. Liu, G., et al.: Myelin sheath as a dielectric waveguide for signal propagation in the mid-infrared to terahertz spectral range. Adv. Funct. Mater. **29**(7) (2019)
4. Lou, J., et al.: Calibration-free, high-precision, and robust terahertz ultrafast metasurfaces for monitoring gastric cancers. Proc. Natl. Acad. Sci. U.S.A. **119**(43), e2209218119 (2022)
5. Zhang, Z., et al.: Advanced terahertz refractive sensing and fingerprint recognition through metasurface-excited surface waves. Adv. Mater. e2308453 (2024)

6. Wang, R., et al.: Electric Fano resonance-based terahertz metasensors. Nanoscale **13**(44), 18467–18472 (2021)
7. Chang, C., et al.: High-gain Thompson-scattering x-ray free-electron laser by time-synchronic laterally tilted optical wave. Phys. Rev. Lett. **110**(6), 064802 (2013)
8. Gao, M., et al.: Enhanced and frequency-selectable terahertz emission from two crossed air-plasmas. Phys. Rev. Res. **5**(2) (2023)
9. Wang, P., et al.: Progress on cutting-edge infrared-terahertz biophysics. IEEE Trans. Microw. Theory Tech. **70**(11), 5117–5140 (2022)
10. Xiang, Z., et al.: A new viewpoint and model of neural signal generation and transmission: Signal transmission on unmyelinated neurons. Nano Res. **14**(3), 590–600 (2020)
11. Xiang, Z., et al.: A primary model of THz and far-infrared signal generation and conduction in neuron systems based on the hypothesis of the ordered phase of water molecules on the neuron surface I: signal characteristics. Sci. Bull. **65**(4), 308–317 (2020)
12. Liu, Y., et al.: Amplification of terahertz/infrared field at the nodes of Ranvier for myelinated nerve. Sci. China Phys. Mech. Astron. **63**(7) (2020)
13. Li, Y., et al.: Physicochemical insights on terahertz wave diminished side effects of drugs from slow dissociation. ACS Nano **16**(5), 8419–8426 (2022)
14. Li, Y., et al.: Terahertz wave enhances permeability of the voltage-gated calcium channel. J. Am. Chem. Soc. **143**(11), 4311–4318 (2021)
15. Tan, X., et al.: Neuromodulation of chemical synaptic transmission driven by THz photons. Research 2022 (2022)
16. Wu, K., et al.: Terahertz wave accelerates DNA unwinding: a molecular dynamics simulation study. J. Phys. Chem. Lett. **11**(17), 7002–7008 (2020)
17. Yin, J., et al.: Terahertz photons inhibit cancer cells long term by suppressing nano telomerase activity. ACS Nano (2024)
18. Zhang, Q.L., et al.: Ultrahigh-flux water nanopumps generated by asymmetric terahertz absorption, Phys. Rev. Lett. (2024)
19. Zhu, Z., et al.: Characteristic fingerprint spectrum of neurotransmitter norepinephrine with broadband terahertz time-domain spectroscopy. Analyst **144**(8), 2504–2510 (2019)
20. Zhu, Z., et al.: Tunable surface wettability via terahertz electrowave controlled vicinal subnanoscale water layer. Nano Lett. **24**(10), 3243–3248 (2024)
21. Zhang, J., et al.: Non-invasive, opsin-free mid-infrared modulation activates cortical neurons and accelerates associative learning. Nat. Commun. **12**(1), 2730 (2021)
22. Liu, X., et al.: Nonthermal and reversible control of neuronal signaling and behavior by midinfrared stimulation. Proc. Natl. Acad. Sci. USA **118**(10) (2021)
23. Yu, Y., et al.: Terahertz photons improve cognitive functions in posttraumatic stress disorder. Research (Wash D C) **6**(0278) (2023)
24. Peng, W., et al.: High-frequency terahertz waves disrupt Alzheimer's β-amyloid fibril formation. eLight **3**(1) (2023)
25. Eshun-Wilson, L., et al.: Effects of alpha-tubulin acetylation on microtubule structure and stability. Proc. Natl. Acad. Sci. U.S.A. **116**(21), 10366–10371 (2019)
26. Humphrey, W., et al.: VMD: visual molecular dynamics. J. Mol. Graph. **14**(1), 33–38 (1996)
27. Lu, T., Chen, F.: Multiwfn: a multifunctional wavefunction analyzer. J. Comput. Chem.Comput. Chem. **33**(5), 580–592 (2012)

Quantum Mechanism of ATP/ADP Hydrolysis: A Journey into Biophotons and THz Biophysics

Xiangyuan Pang(✉)

Tsinghua University, Haidian District, Beijing 100084, People's Republic of China
pangxy20@mails.tsinghua.edu.cn

Abstract. ATP/ADP hydrolysis serves as a fundamental driving force for numerous life processes. From the traditional thermal perspective to the involvement of biophotons, the researches outlined here open avenues for further investigations into the biological mechanisms of energy conversion in the processes of ATP/ADP hydrolysis. This emerging field invites future studies to unravel the full spectrum of interactions shaping bioenergetics.

Keywords: ATP/ADP Hydrolysis · Biophotons · Quantum Chemical Computation · THz Biophysics

1 Introduction

The Terahertz science and technology have broad applications in THz bioeffects [1–7] and bio-sensors [8–14]. Recently, the information transmitted in neurons is thought as Terahertz photons with specific frequencies [15–28]. Based on the research mentioned above, biophotons many also play a role in the process of ATP/ADP hydrolysis, contributing to part of the release of energy.

The hydrolysis of ATP/ADP, conventionally viewed as a thermal process, has been a topic of extensive discussion. However, the biological mechanism underlying the high efficiency of ATP-driven biosynthesis remains unclear, compared with the light harvesting of photosynthesis, which is known to reach an efficiency of ~ 95% through quantum energy transfer of photons. This article discussed this research field and proposed the hypothesis that the ATP/ADP hydrolysis may also involve a quantized process. Further research could provide valuable insights into verifying this supposition.

2 Research Review on ATP/ADP Hydrolysis

2.1 ATP/ADP Hydrolysis Regarded as Thermodynamic Process

There have been several researches attempting to explain the energy convention in ATP/ADP hydrolysis. Kamerlin and Warshel [29] employed Gaussian03 software and QM/MM-FEP methods to explore the energy conversion in ATP hydrolysis. Factors

C. Chang et al. (Eds.): YTHZ 2024, SPPHY 400, pp. 289–292, 2024.
https://doi.org/10.1007/978-981-97-3786-4_59

such as cavitation energy, configurational entropy, and cage entropy were systematically examined. Their findings indicated an average free energy change of -6.5 kcal/mol (27.196 kJ/mol), though the accuracy of their calculations is sensitive to the theoretical approach used. This initial study laid the groundwork for subsequent discussion on the topic.

In a study by Bergman, Kashiwaya, and Veech [30], the impact of pH and free Mg^{2+} on ATP-linked enzymes and the Gibbs free energy of ATP hydrolysis was analyzed. As pH levels and concentrations of $[Mg^{2+}]$ varied, the Gibbs free energy would change within the range of 59 kJ/mol to 62 kJ/mol. Additionally, Frey and Arabshahi [31] calculated the standard free energy change associated with the hydrolysis of the alpha-peta-phosphoanhydride bridge in ATP. Their findings indicated a free energy change of approximately -10.9 kcal/mol during the hydrolysis of ATP to AMP and PP_i, which is considerably more negative than that in the hydrolysis of ATP to ADP and P_i.

2.2 Intrinsic Biophotons and Biological Process

Biophotons, considered carriers of bioinformation in biological systems, still play a mysterious role in biological process. W.P. Mei's seminal work in 1993 [32] concluded the inseparability of biophoton emission from basial biological phenomena, including cell division, apoptosis, oxidative metabolism, and carcinogenesis. Experiments conducted by Scott et al. [33] provided evidence confirming the close connection between biophoton emission and the biochemical and biophysical process of cell multiplication and successive growth.

2.3 DNTP Hydrolysis Release PB-photons

Connecting these findings, Li et al. [34] conducted an AuNP-PCR experiment, demonstrating that energy released during phosphoanhydride-bond (PB) hydrolysis of dNTPs is carried by photons. Their calculations based on density functional theory (DFT) indicated a quantized process with a PB-photon of 34THz, mainly released by the formation of the P-OH bond. Given that identical bonds break in the hydrolysis of both ATP and dNTP, their work provided an example for employing similar simulation methods to calculate the photons released during ATP/ADP hydrolysis.

3 Conclusion

Based on the previous research, we propose the hypothesis that the ATP/ADP hydrolysis also involves a photon transfer process, with the frequency of ~ THz. The energy released by ATP/ADP hydrolysis is mainly carried and transferred by these THz biophotons, while the rest of energy may dissipate as thermal energy. Further quantum chemical computation based on DFT or other relevant theoretical frameworks, coupled with experimental investigations, may explain the micro process of hydrolysis and photon transfer.

Despite the calculation of photon frequency, there are certain issues worth discussion:

- Not every photon generated by the break/formation of bonds may necessarily be released. The release of photons could be influenced by factors such as the specific molecular environment and energy pathways involved.
- The process and photon frequency may vary in cells or when enzymes are involved. Cellular conditions and the catalytic influence of enzymes can introduce additional complexity.
- The purpose or utilization of the photons released during ATP/ADP hydrolysis is an intriguing question. It remains an area of investigation to determine how these photons, particularly in the THz frequency range, might be involved in cellular processes or signaling.
- Whether a similar process involving photon release occurs in the biosynthesis of ATP/ADP is an open question. Investigating biosynthetic pathways could provide insights into whether similar photon transfer processes are inherent in the production of these cellular energy carriers.

References

1. Li, Y., et al.: Terahertz wave enhances permeability of the voltage-gated calcium channel. J. Am. Chem. Soc. **143**, 9 (2021)
2. Liu, X., et al.: Nonthermal and reversible control of neuronal signaling and behavior by midinfrared stimulation. Proc. Natl. Acad. Sci. **118**(10), e2015685118 (2021)
3. Zhang, J.X., et al.: Non-invasive, opsin-free mid-infrared modulation activates cortical neurons and accelerates associative learning. Nat. Comm. **12**, 2730 (2021)
4. Yu, Y., et al.: Terahertz photons improve cognitive functions in posttraumatic stress disorder. Research **6**, 0278 (2023)
5. Peng, W., et al.: High frequency terahertz waves disrupt Alzheimer's amyloid fibril formation. eLight **3**, 18 (2023)
6. Tan, X., et al.: Neuromodulation of chemical sysnaptic transmission driven by THz photons. Research **2022**, 0010 (2022)
7. Li, Y., et al.: Physicochemical insights on terahertz wave diminished side effects of drugs from slow dissociation. ACS Nano **16**(5), 8419 (2022)
8. Wang, R., et al.: Electric Fano resonance-based terahertz metasensors. Nanoscale **13**, 18467–18472 (2021)
9. Lou, J., et al.: Optically controlled ultrafast terahertz metadevices with ultralow pump threshold. Small **17**, 2104275 (2021)
10. Zhang, Z., et al., Advanced terahertz refractive sensing and fingerprint recognition through metasurface-excited surface waves. Adv. Mater., 2308453 (2024)
11. J. Lou et al., Calibration-free, high-precision, and robust terahertz ultrafast metasurfaces for monitoring gastric cancers. PNAS **119**(43) (2022)
12. Wang, R., et al.: Ultrasensitive terahertz biodetection enabled by quasi-BIC-based metasensors. Small **19**, 2301165 (2023)
13. Jiao, Y., et al.: Photoactive terahertz metasurfaces for ultrafast switchable sensing of colorectal cells. Mater. Horiz. **9**, 2984–2992 (2022)
14. Sun, L., et al.: A pixelated frequency-agile metasurface for broadband terahertz molecular fingerprint sensing. Nanoscale **14**, 9681–9685 (2022)

15. Liu, G., et al.: Myelin sheath as a dielectric waveguide for signal propagation in the mid-infrared to terahertz spectral range. Adv. Funct. Mater. **29**(7), 1807862 (2019)
16. Wang, P., et al.: Progress on cutting-edge infrared-terahertz biophysics. IEEE Trans. Microw. Theory Tech. **70**(11), 5117–5140 (2022)
17. Xiang, Z., et al.: A primary model of THz and far-infrared signal generation and conduction in neuron systems based on the hypothesis of the ordered phase of water molecules on the neuron surface. Sci. Bull. **65**, 308 (2020)
18. Liu, Y., et al.: Amplification of terahertz/infrared field at the nodes of Ranvier for myelinated nerve. Sci. China Phys. Mech. Astron. **63**(7), 274211 (2020)
19. Yin, J., et al.: Terahertz photons inhibit cancer cells long term by suppressing nano telomerase. ACS Nano **18**(6), 4796–4910 (2024)
20. Wu, K., et al.: Terahertz wave accelerates DNA unwinding: a molecular dynamics simulation study. J. Phys. Chem. Lett. **11**(17), 7002–7008 (2020)
21. Zhang, C., et al.: Driving DNA origami assembly with a terahertz wave. Nano Lett. **22**, 468–475 (2022)
22. Yuan, Y., et al.: Midinfrared radiation accelerates DNA unwinding. ACS Photonics (2024). https://doi.org/10.1021/acsphotonics.3c01548
23. Xiao, T., et al.: Sensory input dependent gain modulation of the optokinetic nystagmus by mid-infrared stimulation in pigeons. Elife **12**, e78729 (2023)
24. Sun, L., et al.: Inhibition of cancer cell migration and glycolysis by terahertz wave modulation via altered chromatin accessibility. Research **2022**, 9860679 (2022)
25. Zhang, Q.L., et al.: Ultrahigh-flux water nanopumps generated by asymmetric terahertz absorption. Phys. Rev. Lett. (2024)
26. Zhu, Z., et al.: Characteristic fingerprint spectrum of neurotransmitter norepinephrine with broadband terahertz time-domain spectroscopy. Analyst **144**(8), 2504–2510 (2019)
27. Tang, C., et al.: Integrating terahertz metamaterial and water nanodroplets for ultrasensitive detection of amyloid β aggregates in liquids. Sens. Actuators B. Chem. **329**, 129113 (2021)
28. Zhu, Z., et al.: Tunable surface wettability via terahertz electrowave controlled vicinal subnanoscale water layer. Nano Lett. **24**(10), 3243–3248 (2024)
29. Kamerlin, S.C.L., Warshel, A.: On the energetics of ATP hydrolysis in solution. J. Phys. Chem. B **113**(47), 15692–15698 (2009)
30. Bergman, et al.: The effect of pH and Free Mg2+ on ATP linked enzymes and the calculation of Gibbs free energy of ATP hydrolysis. JPCB **114**, 16137 (2010)
31. Frey, P.A., Arabshahi, A.: Standard free energy change for the hydrolysis of the alpha, beta-phosphoanhydride bridge in ATP. Biochemistry **34**(36), 11307–11310 (1995)
32. Mei, W.P.: About the nature of biophotons. J. Biol. Syst. **02**, 25–42 (1994)
33. Scott, R., et al.: Two-dimensional detection of ultraweak bioluminescence using a single-photon image acquisition system. Ann. Meeting Optical Soc. Am. (1987)
34. Li, N., et al.: Demonstration of biophoton-driven DNA replication via gold nanoparticle-distance modulated yield oscillation. Nano Res. **14**(1), 40–45 (2021)

Author Index

C. Chang et al. (Eds.): YTHZ 2024, SPPHY 400, pp. 293–295, 2024.
https://doi.org/10.1007/978-981-97-3786-4

The manufacturer's authorised representative in the EU is Springer Nature Customer Service Centre GmbH, Europaplatz 3, 69115 Heidelberg, Germany. If you have any concerns regarding our products, please contact ProductSafety@springernature.com

Printed and bound by CPI Group (UK) Ltd, Croydon, CR0 4YY
01/10/2026
02225862-0008